Engineering Measurements

C. V. COLLETT, C.Eng., M.I.Mech.E., M.I.Prod.E., received his basic engineering training in the machine tool overhaul department of J. Lucas Ltd., where he gained considerable experience in machine tool construction and in the requirements for accurate alignment and the relationships of the various elements of machine tools. He was later employed by Harold Gauges Ltd. where he increased his experience of the techniques of metrology and, prior to entering education, was a designer in the gravity and forge die design office of Alcan in Birmingham. His experience also includes jig and fixture design for the machining of aluminium castings. Since 1968 he has been a Lecturer in the Department of Mechanical and Production Engineering at the City of Birmingham Polytechnic.

A. D. HOPE graduated from Leeds University in 1964 with a B.Sc.(Hons.) in fuel technology. He then spent the following four years in industry with the aerospace division of J. Lucas Ltd. as a graduate apprentice and as a development engineer on aircraft fuel systems and controls. He joined the City of Birmingham Polytechnic in 1968 as a Lecturer in mechanical engineering specializing in measurement and control engineering.

ENGINEERING MEASUREMENTS

C. V. Collett

A. D. Hope

PITMAN

PITMAN PUBLISHING LIMITED
39 Parker Street, London WC2B 5PB

Associated Companies
Copp Clark/Pitman, Toronto
Fearon Pitman Publishers Inc, San Francisco
Pitman Publishing New Zealand Ltd, Wellington
Pitman Publishing Pty Ltd, Melbourne

First published in Great Britain 1974
Reprinted 1979

Text set in 10/11 pt Monotype Times Roman
printed by photolithography and bound in Great Britain
by Biddles Ltd, Guildford, Surrey

ISBN 0 273 40423 7 (cased edition)
ISBN 0 273 40419 9 (paperback edition)

Contents

Preface

Measurement is an essential part of the development of technology and as technology becomes more complex so the techniques of measurement become more sophisticated.

Some years ago on the representation of the Higher National Certificate, a new subject, "Engineering Measurement", was introduced in recognition of the increasing importance to the technician engineer of measuring methods and techniques.

This volume was originally conceived to cover the engineering measurement requirement of the H.N.C. in Engineering, but during the course of its preparation, it became apparent to the authors that, while fully covering the requirements of the H.N.C. in this regard, the appeal of the subject matter was more broadly based and they suggest that it will be of use and interest to technician engineers taking the more advanced City and Guilds courses and the H.N.D., and as introductory reading for first-year undergraduates and for practising engineers.

Although the change to SI units is now substantially completed, some British Standards are still in the course of revision and have not yet been published in the new units. Consequently standards are sometimes quoted which are still in Imperial units. Where this has been done the approximate metric equivalent has been placed in parenthesis for the reader's convenience. Apart from these considerations, no concession has been made in the text to the former units.

We wish to acknowledge the generous help and advice of colleagues and the many industrial organizations who have kindly provided us with information.

We dedicate this book to our respective wives, without whose monumental patience the enterprise would not have been possible.

C.V.C.
A.D.H.

1 The Characteristics of Measuring Systems

1.1 Measuring Systems

All measuring systems include three basic elements:

(*a*) A detecting and measuring element (transducer) which detects the physical variable to be measured, e.g. pressure, temperature, rate of flow, and converts the signal into a more usable form. In practice the physical variable is usually transformed into a mechanical or electrical signal.

(*b*) An intermediate stage which modifies the signal from the transducer so that a desirable output is available. For example, in the case of an electrical output signal amplification may be required before the signal can be indicated or recorded.

(*c*) An indicating or recording stage.

The simple Bourdon-tube pressure gauge discussed in Chapter 10 can be considered as an example of the general measurement system. The Bourdon-tube itself represents the transducer as it converts the pressure signal into a mechanical displacement of the tube. The gearing constitutes the intermediate stage as a relatively small displacement at the end of the Bourdon-tube is amplified to give a relatively large displacement of the centre gear. Finally the indicator stage consists of the pointer and dial. The dial is calibrated in pressure units and thus the pointer gives an indication of the pressure acting on the Bourdon-tube.

A second example is the piezoelectric pressure transducer also discussed in Chapter 10. In this case the crystal itself converts the pressure signal into an output voltage proportional to the pressure. Since it is necessary to indicate or record this voltage an amplification stage is required. Also electrical impedance matching between the transducer and the readout circuitry is usually a critical matter requiring careful design considerations (see Chapter 5). The final stage of the system may then be a voltmeter or a recorder.

Measuring instruments may broadly be divided into two groups:

(*a*) *Indicating instruments* in which the value of the measured quantity is visually indicated, but is not recorded. For example, ammeter, Bourdon-tube pressure gauge, mercury-in-glass thermometer.

(*b*) *Recording instruments* in which the values of the measured quantity are recorded on a chart. For example, barograph, surface-finish recorder.

1.2 Calibration and Graduation of Measuring Instruments

Calibration is defined as the process of determining the characteristic relationship between the values of the physical quantity applied to the instrument and the corresponding positions of the index, i.e. a chart of measured quantity versus instrument readings is required.

Calibration may be achieved by comparison with

(i) a primary standard
(ii) a secondary standard which has a higher accuracy than the instrument to be calibrated, or
(iii) a known input source.

As an example, a flowmeter may be calibrated by comparing it with a standard flow measurement device, another flowmeter of known accuracy, or by direct calibration with a primary measurement. The third method may easily be achieved by weighing a quantity of liquid in a tank and then recording the time taken for the liquid to flow through the meter.

The first step in the graduation of an instrument scale is to locate certain fixed points. The division of the scale into a specified number of increments is then made with reference to the fixed points. As an example consider a simple liquid in glass thermometer. For most purposes the scale can be assumed to be linear and has 100 Celsius units between the freezing point and the boiling point of water at standard atmospheric pressure. The fixed points 0°C and 100°C are obtained by standing the thermometer in melting ice and steam from boiling water respectively, the distance between the fixed points then being divided into 100 equal parts.

For an instrument having an arbitrary scale, the factor by which the indication has to be multiplied in order to obtain the nominal value of the quantity measured is known as the *scale factor*.

It is essential that, after a measuring instrument has been graduated and calibrated, the instrument be tested for the purpose of assessing the instrument errors. This process is known as verification and is employed to determine whether an instrument does in fact comply with a prescribed specification.

1.3 Sources of Error in Measuring Instruments

The main sources of error in a measuring system may be summarized as follows:

(*a*) The *observation error*, which is the error committed by the observer when reading the indication of an instrument. The main reasons for errors of this type are simple misreading, errors due to parallax, or faulty estimation of the fractional part of a scale interval.
(*b*) The *scale error*, due to the difference between the actual position of the scale mark and its theoretical position on a scale correctly graduated in accordance with the assumed law of operation of the instrument, i.e. the instrument scale may not be perfect.
(*c*) The *indication error* which is defined as the error obtained by subtracting the true value of the quantity measured from the indicated value, due regard being paid to the sign of each. When the instrument is in the

specified conditions of use, and a physical quantity is presented to it for the purpose of verifying the setting, the indication error may be referred to as datum error.

(*d*) *Zero-error* which is the indication when the instrument is in the specified conditions of use and the magnitude of the physical quantity presented to it is zero.

The limits of error for a measurement system are the positive and/or negative values of the errors which must not be exceeded under test. In the case of indication errors the limits may be expressed in the following ways:

(*a*) Directly in units of the measured quantity.
(*b*) As a percentage of the full-scale deflection (e.g. error 10% F.S.D.).
(*c*) As a percentage plus or minus of the expected reading (e.g. $\pm 1\%$ error at 30°C). Errors are often expressed in this form when the error varies over the working range.

The interval error is the difference between two indication errors. This is obtained by subtracting the indication error at one of two scale marks from that at the other, due regard being paid to the signs of each. The interval error is thus expressed as an error over the working range (e.g. error $\pm 1\%$ between 20°C and 50°C).

1.4 Characteristics of Measuring Systems under Dynamic Conditions

In many measurement applications ample time is available for the system to reach a steady state and hence the transient characteristics of the system response need not be considered. In other situations however it may be required to examine the behaviour of a physical variable over a period of time. The measurement problem will usually become more complicated when the transient characteristics are considered, and some of the more important factors affecting the transient response will be discussed during the remainder of this chapter.

1.5 System Response

The most important factor in the performance of a measuring system is that the full effect of an input signal (i.e. change in measured quantity) is not

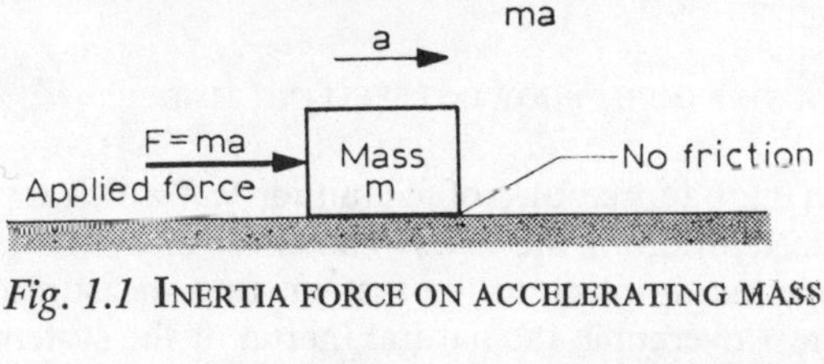

Fig. 1.1 INERTIA FORCE ON ACCELERATING MASS

immediately shown at the output but is almost inevitably subject to some lag or delay in response. This is a delay between cause and effect due to the natural inertia of the system and is known as *measurement lag*.

If a force is applied to the frictionless mass shown in Fig. 1.1 so as to produce an acceleration *a* to the right, then the magnitude of the force is *ma*, where *m* is the mass. The force on the mass acts in the direction in which the mass is required to accelerate, that is, to the right.

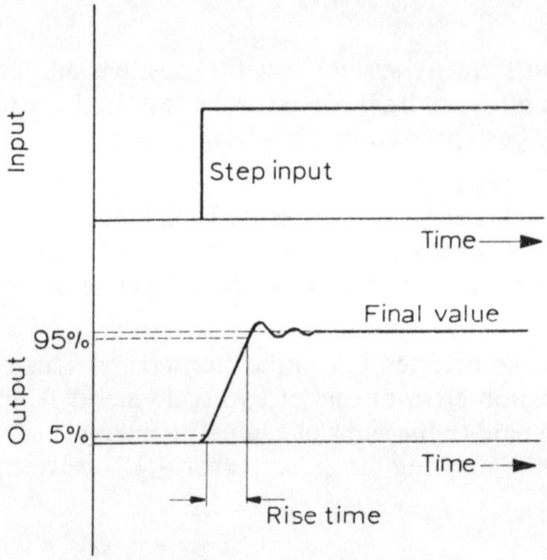

Fig. 1.2 RESPONSE TO A STEP INPUT SHOWING MEASUREMENT LAG

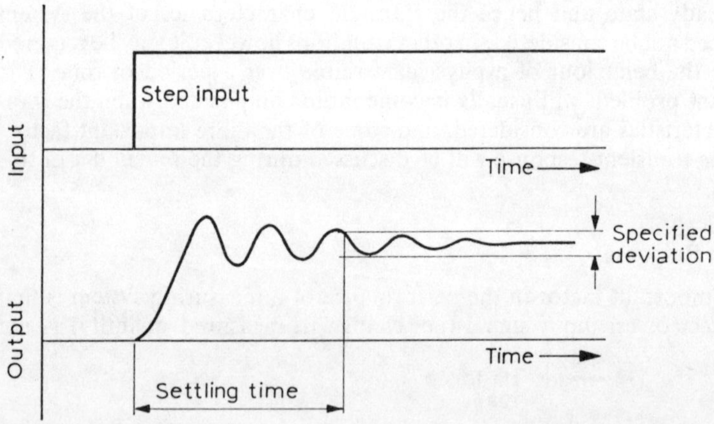

Fig. 1.3 OUTPUT RESPONSE TO A STEP INPUT SHOWING SETTLING TIME

The mass however applies an inertia force, also of magnitude *ma*, which acts to the left and opposes the acceleration of the mass. The effect of inertia is therefore to oppose any change taking place in the system. In a measuring system therefore the output must overcome the natural inertia of the system which results in a time lag between input and output. If an abrupt or step change in the measured quantity takes place then a delay occurs in the output response as shown in Fig. 1.2.

The rise time is usually defined as the time taken for a system to change from 5% to 95% of its final value in the initial part of the curve, but 10% and 90% are also commonly used. The rise time is a measure of the speed of response of a measuring system and in general a system with a short rise time is required.

The settling time is defined as the time taken for the index of an instrument to attain, and remain within, a specified deviation (typically ±2%) from its desired value, after an abrupt change in the measured quantity (Fig. 1.3).

1.6. Periodic Time and Damping

Damping may be an inherent factor in the operation of a measuring instrument or it may deliberately be introduced as a feature in its design. An instrument is

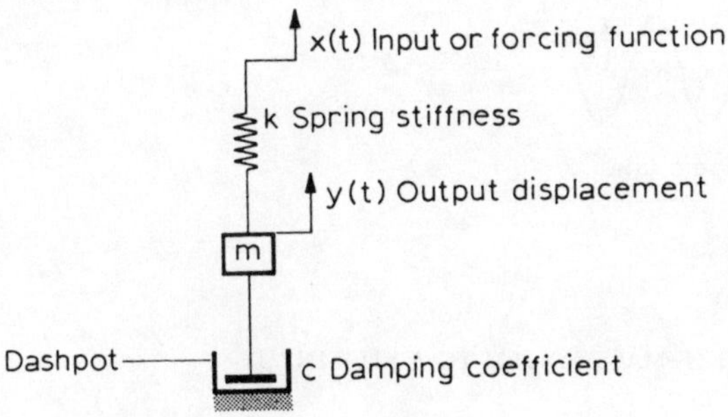

Fig. 1.4 SPRING–MASS DAMPER SYSTEM

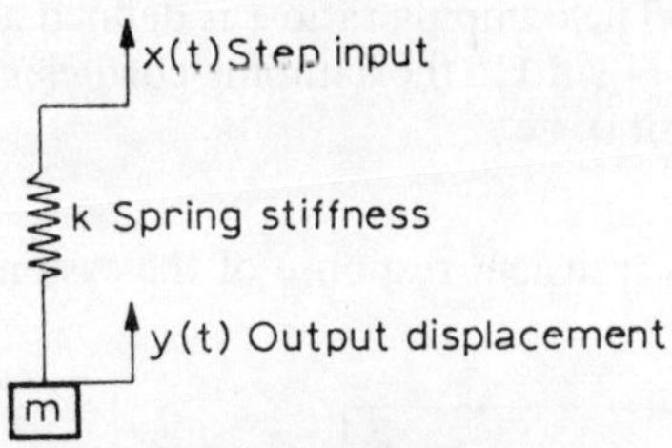

Fig. 1.5 SPRING–MASS SYSTEM

said to be damped when there is a progressive reduction in the amplitude, or complete suppression of successive oscillations of the index after an abrupt change in the value of the measured quantity.

As an example consider the simple spring–mass damper system shown in Fig. 1.4. $x(t)$ is an input displacement variable which is a function of time and acts through the spring–mass damper system to produce an output displacement of the mass $y(t)$. It is assumed that the damping force is proportional to velocity where c, the damping coefficient, is defined as the damping force per unit velocity (N s/m). Consider first the case when the damping force is zero and the system is subjected to a step input displacement. The system may be represented as shown in Fig. 1.5.

The output displacement of the mass $y(t)$ oscillates with simple harmonic motion about the desired value of the output as shown in Fig. 1.6. The amplitude of the motion is governed by the sensitivity of the system. A highly sensitive system tends to produce oscillations of greater amplitude than a system of low sensitivity. The periodic time of the resulting simple harmonic motion is $2\pi/\omega_n$ seconds where ω_n is the natural undamped radiancy of the system in radians

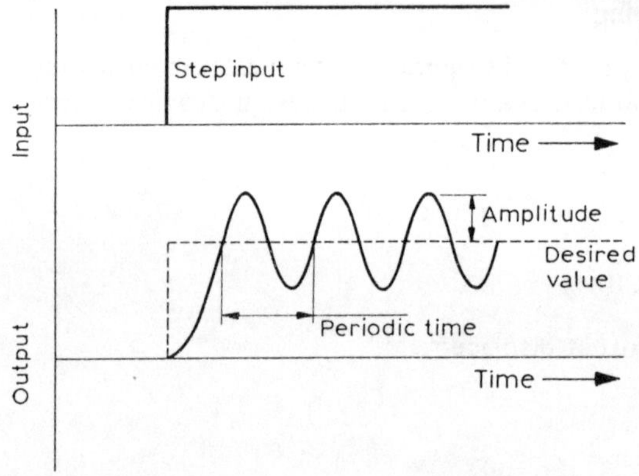

Fig. 1.6 RESPONSE OF A SPRING–MASS SYSTEM TO A STEP INPUT

per second. The natural undamped frequency of the system is given by $\omega_n/2\pi$ hertz.

When damping is present in the system (Fig. 1.4) the resulting motion depends largely on the amount of damping. The damping ratio ξ is defined as the ratio of the actual damping coefficient c (N s/m) to the damping coefficient required to produce critical damping c_c (N s/m), i.e.

$$\xi = c/c_c$$

The magnitude of the damping ratio affects transient response of the system as shown below.

Magnitude of damping ratio	*Transient response*
Zero	Undamped simple harmonic motion
Greater than unity	Overdamped motion
Unity	Critical damping
Less than unity	Underdamped, oscillatory motion

The case of undamped simple harmonic motion when the magnitude of the damping ratio is zero has already been discussed but the other three cases will now be considered.

(*a*) *Overdamped Motion* (when ξ is greater than unity).
An instrument is said to be overdamped when the degree of damping is more than sufficient to prevent oscillation of the index after an abrupt change in the value of the measured quantity. Referring to the spring–mass damper system shown in Fig. 1.4, if the system is given a step input displacement the output

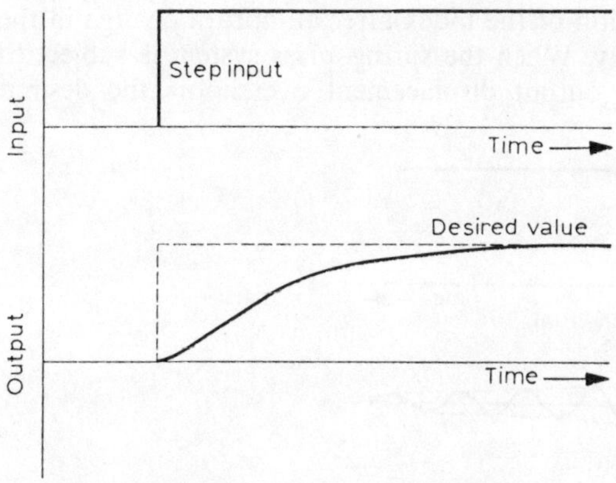

Fig. 1.7 OVERDAMPED RESPONSE TO A STEP INPUT

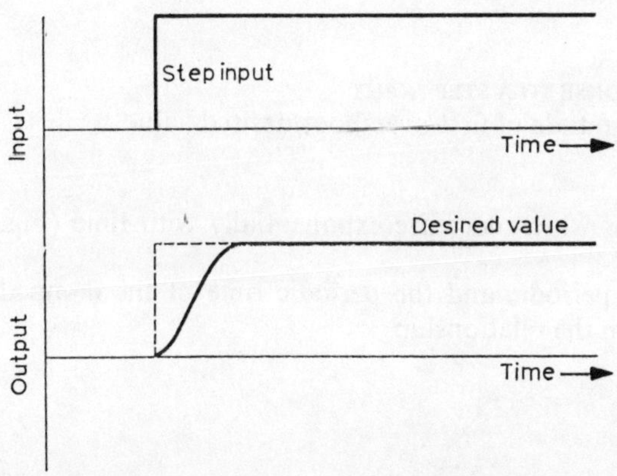

Fig. 1.8 CRITICALLY DAMPED RESPONSE TO A STEP INPUT

will approach the desired value exponentially and without oscillation as shown in Fig. 1.7. The motion is said to be "dead-beat" under these conditions.

(*b*) *Critically damped motion* (when ξ is equal to unity).
An instrument is said to be critically damped when it is subject to the minimum degree of damping which will suffice to prevent oscillation of the index after an abrupt change in the value of the measured quantity. The spring–mass system under critically damped conditions behaves in a similar manner to the previous overdamped case with the exception that the output increases much more

rapidly. This represents the transition state between "dead-beat" and oscillatory motion (Fig. 1.8). An instrument is referred to as aperiodic when the motion of its index is either critically damped or overdamped.

(*c*) *Underdamped motion* (when ξ is less than unity).
An instrument is said to be underdamped when the degree of damping is insufficient to prevent oscillation of the index after an abrupt change in the value of the measured quantity. When the spring–mass system is subject to underdamped conditions the output displacement overshoots the desired

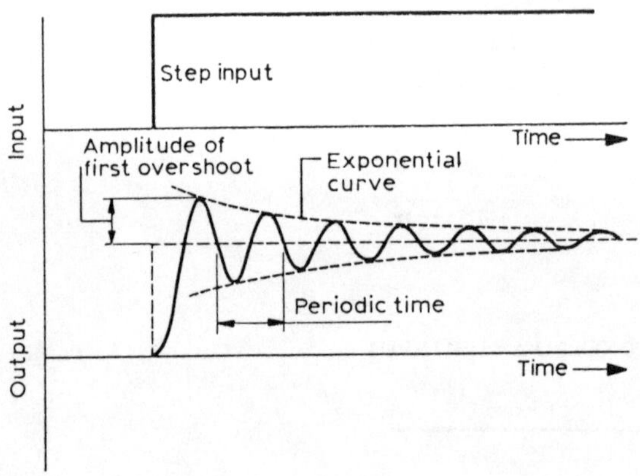

Fig. 1.9 UNDERDAMPED RESPONSE TO A STEP INPUT
% overshoot = magnitude of first overshoot/desired value

value resulting in oscillations which decrease exponentially with time (Fig. 1.9).

The oscillatory motion is periodic and the periodic time of the damped oscillations can be found from the relationship

$$t_{pd} = \frac{2\pi}{\omega_n\sqrt{(1 - \xi^2)}}$$

where t_{pd} is the periodic time of the damped oscillations (s)
ω_n is the natural undamped radiancy of the system (rad/s)
ξ is the damping ratio.

The percentage overshoot and number of oscillations of the system depend largely on the damping ratio. For example a low damping ratio produces a large percentage overshoot and a large number of oscillations, whereas an increased damping ratio reduces both the percentage overshoot and the number of oscillations. In practice sufficient damping is required to prevent excessive oscillation but unfortunately an increase in the damping ratio also increases the rise time. An optimum damping ratio is therefore required which will produce a short rise time with a small number of oscillations.

1.7. Sensitivity, Accuracy and Precision

The *sensitivity* of an instrument at any indicated value is the relationship between the movement of the index and the change in the measured quantity that produces it. A highly sensitive instrument therefore produces a large movement of the index for a small change in the measured quantity. Such an instrument tends to produce a large overshoot of the index with resulting oscillations. The oscillations can easily be reduced by increasing the damping ratio but again the rise time will be increased as discussed in the previous section. In practice therefore an optimum setting is required which will produce a high instrument sensitivity together with a small number of oscillations and a reasonably short rise time.

Accuracy is the closeness with which the indications of an instrument approach the true values of the quantities measured. A term frequently used to specify the precision of an instrument is repeatability which is defined as the ability of the instrument to reproduce readings when a series of tests is carried out in a short interval of time and under fixed conditions of use.

Two further terms used to define reproducibility are:

(*a*) *Stability*. The reproducibility of the mean readings of an instrument when tested under defined conditions of use, repeated on different occasions separated by intervals of time which are long compared to the time of taking a reading.

(*b*) *Constancy*. The reproducibility of the mean reading of an instrument over a period of time, when the quantity to be measured is presented continuously, and the conditions of test are allowed to vary within specified limits.

In assessing stability the readings are corrected for any departure of the actual test conditions from the defined conditions, constancy however includes effects due to variations from the defined conditions.

1.8 Hysteresis in Measuring Systems

A system free from hysteresis will produce the same reading irrespective of whether the reading has been achieved by increasing from a lesser value or

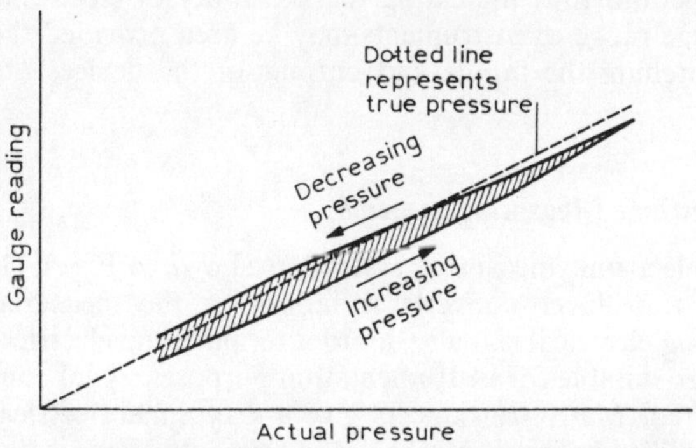

Fig. 1.10 TYPICAL HYSTERESIS CURVE FOR A PRESSURE GAUGE

decreasing from a higher value. Owing to hysteresis however many instruments do not reproduce the same reading under these conditions and a typical hysteresis curve for a pressure gauge is shown in Fig. 1.10. Hysteresis arises because of strain energy stored in the system, slack motion in bearings and gears, and bearing friction, etc.

If the hysteresis band is appreciable it may well be necessary to use average values of the gauge reading up and down. A correction curve may be drawn for the gauge showing the error between the true pressure and the average gauge pressure. When the gauge is used the true pressure may be found by referring to the correction curve by adding or subtracting the appropriate correction to the average gauge reading.

1.9 Electronic Methods and Mechanical Measurements

The particular advantages offered by the use of electronics in mechanical measurements may be summarized as follows:

(*a*) Speed of response is high.
(*b*) The instruments are versatile and convenient to use.
(*c*) The instruments may be adapted to difficult conditions of measurement.
(*d*) It is not always necessary to have a physical connection between the measuring device and the indicating or recording instrument.

The most significant advantage of electronic devices over other measuring systems is the exceptionally high speed of operation. Electronic instruments may be used for the measurement of rapidly changing functions of time where a mechanical system would be completely unsuitable.

With the application of radiotelemetry there is virtually no limit to the distance which may exist between the measuring device and the indicating or recording instrument. Thus measurements may be made in dangerous and hostile environments or inaccessible places. A good example of this application is that of space research.

Once the measured quantity has been converted into an electrical signal, the signal can be applied to indicate, record, or in the case of control systems perform a controlling function, thus indicating the versatility of electronic systems. Any one of a large range of instruments may be used provided the basic requirements of matching the inputs and outputs of the devices are carried out.

1.10 The Elements of Electronic Measuring Systems

The basic elements of an electronic measuring system are shown in Fig. 1.11. The sensing element or transducer converts variations in the measured quantity into corresponding electrical signals. In order to make the electrical output from the transducer suitable for instrumentation purposes signal conditioning is usually required. Many transducers give a very small electrical output which must be amplified before it can be used to operate an indicator or recorder. In other applications it may be necessary to convert a d.c. signal into an a.c. signal or convert the signal into pulses for digital indication, transmission of information, or computer processing.

The third stage of the measuring system is that of indication or recording and a number of methods are available depending on the type and accuracy of the information required. Indication may be achieved by means of a pointer and scale, digital display, teleprinting, or a cathode ray oscilloscope. Recording

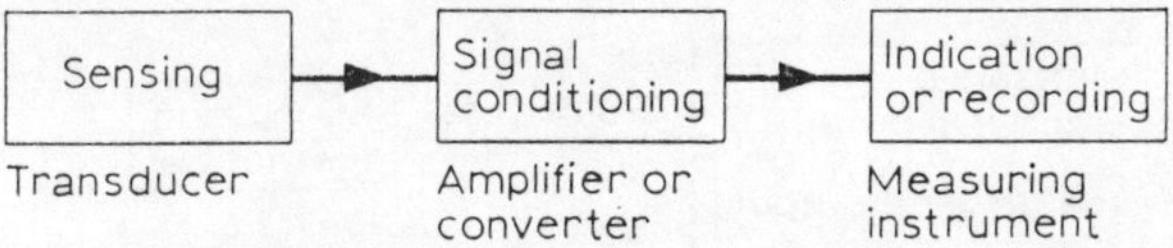

Fig. 1.11 BASIC ELEMENTS OF AN ELECTRONIC MEASURING SYSTEM

of slowly changing variables may be achieved by means of an ink trace on a paper chart. Ultra-violet light beam traces are normally used for recording rapidly changing variables but other methods available are:

(i) A camera used to photograph a cathode ray oscilloscope trace.
(ii) Punched tape.
(iii) Magnetic tape.
(iv) Videotape.

1.11 Electronic Control

In many applications a controlling function is required in addition to, or instead of, measurement. The block diagram (Fig. 1.12) shows how the controller uses the electrical output of the transducer to operate a servo system which controls the measured variable.

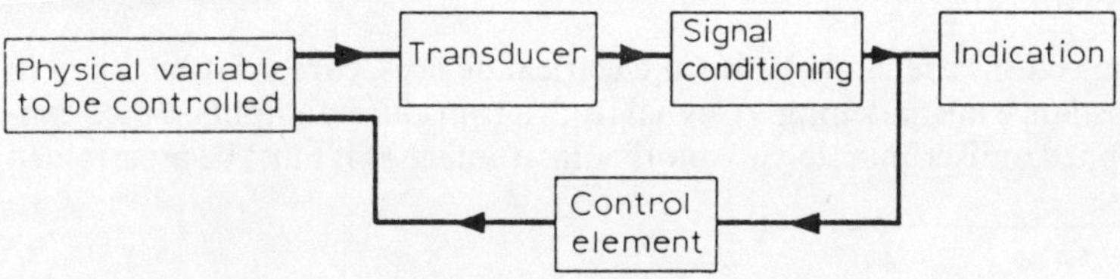

Fig. 1.12 BASIC ELEMENTS OF AN ELECTRONIC CONTROL SYSTEM

The controller compares the actual electrical output with the desired (or reference) electrical output. The difference between the actual or measured output and the desired output is known as the error or deviation. This error or deviation is used to operate a valve and adjust the measured variable.

Consider as an example the temperature control of a furnace shown in Fig. 1.13. If the temperature of the furnace is higher than the desired temperature then the measured electrical output from the thermocouple will be greater than the desired output and a positive deviation or error will occur. This error is used to close the main gas inlet valve to the furnace by an amount proportional to the error thus causing the temperature to fall. If, however, the temperature falls below the desired value a deviation or error of opposite sign is produced which is used to open the gas valve and thus increase the temperature.

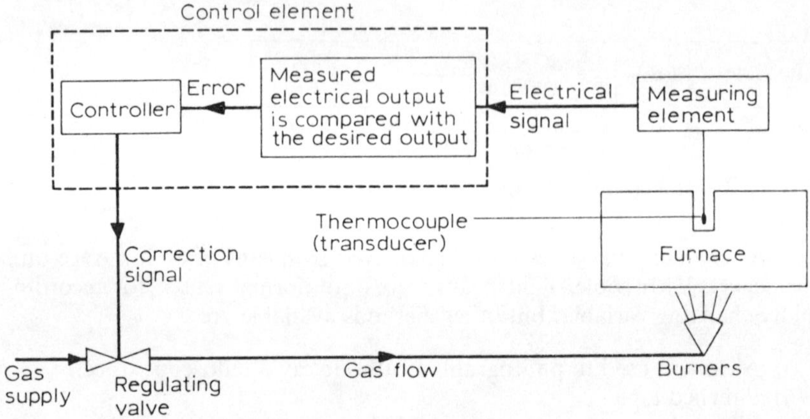

Fig. 1.13 CONTROL OF FURNACE TEMPERATURE

1.12 Practical Considerations for Electrical Instrumentation

It is often necessary to connect various items of electrical equipment such as a transducer, pre-amplifier, and an indicating or recording instrument in order to obtain a complete measurement system. In practice certain precautions must be taken as shown below.

Impedance Matching

When connections are made between electrical devices, care must be taken to avoid impedance mismatching. A signal from a transducer, amplifier, etc., cannot be applied indiscriminately to another instrument as it must be ascertained

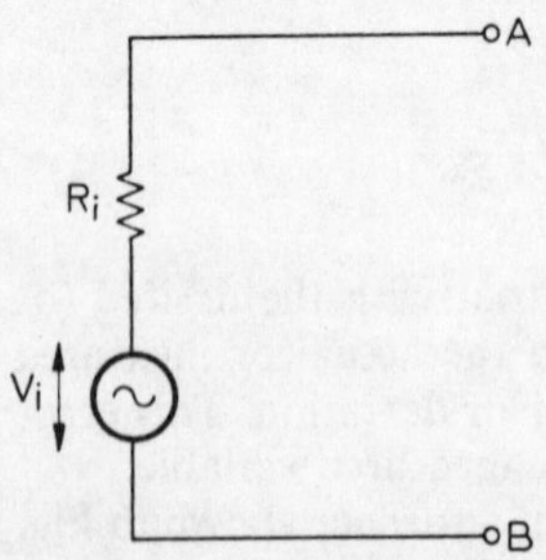

Fig. 1.14 INPUT IMPEDANCE OF A TWO-TERMINAL RESISTIVE DEVICE
A and B are the connecting terminals for the instrument

that the signal is not excessively loaded by the input circuit of the following instrument. Considering a resistive circuit the input impedance of a two-terminal device behaves as if the internal resistance R_i were connected in series with the internal voltage source V_i as shown in Fig. 1.14.

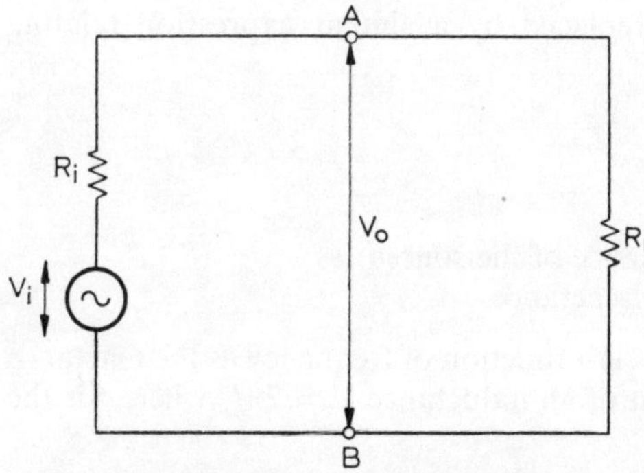

Fig. 1.15 TWO-TERMINAL DEVICE CONNECTED TO CIRCUIT OF INPUT RESISTANCE

If the device is connected to a circuit of input resistance R_L then the voltage at the output terminals A and B will depend on the value of R_L (Fig. 1.15). The potential difference V_o across R_L is given by

$$V_o = \frac{R_L}{R_i + R_L} V_i \tag{i}$$

There are two matching conditions:

(*a*) For maximum voltage transmission the internal voltage source V_i must equal the output voltage V_o at the terminals A and B.

Now $V_i \simeq V_o$ when $R_i \ll R_L$.

Thus for maximum voltage transmission the resistance of the load must be very much greater than that of the signal source.

(*b*) If power is to be delivered from the device to the external load R_L the power dissipated in the load is given by

$$P = V_o^2/R_L$$

Using (i)

$$P = \frac{R_L}{(R_i + R_L)^2} V_i^2$$

Differentiating P with respect to R_L

$$\frac{dP}{dR_L} = \frac{(R_i - R_L)}{(R_i + R_L)^3} V_i^2$$

for the power to be a maximum $dP/dR_L = 0$ and

$$R_L = R_i$$

Thus for maximum power transmission the resistance of the external load must equal the internal resistance.

The internal impedance and external load of a complicated electronic device may contain inductive and capacitive components that will be important in a.c. transmission and dissipation. In the case of voltage transmission the expression given above may be replaced by a similar expression relating impedances:

$$V_o = \frac{Z_L}{Z_i + Z_L} V_i$$

where Z_i is the impedance or reactance of the source
Z_L is the load impedance or reactance.

Impedance matching in a.c. circuits is a function of frequency as the reactance of a capacitor $C = 1/2\pi f_C$ and that of an inductance $L = 2\pi f_L$ where f is the frequency in hertz.

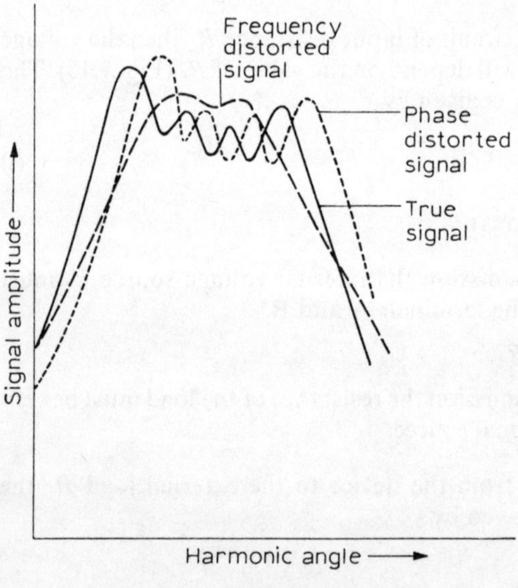

Fig. 1.16 SIGNAL DISTORTION PRODUCED BY POOR FREQUENCY RESPONSE AND POOR PHASE SHIFT RESPONSE

Distortion

Distortion is a term which is used to specify the variation of a signal from its true form. Distortion of the signal may occur as a result of poor frequency response or poor phase shift response. Various circuits are employed in electronic devices to reduce the effect of distortion to very small values (Fig. 1.16).

1.13 Basic Standards of Measurement

The fundamental units of the SI system are mass, length, time, temperature, electrical current, luminous intensity, and the mole.

MASS The mass of a body is defined as the quantity of matter of which the body consists and can only be changed if matter is added to or subtracted from the body.

The fundamental unit of mass is defined as the kilogramme (kg). The international prototype kilogramme was legalized in 1889 at the First General Conference of Weights and Measures and remains the primary standard. It is housed at the International Bureau of Weights and Measures at Sèvres, France, and takes the form of a 90% platinum–10% iridium alloy cylinder of equal height and diameter. Duplicates are held by various countries including Great Britain, where it is in the custody of the National Physical Laboratory (N.P.L.) and may be compared with the prototype on request. The comparison is made by means of accurate scales to a precision of at least 1 in 10^8.

LENGTH The fundamental unit of length is the metre (m). The first metric standard was defined at the time of the French Revolution as the distance between the ends of a platinum bar at 0°C. The International Prototype metre which replaced the original standard was defined in 1889 as the distance between two scribed lines on a 90% platinum–10% iridium bar at 0°C.

Because of the unsatisfactory nature of this and other material standards such as the Imperial Standard Yard, also a line standard, the metre was defined in 1960 as a length equal to 1 650 763·73 wavelengths in vacuum of a particular radiation band emitted by the Krypton 86 atom. This has the advantage of being easily reproducible to a precision of 1 part in 10^8 and avoids the difficulties of intercomparison between material standards used in the past, which suffered from progressive changes in length with time.

TIME The fundamental unit of time is the second. The second was originally defined as 1/86 400 part of the mean solar day but because of the variation of the solar day due to irregularities in the rotation of the earth at different periods of the year, and variations due to the movement of the earth's pole, a definition based purely on the mean solar day means that some days will consist of more or less parts of a second than the mean solar day. Such a definition is not sufficiently accurate for many scientific purposes. By 1967, experimental work had shown that an atomic time standard could provide a very precise definition of the second independently of fluctuations in the earth's rotation, etc., and thus could fulfil the requirements of advanced metrology.

The 1967 General Conference of Weights and Measures adopted a definition, which may be understood as follows:

> The second is the duration of 9 192 631 770 periods of oscillation of the radiation emitted by the caesium 133 atom under precisely defined conditions of resonance.

Such precise oscillations can be reproduced electronically to an accuracy of 1 part in 10^{11} and such instruments are commercially available.

Time signals conforming to a scale known as Co-ordinated Universal Time (UTC) are broadcast by radio. In order that a day may contain 86 400 seconds (as defined above) and be approximately equal to the period of one rotation of the earth, the second as defined by the Co-ordinated Universal Time scale is shorter than that defined by the atomic time scale.

In 1966 the following offset was agreed:

$$\frac{1 \text{ second (UTC)}}{1 \text{ second (true)}} = 1 - 300 \times 10^{-10}$$

As the number of seconds in the day is constant at 86 400 and the period of rotation of the earth varies slightly, it may sometimes be necessary to increase or decrease the first minute of each month by one or more tenths of a second.

Both UTC and true time signals are broadcast by radio so that precision clocks may be calibrated.

THERMODYNAMIC TEMPERATURE In 1954, the General Conference of Weights and Measures selected the equilibrium point of water, water vapour, and ice—the triple point—as the fundamental fixed point of the thermodynamic temperature scale.

In 1967 the name kelvin (K) was given to the unit of thermodynamic temperature (T) and the temperature 273·16 K was assigned to the triple point of water. The kelvin is therefore defined as the fraction 1/273·16 of the thermodynamic temperature of the triple point. The unit kelvin (K) is used to express an interval or difference in temperature.

The Celsius Scale (t) is distinguished from the kelvin scale (T) in the International Practical Temperature Scale as follows:

$$t = T - 273{\cdot}15 \text{ K}$$

where the unit of t is the degree Celsius (°C), which is equal to the kelvin (K). A temperature difference expressed in kelvins may therefore be expressed in degrees Celsius.

$$\text{Triple point of water} = 273{\cdot}16 \text{ K}$$
$$t = 273{\cdot}16 - 273{\cdot}15 = 0{\cdot}01\,^{\circ}\text{C}$$

(Full details of the I.P.T.S. appeared in *Metrologia*, vol. 5, no. 35, 1969.)

ELECTRIC CURRENT The General Conference of Weights and Measures in 1948 adopted the ampere as the fundamental unit of electric current according to the following definition:

The ampere (A) is that constant current which, if maintained in two straight parallel conductors of infinite length, of negligible circular cross-section, and placed one metre apart in a vacuum, would produce between these conductors a force equal to 2×10^{-7} newton per metre of length.

LUMINOUS INTENSITY The fundamental unit of luminous intensity is the candela (cd), which is defined as follows. The candela is the luminous intensity, in the perpendicular direction, of a surface of 1/600 000 square metres of a black body at the temperature of freezing platinum under a pressure of 101 325 newtons per square metre.

In practice groups of specially designed direct current lamps are colour matched against the primary black body standard so that a known luminous intensity may be more easily produced for calibration purposes.

THE MOLE The unit of amount of substance is the mole (mol) which is defined as follows. The mole is the amount of substance of a system which contains as many elementary entities as there are atoms in 0·012 kg of carbon 12.

The term "entity" must be specified for this purpose, and may be atoms, molecules, electrons, etc.

In practice, the mass corresponding to one mole of a substance, i.e. the molar mass, is calculated as follows:

$$M(X) = \frac{m(X)}{m(^{12}\mathrm{C})} \times 0{\cdot}012 \text{ kg/mol}$$

where $M(X)$ = the molar mass of substance X
$m(X)$ = the atomic mass of the substance X
$m(^{12}\mathrm{C})$ = the atomic mass of carbon 12
0·012 kg/mol = the molar mass of carbon 12.

For example, the atomic mass of oxygen (O_2) is 16.

$$M(O_2) = \frac{2 \times 16}{12} \times 0{\cdot}012 \text{ kg/mol} = 0{\cdot}032 \text{ kg/mol}$$

SUPPLEMENTARY UNITS The two supplementary units of the SI system are the radian (rad) for the plane angle and the steradian (sr) for the solid angle.

DERIVED UNITS Derived units are expressed in terms of the fundamental and/or supplementary units. Some important derived units are as follows.

FORCE Force is defined as the product of mass and acceleration.

If F = force, m = mass, a = acceleration, then

$$F = ma$$

The unit of force is the newton which is defined as the force required to produce an acceleration of 1 m/s² when applied to a mass of 1 kg. Therefore

$$1 \text{ N} = 1 \text{ kg} \times 1 \text{ m/s}^2$$

WEIGHT The force to which a mass is subjected by the presence of a gravitational field is its weight.

Let the mass of the body = m kg
The acceleration due to gravity = g m/s²

Then the force on the body due to gravity (weight) = mass × acceleration due to gravity. Therefore

$$F = m \text{ kg} \times g \text{ m/s}^2 = mg \text{ kg m/s}^2$$

$$F = mg \text{ newtons}$$

PRESSURE AND STRESS Pressure and stress are defined as force per unit area. Thus the derived unit in both cases is the newton per square metre and 1 N/m^2 has been assigned the name Pascal (Pa).

MULTIPLES OF SI UNITS The prefixes for use with SI units are given in the table.

Multiplying Factor	*Prefix* *Name*	*Symbol*
10^{12}	tera	T
10^{9}	giga	G
10^{6}	mega	M
10^{3}	kilo	k
10^{2}	hecto	h
10	deca	da
10^{-1}	deci	d
10^{-2}	centi	c
10^{-3}	milli	m
10^{-6}	micro	μ
10^{-9}	nano	n
10^{-12}	pico	p
10^{-15}	femto	f
10^{-18}	atto	a

2 Interferometry

The length of standards used in precision measurement must be found to the highest possible accuracy and the technique of interferometry in which light is used as a means of measurement provides the degree of accuracy required for both length measurement and flatness testing. Mercury vapour and sodium lamps designed to radiate light of closely defined wavelengths are used as light sources for these purposes.

Certain aspects of the behaviour of light, sufficient to provide an understanding of absolute length measurement, are outlined in the following paragraphs.

2.1 White Light

Light is a form of energy which can be considered to travel from one place to another in the form of electromagnetic sinusoidal waves. The wavelength λ determines the colour, and the amplitude a determines the intensity of the light. White light is a combination of all the colours of the visible spectrum, red, orange, yellow, green, blue, indigo and violet, each colour band consisting of a group of similar wavelengths. Because white light is a combination of light of all colours and is therefore a combination of all the wavelengths of the visible spectrum, it is not in a form suitable for length measurement by interferometry. To overcome this difficulty a substantially monochromatic light source such as a mercury isotope 198 discharge lamp is used, the light produced having an average wavelength of 0·53(4) μm to an accuracy of about one part in one hundred million.

2.2 Monochromatic Light

A beam of monochromatic light can be considered as an infinite number of rays all having the same wavelength. When the waves are emitted from the lamp they are in phase and the intensity of the light, which increases as the square of the amplitude of the waveform, depends upon the power of the light source. If one half of the waves are displaced a distance equal to one half wavelength, the algebraic sum of the amplitudes, assuming that they are all of equal value, will be zero and any object in the path of light in this condition will not be illuminated. This effect is known as interference. At any intermediate condition the interference effect will be lessened and the light will be of reduced intensity (Fig. 2.1).

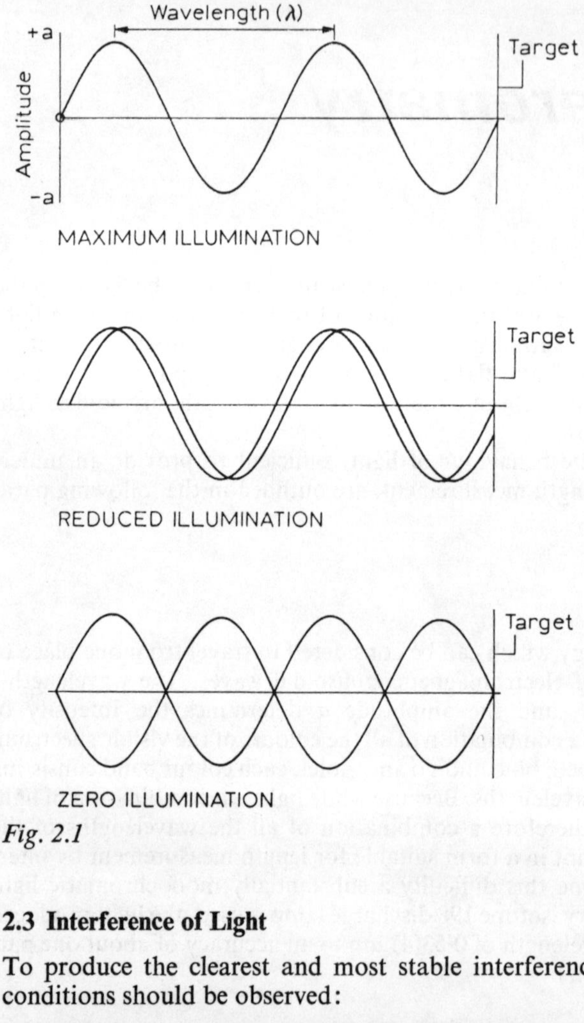

Fig. 2.1

2.3 Interference of Light

To produce the clearest and most stable interference effects the following conditions should be observed:

(1) The light must be of uniform frequency and wavelength.
(2) Any phase differences existing should not vary with time.
(3) The amplitudes should be equal or nearly equal.

A method for producing bands of interference is shown in Fig. 2.2. Light passes through the very narrow slit A and thence through the slits B and C which are close together. Thus two separate beams of light are formed which for the purposes of this explanation may be assumed to be in phase. If the ray paths BO and CO are exactly equal then the waves on these paths will be in phase at the screen and a band of light will occur at point O. At some point M the ray path difference will be equal to one half wavelength, i.e. CM − BM $= \frac{1}{2}\lambda$. Thus, at the screen, the waves will be 180° out of phase and a band of darkness will occur at M, and similarly at N. At point P, the ray path difference will be one wavelength, and the waves again will be in phase, a band of light appearing at point P and similarly at Q. Thus a series of light and dark bands are produced. The dark bands are known as interference fringes.

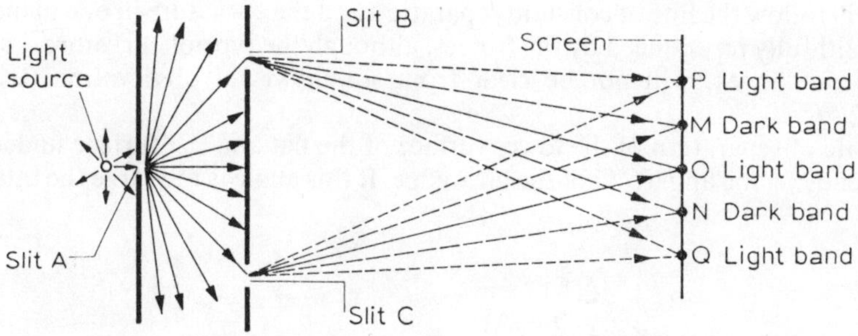

Fig. 2.2 METHOD FOR PRODUCING INTERFERENCE BANDS

2.4 Flatness Testing

Slip gauges are the best known example in engineering requiring surfaces of a very high degree of flatness of the order of 0·25 μm for workshop grades to 0·08 μm for reference grades (see B.S.S. 4311, 1968). The final processes of slip gauge manufacture involve very careful lapping to produce a surface of the flatness required. The surface is highly reflective which makes it suitable for checking by interferometry. The device used to produce ray paths of various lengths is the optical flat.

An optical flat is a circular piece of unstressed glass or quartz usually about 5 cm diameter but which may be 30 cm in diameter, or more. The thickness is sufficient to prevent significant flexure under its own weight. The upper and lower surfaces are ground optically flat so that they become reference planes. Consider an optical flat placed upon a flat metal surface so that a thin wedge film of air is entrapped between them (Fig. 2.3). When suitably illuminated, interference fringes will appear at the metal surface. The behaviour of a ray of light as it passes through the flat to the surface and thence to the observer's eye may be understood as follows. Ray L_1 follows the path AB. At B part of the ray is reflected to follow path BF while the remainder continues along the path BC and is reflected from the metal surface C along the path DE. Both parts of the ray are recombined at the eye, having traversed unequal distances.

If the waves reaching the eye are 180° out of phase a straight dark band or fringe will be seen to lie across the surface. The fringe is straight because the surface is flat as fringes always follow a line of constant distance between the surface and the lower face of the flat. They are, in a sense, contour lines. Straight fringes always lie parallel to the line of contact or line of minimum displacement of the optical flat from the surface.

Referring again to Fig. 2.3, at some point J to the right of C, the gap between the two surfaces has increased such that the difference between the distance traversed by the two portions of ray L_2 has increased by one half wavelength. The waves will then be in phase and a bright band will be seen. A further increase of one half wavelength will produce a second dark band at Q. In this way successive light and dark bands are produced which follow the contour of the surface, each band indicating a path of constant separation between the optical flat and the surface under examination. If the surface is curved then the bands will follow the line of constant separation and the curvature in one plane will be faithfully reproduced by the fringes, although the type of curvature, i.e. concave or convex, will not be clear from simple visual observation (see section 2.5).

The rate of separation of the lower surface of the flat and the surface under test depends on the angle ($\delta\theta$) of the air wedge. If this angle is too large the rate

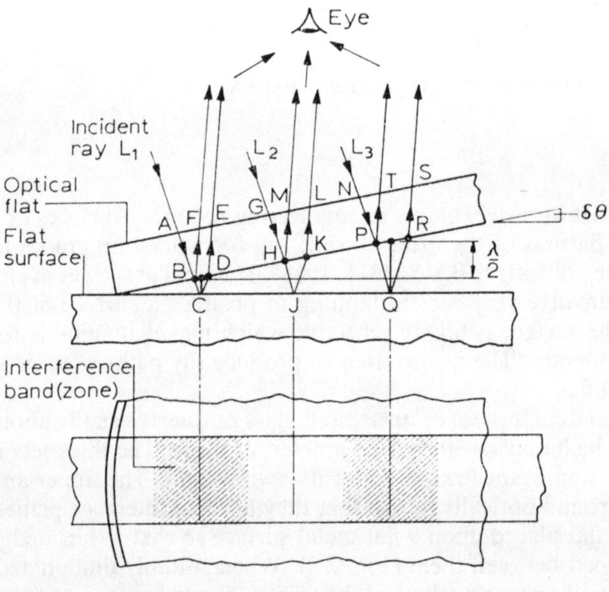

Fig. 2.3 USE OF OPTICAL FLAT

of separation is correspondingly too great and the fringes will be so close together that they cannot be distinguished one from another. If the angle is too small the fringes will be so far apart as not to appear in sufficient numbers or not at all. Thus the pitch of the fringes varies inversely as the angle $\delta\theta$, increasing as $\delta\theta$ decreases and decreasing as it increases. Consequently it may be necessary to perform a number of trial placings with the optical flat before satisfactory results are obtained. Since the path difference from one fringe to the next similar fringe is one whole wavelength, then PQR − BCD = λ. When the angle $\delta\theta$ is very small the vertical displacement of the surface from the points Q and C are seen to be very nearly $\frac{1}{2}$PQR and $\frac{1}{2}$BCD respectively, thus

$$\tfrac{1}{2}\text{PQR} - \tfrac{1}{2}\text{BCD} = \tfrac{1}{2}\lambda$$

The change in elevation between the optical flat and the surface can be calculated by counting the number of fringes and multiplying by the length of one half wavelength of the light used (Fig. 2.4).

Incident light: wavelength $\lambda = 0{\cdot}5\ \mu m$, i.e. $\frac{1}{2}\lambda = 0{\cdot}25\ \mu m$
Number of fringes = 3. Thus, separation $= 3 \times \frac{1}{2}\lambda = 0{\cdot}75\ \mu m$

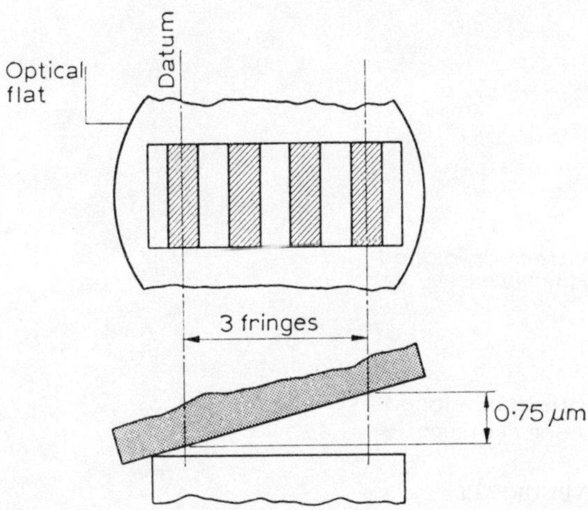

Fig. 2.4

2.5 Contour Testing

Consider a surface which is uniformly convex in one plane, and upon which is mounted an optical flat as shown in Fig. 2.5*a*.

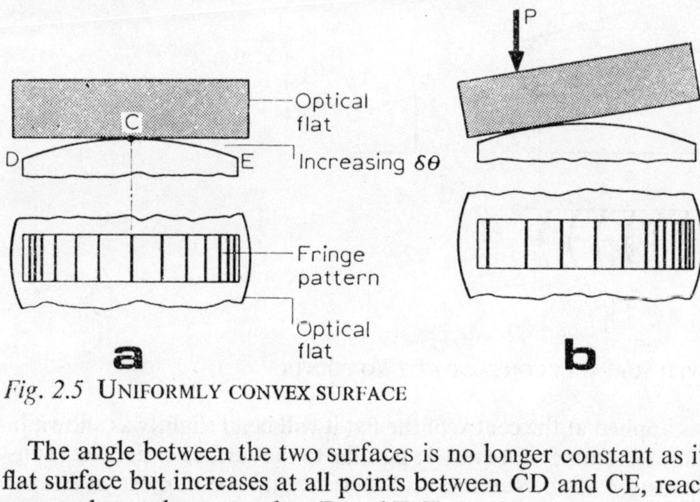

Fig. 2.5 UNIFORMLY CONVEX SURFACE

The angle between the two surfaces is no longer constant as it was for the flat surface but increases at all points between CD and CE, reaching a maximum value at the outer edges D and E. From the previous discussion it will be appreciated that the fringes will be closer together towards the outer edges. To test whether the surface is concave or convex, a light pressure P is applied towards one edge of the flat. If the surface is convex then the flat will roll over the surface to a new line of contact the bands taking up a new configuration similar to that shown in Fig. 2.5*b*.

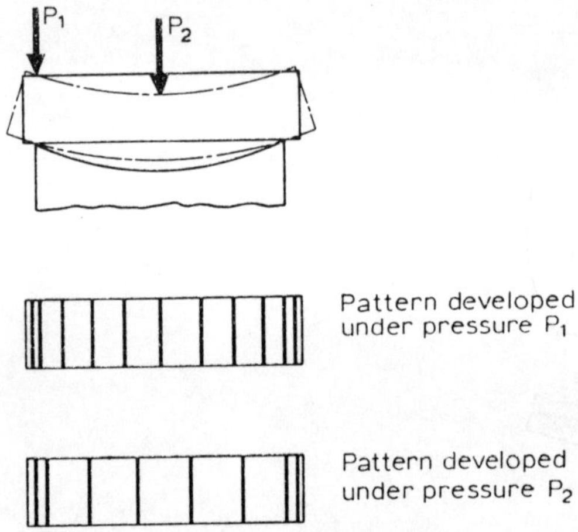

Fig. 2.6 UNIFORMLY CONCAVE SURFACE

The pattern for a uniformly concave surface in one plane is shown in Fig. 2.6.

If a light pressure is applied at point P_1 then little or no change in fringe pattern will occur as there is virtually no distortion of the optical flat. If now

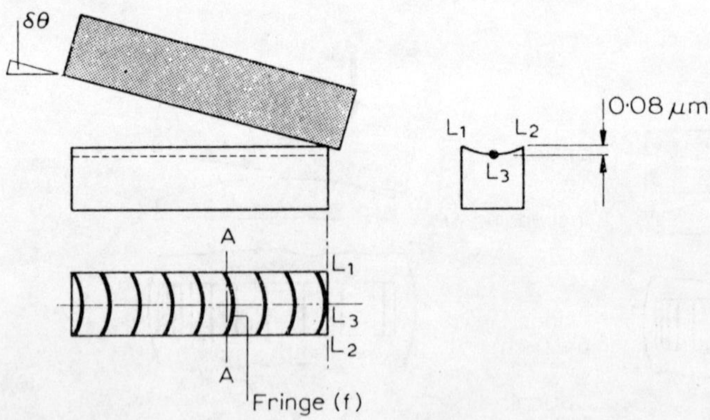

Fig. 2.7 CURVED SURFACE: CONTACT AT TWO POINTS

a pressure P_2 is applied at the centre of the flat it will bend slightly as shown in chain dot outline becoming more nearly parallel to the surface under test. Thus the rate of separation of the surfaces will decrease and the number of fringes appearing will be less numerous. If the surface tested is spherically concave or convex then a "bull's eye" fringe pattern will appear. The same method of testing is used.

Fig. 2.7 shows a curved surface, with contact at points L_1 and L_2. The fringes curve away from a line through the points of contact and, keeping in mind that the fringes are contour lines which indicate that the surface is at a constant distance from the flat along their path, then it will be seen that a concave surface will produce this fringe pattern. A convex surface would reveal fringes curving towards the line or points of contact. The distance from the

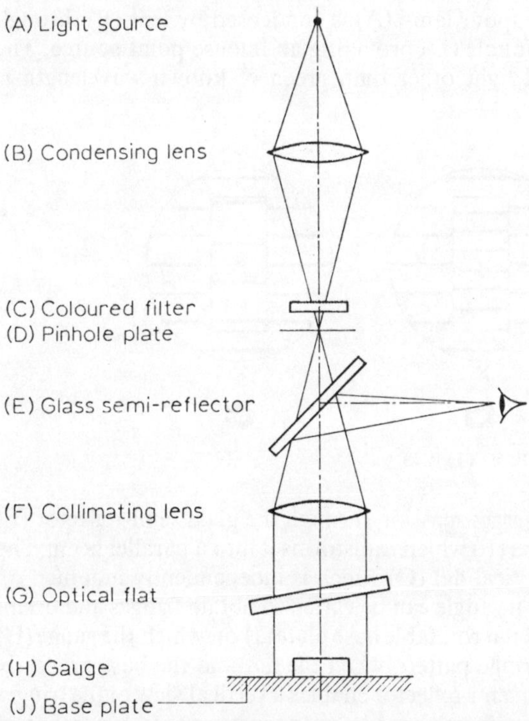

Fig. 2.8 FLATNESS INTERFEROMETER (N.P.L. TYPE): OPTICAL SYSTEM

centre of one of the fringes (f) to the line A-A suggests that its curvature is approximately equal to one third of one fringe pitch, indicating that the curvature is equal to $\frac{1}{3} \times \frac{1}{2}\lambda$ from centre to edge. Thus the centre is approximately 0·08 μm below the edges, assuming a half wavelength equal to 0·25 μm.

2.6 Optical Interferometers

When an optical flat is placed on a surface there is no control over the position of the air wedge, and it is often very difficult to adjust the flat so as to orientate the fringes to the best advantage. Interferometers, while overcoming this difficulty, can also be used for checking the parallelism of opposed gauge faces and the gauge length. There are two basic types of interferometer, the flatness interferometer and the length interferometer. The fundamental difference between them is that the latter incorporates a constant deviation prism which splits the light into a number of parallel beams each having a different and closely defined wavelength of known value.

2.7 The Flatness Interferometer

The optical layout of a flatness interferometer of the N.P.L. type is shown in Fig. 2.8.

Light from the mercury vapour lamp (A) is condensed by lens (B) through the green filter (C) to the pinhole (D) providing an intense point source. The green filter ensures that all light other than green of known wavelength is

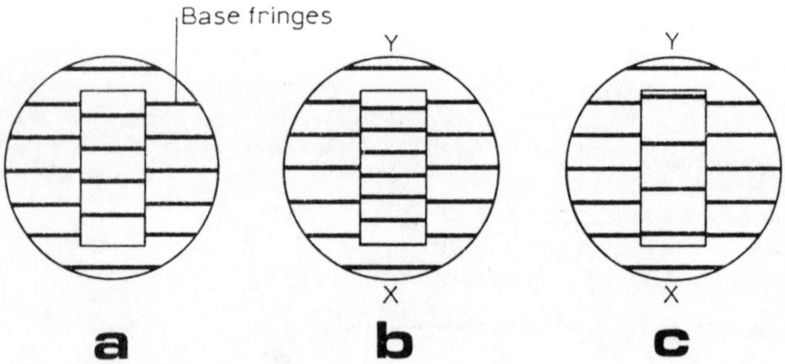

Fig. 2.9 EXAMPLES OF FRINGE PATTERNS

excluded. The light then diverges, passing through the glass semi-reflector (E) and on to the collimating lens (F) which transforms it into a parallel beam. The beam passes through the optical flat (G) which is independently mounted on three adjustable legs so that its angle can be varied to obtain fringes and orientate them at will. The lapped flat rotatable base plate (J) on which the gauge (H) is wrung will also show a fringe pattern when the flat and the base are fairly near to each other. The glass semi-reflector enables a vertical view of the fringes to be obtained while viewing from the side of the instrument. A number of examples of fringe patterns are shown in Fig. 2.9*a* to *c*.

At 2.9*a*, the upper surface of a gauge block which is both flat and parallel to the base is shown. The fringes are straight, parallel to the base fringes, and equi-spaced, and their spacing is the same as the base fringe spacing. Thus the angle $\delta\theta$ is the same for both the gauge and the base. The displacement of the gauge fringes relative to the base fringes is entirely a function of the gauge length and the wavelength of the light used and is of no concern in flatness testing.

The fields of view at 2.9*b* and *c* show a gauge which is flat but not parallel to the base in the direction X-Y. The difference between the number of gauge and base fringes multiplied by $\frac{1}{2}\lambda$ will give the amount of taper present.

Beyond a gauge length of about 25 mm the fringe pattern on the base plate becomes more difficult to observe and the following method may then be used to estimate the degree of taper present.

The number of fringes on the workpiece are counted, the base is rotated through 180° without disturbing the instrument setting, and another fringe count is taken. Half the difference between the two counts multiplied by $\frac{1}{2}\lambda$ will give the amount of taper present (see Fig. 2.10). The need to halve the difference is caused by the rotation of the base through 180° and is similar to the effect on the bubble reading when using the level comparator (see Section 3.12).

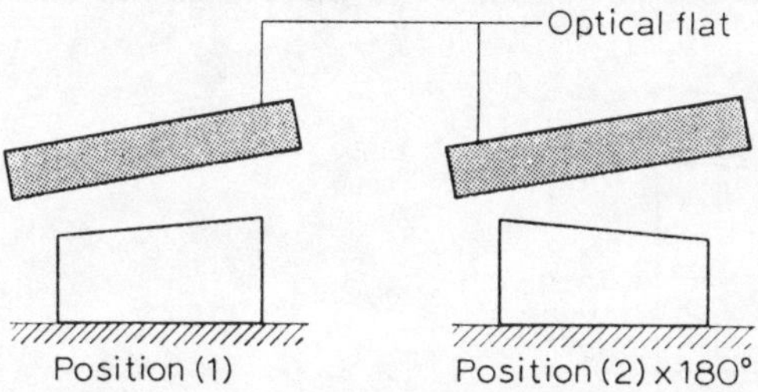

Fig. 2.10

Fringe count position (1) = 5
Fringe count system (2) = 11
Difference ÷ 2 = 3 and $\frac{1}{2}\lambda = 0{\cdot}25\ \mu m$
Gauge taper = $3 \times 0{\cdot}25 = 0{\cdot}75\ \mu m$

At Fig. 2.11*a* a gauge ABCD is shown whose upper surface is flat but makes a compound angle with the base and thus with the flat. If the air wedge is increasing in the direction X-Y then the direction of the taper is indicated by the arrow which is perpendicular to the fringe direction. The fringe spacing on the gauge is less than the base fringe spacing, thus corner D is the lowest point on the gauge. Because the surface is flat, corner B is the highest point. The intermediate points A and C are determined by the fringe count along AB and CD.

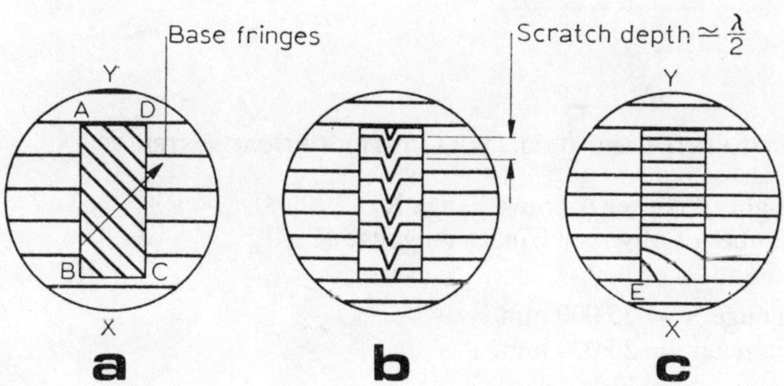

Fig. 2.11 FRINGE PATTERNS FOR A REFERENCE GAUGE

Fringes along AB = 7 Fringes along CD = 3

Thus A is lower than C. The order of descending height from B is C, A, D.

The effect of a scratch on the gauge surface is shown at 2.11*b* and an estimation of its depth can be made. A gauge which is flat but not parallel to the base at one end and curved towards point E is shown at 2.11*c*.

EXAMPLE 2.1

A reference gauge X of known length is wrung to a surface and is in close contact with a cylinder of known size which in turn abuts gauge Y. The measuring surfaces of gauge Y are flat but not parallel and the optical flat makes contact at corner A (Fig. 2.12).

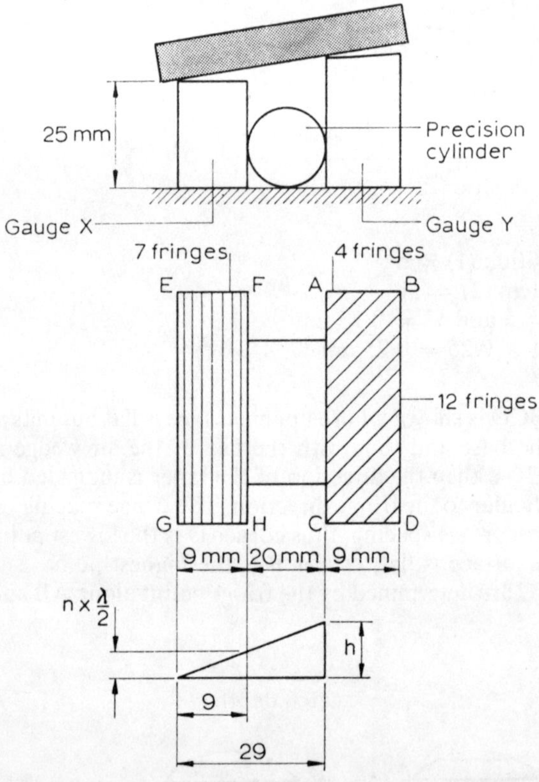

Fig. 2.12 LENGTH INTERFEROMETER (N.P.L. TYPE): OPTICAL SYSTEM

Let h = height of corner A above gauge X
n = number of observed fringes on gauge X
$\lambda = 0{\cdot}5\ \mu m$

Length of gauge X = 25·000 mm.
Cylinder diameter = 20·000 mm.

Thus,

$$\frac{h}{29} = \frac{7 \times \lambda}{2 \times 9}$$

$$h = 5{\cdot}63\ \mu m$$

Height of corner A above reference gauge = $+5{\cdot}63\ \mu m$.

Since there are three fringes less on AB than there are on EF, B is higher than A by

$$(7 - 4)\ \lambda/2 = 3\lambda/2 = +0{\cdot}75\ \mu m$$

Point C is lower than A by

$$(0 - 12)\ \lambda/2 = -6\lambda = -3\ \mu m$$

Point D is higher than C by

$$(7 - 4)\ \lambda/2 = 3\lambda/2 = +0{\cdot}75\ \mu m$$

Therefore, height of D relative to A = $-3 + 0{\cdot}75 = 2{\cdot}25\ \mu m$.

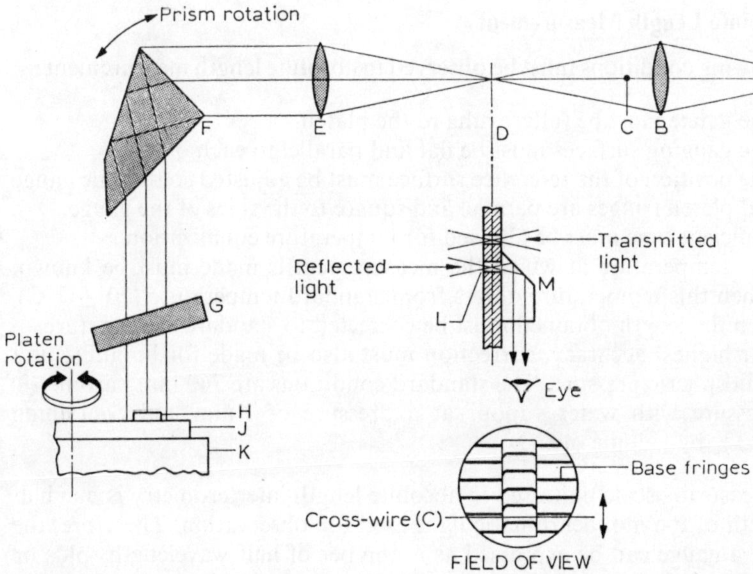

Fig. 2.13

2.8 The N.P.L.-type Length Interferometer

The optical system of this instrument is shown in Fig. 2.13. Light from the hot cathode cadmium lamp (A) is focused by the condensing lens (B) at the plane of the pinhole plate (D) thus giving an intense point source of light. An adjustable cross-wire is located at (C) and its reflected image will be seen in the field of view. This wire is a useful aid for estimating the displacement of the gauge fringes relative to the base fringes (see section 2.9). The collimating lens (E) projects a parallel beam of light on to the constant deviation prism (F) which splits the light into its component colours; these, on leaving the prism, are in the form of separated parallel beams. The prism can be rotated so that beams of different wavelengths (colours) may be brought to bear on the reference plane (optical flat G), the gauge (H), and the rotatable platen (J). The light is reflected through the system by the gauge and platen surfaces and is brought to focus at the plane of the pinhole but slightly displaced to one side. This displacement is caused by the angle of tilt of the reference plane which determines the angle of reflection. A second pinhole (L) allows the light to pass on to a 45° reflecting prism (M) which enables observations to be made from the side of the instrument. For convenience the reflecting prism is shown in plan but in the instrument it lies in the horizontal plane. A further point to note is that focusing at (L) reverses the image; thus the reference plane position in the diagram will produce an image of the gauge as shown. The slight curvature on the gauge edges results from carefully stoning a small radius to give protection from damage.

2.9 Absolute Length Measurement

The following conditions must be observed in absolute length measurement:

(1) The gauge must be fully wrung to the platen.
(2) The gauging surfaces must be flat and parallel to each other.
(3) The position of the reference surface must be adjusted so that the gauge and platen fringes are parallel and square to the sides of the gauge.
(4) Sufficient time must be allowed for temperature equalization.
(5) The temperature at which the measurement is made must be known. When this temperature differs from standard temperature ($20 \pm \frac{1}{2}$°C), then the length obtained must be corrected to standard temperature.
(6) For highest accuracy, correction must also be made for humidity and atmospheric pressure. The standard conditions are 760 mm barometric pressure with water vapour at a pressure of 7 mm and containing 0·03% by volume of CO_2.

The measuring scale unit used in absolute length interferometry is one half wavelength of the monochromatic light used for observation. Therefore, the length of a gauge can be expressed as a number of half wavelengths plus or minus a fraction of one half wavelength. For example, the length of a slip gauge exactly 2 mm long when observed under light of 0·5 μm wavelength is

$$(2 \times 10^3 \times 2)/0{\cdot}5 = 8000 \text{ half wavelengths}$$

where 2×10^3 is the gauge length expressed in micrometers. The same gauge observed under a wavelength of 0·51 μm is approximately 7843·13 half wavelengths long.

A little thought will indicate that when the gauge is an exact number of half wavelength units long the fringes observed on the gauge should be exactly in line with the fringes observed at the platen (Fig. 2.14*a*). It follows that the fringes observed on any gauge will be displaced relative to the platen fringes when the gauge length is any number of half wavelengths plus a fraction and that such displacement will be directly proportional to the fraction. The displacement fraction expressed as a proportion of one platen fringe spacing may be determined as shown at 2.14*b*, where it is seen to be approximately 0·7 when using platen fringe P, as a datum.

Although the nominal length of a gauge can be expressed in terms of half wavelength units, its true length cannot be determined from observation under one beam of monochromatic light alone.

If a gauge of a given nominal length, say 2 mm, is exactly 2 mm long then the half wavelength fractions successively observed when the gauge is viewed under a number of monochromatic beams will be equal to the calculated fractions in each case, but, if the observed fractions differ from the calculated fractions, an error in gauge length is indicated. The fractions observed, taken as a group, can be assumed to be unique to the length of the gauge under observation, that is, no other gauge error can produce the same group of fractional values within

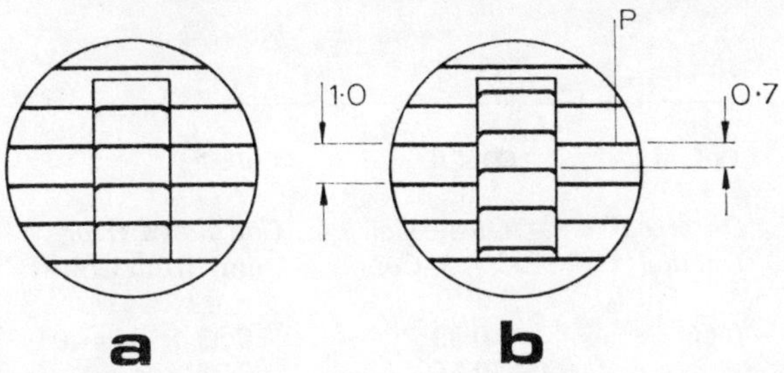

Fig. 2.14

the region of the expected error in gauge length. This is a valid assumption provided that a sufficient number of observations are made (three or four), as absolute length interferometry is concerned only with small deviations from nominal length and under these circumstances the probability of more than one result from this number of readings is virtually zero.

The method outlined in thc following example demonstrates how an error in length may be determined when measurement is made using a number of different scales. For the purpose of the example the error is assumed, as the "observed" fraction cannot otherwise be stated.

EXAMPLE 2.2

Nominal length = 31 mm
Actual length = 32 mm
Error = +1 mm

Measuring scales

Scale (1)	3 mm units
Scale (2)	4 mm units
Scale (3)	5 mm units
Scale (4)	7 mm units

Nominal Length		*Scale fraction*
Scale (1)	31/3 = 10·33	0·33
Scale (2)	31/4 = 7·75	0·75
Scale (3)	31/5 = 6·20	0·20
Scale (4)	31/7 = 4·43	0·43

	Actual Length	*"Observed" fraction*
Scale (1)	32/3 = 10·66	0·66
Scale (2)	32/4 = 8·00	0·00
Scale (3)	32/5 = 6·40	0·40
Scale (4)	32/7 = 4·57	0·57

The difference between the two groups of fractions when added algebraically to the nominal length yields the actual length.

The evaluation of the error from the calculated and the "observed" fractions is shown in Table 2.1. The results obtained from scales 1, 3 and 4 shown in

TABLE 2.1

COL. 1	COL. 2	COL. 3	COL. 4	COL. 5
Scale	*Calculated fraction*	*"Observed" fraction*	*Obs.—Cal. Col. 3–Col. 2*	*Calculated error* (mm) (from Col. 4)
1	0·33	0·66	0·33	0·33 × 3 = +1
			−0·75	−0·75 × 4 = −3
2	0·75	0·00	or	or
			$\bar{1}$·25	0·25 × 4 = +1
3	0·20	0·40	0·20	0·20 × 5 = +1
4	0·43	0·57	0·14	0·14 × 7 = +1

column 5 directly indicate a +1 mm error. The coincidence of these results alone is sufficient to confirm the error, but an analysis of the result from scale 2 (column 4), also yields a coincident value (Fig. 2.15).

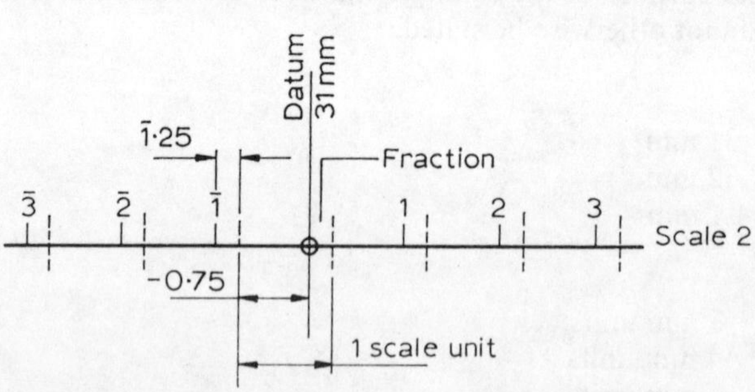

Fig. 2.15

By adding one scale 2 unit, the position of the fraction relative to the datum is seen to coincide with the positions obtained from scales 1, 3 and 4, i.e. its position is inferred from the other results. A diagram of the scales and the fraction coincidences is shown in Fig. 2.16.

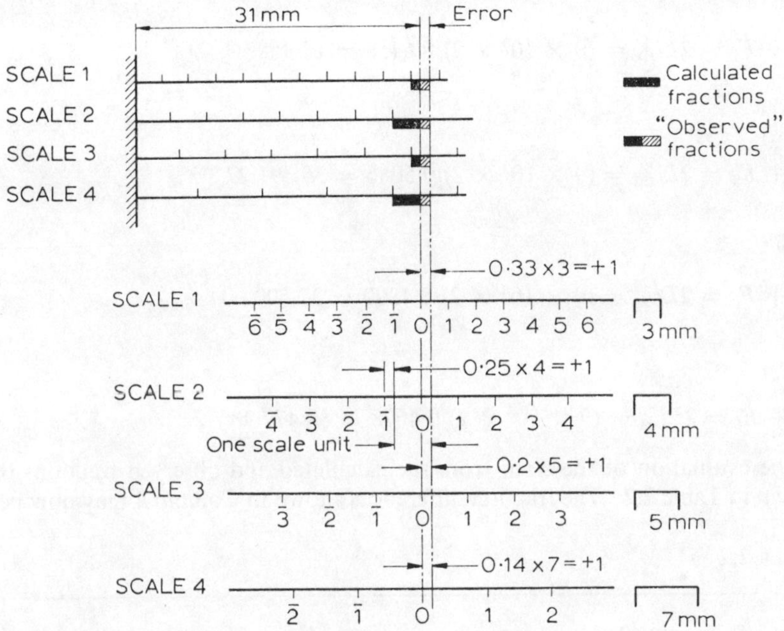

Fig. 2.16

EXAMPLE 2.3

This example illustrates the measurement of a 9 mm block gauge using the N.P.L. type interferometer shown in Fig. 2.13. In practice the calculated fraction would be obtained from a table, but for the purpose of this example the fractions are obtained by calculation.

Light source is a cadmium–mercury discharge lamp.

Colour	*Radiation wavelength* (μm)	*Observed fraction*
Red (λ_1)	0·6438	0·0
Green (λ_2)	0·5086	0·5
Blue (λ_3)	0·4800	0·3
Violet (λ_4)	0·4678	0·4
Nominal slip gauge length = 9 mm		

Nominal gauge length in terms of $\frac{1}{2}$ wavelength units

Let N = number of half wavelengths, F = fractional half wavelengths, and L = nominal gauge length (μm).

Red

$$N_1 + F_1 = 2L/\lambda_1 = (9 \times 10^3 \times 2)/0{\cdot}6438 = 17\,958{\cdot}\underline{99} \quad \lambda_1$$

Green

$$N_2 + F_2 = 2L/\lambda_2 = (9 \times 10^3 \times 2)/0{\cdot}5086 = 35\,391{\cdot}\underline{27} \quad \lambda_2$$

Blue

$$N_3 + F_3 = 2L/\lambda_3 = (9 \times 10^3 \times 2)/0{\cdot}4800 = 37\,500{\cdot}\underline{00} \quad \lambda_3$$

Violet

$$N_4 + F_4 = 2L/\lambda_4 = (9 \times 10^3 \times 2)/0{\cdot}4678 = 38\,477{\cdot}\underline{98} \quad \lambda_4$$

The evaluation of the error from the calculated and observed fractions is shown in Table 2.2. The fraction differences shown in Column 4 may now be

TABLE 2.2

COL. 1	COL. 2	COL. 3	COL. 4	COL. 5
W.L.	*Calculated fraction*	*Observed fraction*	*Fraction difference Col. 3–Col. 2*	*Calculated error* (μm)
1	0·99	0·0	−0·99 or $\bar{1}$·01	$1{\cdot}01 \times \frac{1}{2}(0{\cdot}6438) = 0{\cdot}325$
2	0·27	0·5	+0·23	$1{\cdot}23 \times \frac{1}{2}(0{\cdot}5086) = 0{\cdot}313$
3	0·00	0·3	+0·3	$1{\cdot}3 \times \frac{1}{2}(0{\cdot}48) = 0{\cdot}312$
4	0·98	0·4	−0·58 or $\bar{1}$·42	$1{\cdot}42 \times \frac{1}{2}(0{\cdot}4678) = 0{\cdot}332$
				Mean error = 0·32

tested for coincidence on the half wavelength rule designed for this purpose; a simplified diagram of such a rule is shown in Fig. 2.17.

Only the positive part of the fractions need be considered as the negative ($\bar{1}$) part represents one half wavelength unit. A first attempt at finding coincidence is shown under A (Fig. 2.17). In these positions the fraction spread is about 0·1 μm. To test for a closer grouping one half wavelength is added (or subtracted) to each of the fractions. In the case under consideration the addition of one half wavelength is found to provide almost exact coincidence as shown under B. Further additions or subtractions as shown at C and D indicate that the fraction divergence increases. Thus the readings at B represent the true error of about 0·32 μm. The average error is shown in Column 5, Table 2.2.

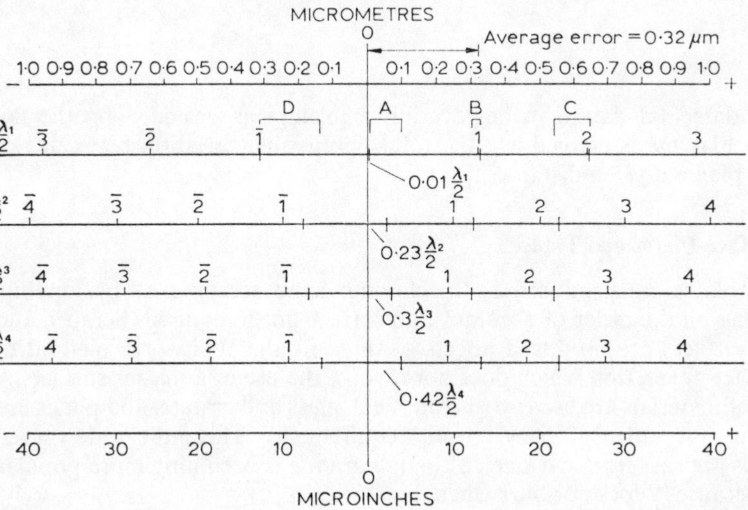

Fig. 2.17

3 Dimensional and Angular Measurement

The fundamental datum in engineering metrology is provided by the flat surface. Flatness is defined as "the total permissible variation between two parallel planes a given distance apart."

3.1 Surface Plates and Tables

Surface plates are produced by machining, hand scraping or by lapping, depending on the order of accuracy and surface finish required. Scraped and lapped surfaces are produced in sets of three by the Whitworth method of flat surface generation which does not involve the use of a master surface. A variety of materials are used; cast iron, steel, glass and granite, and plates are available in a range of sizes to suit requirements. The most widely used materials are cast iron and steel, although granite is becoming more popular partly because it does not burr when damaged.

Steel plates of uniform thickness up to 200 mm diameter are called toolmakers flats or platens. They allow slip gauges to be wrung to their surface and exhibit straight interference fringes when viewed under an optical flat. Lapped cast iron plates are also available up to 450 mm diameter (see B.S. 869: 1959).

Square or rectangular hand-scraped cast iron surface plates are flat to a few micrometers; they are heavily ribbed and deep in section to reduce deflections to a minimum. The same order of accuracy is available in granite plates which are square or rectangular and of uniform thickness. For marking out purposes, etc., planed surface plates and tables are used as the order of accuracy required is less. The present standard for scraped and planed surface plates is B.S. 817: 1957 which provides two grades of accuracy, A and B, for scraped and planed surfaces respectively. This standard is soon to be replaced.

3.2 Straight Edges

The straight edge is the material standard which represents the straight line and the following three types are available:

(*a*) Steel straight edges of deep and narrow rectangular cross-section in a range of lengths up to 6 feet (approximately 2 m). Both narrow faces are straight and parallel to each other to a close tolerance. Two grades A and B are specified in B.S. 863:1939, grade A carrying closer tolerances than grade B.

(*b*) Cast iron straight edges which are in effect long narrow surface plates. They are made in the form of a deep I-section or are strengthened by a deep parabolic web so that resistance to bending along the edge is proportional to the bending moment due to the self-weight of the material. These are manufactured in lengths up to 5 m for use in checking the straightness of long narrow machine slideways, etc. Full details of cast iron edges in B.S. 818:1963.

(*c*) High precision toolmakers' straight edges up to about 200 mm in length which have a "knife" edge of small radius and are finished by grinding and lapping. Such edges are straight to about 1 μm. For further details see B.S. 852:1939. Metric straight edge standards have as yet not been published (1973).

Straight edges three feet long or longer when used edge downwards are subjected to significant deflection owing to their own self-weight. For example, a rectangular edge 4 in × $\frac{1}{2}$ in × 6 ft long when supported at its ends sags at

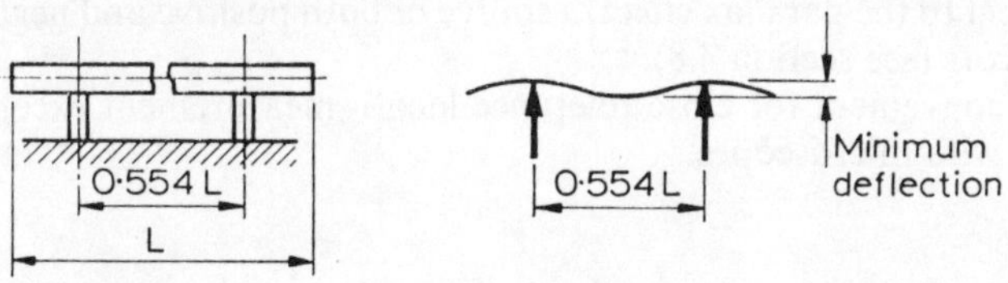

Fig. 3.1 SUPPORTING A UNIFORM CROSS-SECTION BEAM FOR MINIMUM DEFLECTION

its centre approximately 0·0024 in (0.061 mm). A surface having the same curvature would appear to be straight if the edge were placed directly upon it as the edge would follow the contour. The deflection can be reduced in this case to a maximum of 0·000 05 in (0·0013 mm) by supporting the edge at two points equidistant from each end and 0·554L apart where L is the length of the beam (Fig. 3.1).

3.3 The Material Standards of Length

The material standards used for length measurement are available in two forms:

(*a*) Line comparison standards (rules, tapes).
(*b*) End comparison standards (gauge blocks or slip gauges and length or end bars).

Rules may be used for direct measurement of length or indirectly when, for instance, a length is first transferred to dividers and then measured. Similarly, gauge blocks may be used directly to measure the width of a slot, or indirectly when an instrument reading is first standardized using gauge blocks, and the reading over the measured part is then compared with the standardized reading.

3.4 Line Measurement and End Measurement

A summary of the characteristics of line and end standards indicates that line standards do not readily provide high accuracy while end standards are more suited to accuracy requirements of a high order.

Line Standards

(1) Scales can be accurately engraved but it is difficult to take full advantage of this accuracy, e.g. a steel rule can be read to about ±0·2 mm of true dimension.
(2) A scale is quick and easy to use over a wide range since only one is required.
(3) The scale markings are not subject to wear although significant wear on the leading end leads to "undersizing", i.e. a bias towards negative component error.
(4) A scale does not possess a "built in" datum which would allow easy scale alignment with the axis of measurement, this again leads to "undersizing."
(5) Scales are subject to the parallax effect, a source of both positive and negative reading errors (see section 3.8).
(6) Scales are not convenient for close tolerance length measurement except in conjunction with microscopes.

End Standards

These consist of sets of standard blocks or bars and are used to build up a required length.

(1) End standards are highly accurate and are well suited to measurements of close tolerance.
(2) They are time consuming in use and provide only one dimension at a time.
(3) Dimensional variations as small as 0·000 5 mm can be obtained.
(4) They are subject to wear on their measuring faces.
(5) Groups of blocks are "wrung" together to provide a given size, faulty wringing leads to damage.
(6) End standards have a "built in" datum because their measuring faces are flat and parallel and can be positively located on a datum surface.
(7) They are not subject to the parallax effect as their use depends on "feel."

The accuracy of end and line standards is affected by temperature change and both are originally calibrated at $20 \pm \frac{1}{2}$°C. Also, care is taken in manufacture to ensure that change of shape with time, secular change, is reduced to negligible proportions.

3.5 Comparison of Line and End Standards

With the abolition of the Imperial Standard Yard as the British standard of length, there is now no national line standard. Precision scales, however, are still widely used and techniques have been devised for comparing line standards with end standards, and thus indirectly with the ultimate standard of length, light waves.

Precision scales are often made of 58% nickel–iron alloy which has the same coefficient of linear expansion as steel and is highly corrosion resistant. Equally effective but cheaper scales are made of chromium plated mild steel.

3.6 Gauge Blocks and Length Bars

Gauge blocks and length bars provide the material standards of length for industry. Their existence makes possible the manufacture of interchangeable components as they are made to such a high degree of accuracy and stability that, provided they are used correctly, their stated nominal length is accepted as their true length for most purposes.

Gauge Blocks (Slip Gauges)

Metric gauge blocks are through-hardened high carbon steel rectangular blocks. They are 9 mm × 30 mm in sizes up to 10 mm length and 9 mm × 35 mm in sizes greater than 10 mm length. Their lengths range up to 100 mm and they are available in five grades, 00, calibration, 0, I and II.

GRADE 00 This grade is accompanied by a calibration chart which lists the departure from nominal size of each gauge in the set. They are used as standards by gauge block manufacturers and in industries where work of a similar nature is produced and by the National Physical Laboratory and its associated organizations; there is thus little call for such gauges as their grade of accuracy is beyond normal requirements.

CALIBRATION GRADE This grade provides the highest level of accuracy required in normal engineering practice and is intended for calibrating other blocks in conjunction with suitably accurate comparators. They are used where tolerances are 2 μm or less, and are not intended for general gauge inspection.

The essential difference between 00 and calibration gauges lies in their length tolerance, the latter having broader tolerances than 00 gauges whereas the geometrical tolerances, that is flatness and parallelism of measuring faces, are the same in both cases. The values shown in Table 3.1 illustrate the tolerances of a selection from the two grades.

A calibration chart is supplied with the gauges. Thus for nearly all length determination requirements, calibration grade gauges are of sufficient accuracy provided that true gauge length is known. Because of their broader length tolerances calibration gauges are cheaper than 00 gauges.

GRADES 0, I AND II These grades are intended for general use without reference to their calibrated values, grade 0 providing the inspection grade for gauges and work of high precision. Grade I is the general purpose manufacturing gauge used for measurement in gauge, tool and component production, while grade II may be used for rough setting purposes and for checking components having fairly wide tolerances. The values shown in Table 3.1 illustrate the tolerance difference of a selection from the three grades.

The length tolerances of grades I and II are biased towards positive values to allow for wear on the measuring faces which will occur particularly during the early period of use. Grades 00, calibration and 0 have tolerances equally disposed about the nominal size suggesting that they are not expected to suffer

TABLE 3.1 *Block Gauge Tolerances*
Tolerance unit = $0{\cdot}01\ \mu m$

GAUGE LENGTH (mm)		GAUGE TOLERANCES			
Over	*Up to and inc.*	*Flatness*	*Parallelism*	*Length*	*Grade*
—	20	5	5	± 5	00
—	20	5	5	± 25	CAL
80	100	5	10	± 15	00
80	100	5	10	± 50	CAL
—	20	10	10	± 10	0
80	100	10	15	± 25	0
—	20	15	20	+ 20 − 15	0 0
80	100	15	25	+ 60 − 30	I I
—	20	25	35	+ 50 − 25	I I
80	100	25	35	+ 140 − 100	II II

For full details of gauge block tolerances see B.S. 4311:1968, *Metric gauge blocks.*

wear to any significant degree. Grade 0 has closer length tolerances than the calibration grade as they are not intended for use with a list of their calibrated values.

Length Bars

Metric length bars are steel cylinders approximately 22 mm in diameter and through-hardened or hardened only at their ends, depending on their length. They range in size up to 1·2 m and are available in four grades, Reference, Calibration, Inspection and Workshop.

They are manufactured to the same high quality with regard to their length, flatness and parallelism of their measuring faces as block gauges. The bars take one of two forms depending on their grade. Reference and Calibration bars have plain ends whereas Inspection and Workshop grades have screwed holes at one or both ends and can be fixed together by means of loose fitting studs. Plain 25 mm bars may be wrung on if required.

Reference grade This grade is sometimes used in laboratories or standards rooms under temperature controlled conditions (20 ± ½°C) when length determination of the very highest order of accuracy is required. They are used only in conjunction with the most sensitive comparators, e.g. Brookes level comparator. Their function is similar to that of 00 block gauges.

Calibration grade This grade has somewhat larger length tolerances than the reference grade, but is the same in all other respects. Because their length tolerances are larger, calibration bars are cheaper than reference bars and their function is more broadly based, i.e. they may provide the standard of length required in a factory against which inspection grade bars may be checked from time to time using suitably sensitive comparators. Strict regard for temperature control is required if full advantage of their accuracy is to be taken.

Both reference and calibration bars are supplied with an N.P.L. certificate which lists the difference from nominal size of each gauge in the set, thus the larger length tolerance on calibration bars is no disadvantage.

Inspection grade These bars have virtually the same length tolerance as the calibration grade but are biased towards positive values to combat the effect of wear. The flatness tolerances are rather larger than those of the calibration grade and the parallelism tolerance is about the same or larger. They are intended for use in inspection and tool rooms.

Workshop grade This grade is intended for accurate length measurement in the workshop in built-up lengths and in conjunction with block gauges. All tolerances are larger than the inspection grade and the length tolerance is also biased to combat wear. Length tolerances are determined as shown in Table 3.2, suitable rounding off being applied where required.

Table 3.2 *The Basis for Length Bar Tolerance*

Length bar grade	*Part per million*	*Minimum tolerance* (0·01 μm)
Reference	± 2	± 8
Calibration	± 5	±15
Inspection	+ 7 − 3	±18
Workshop	+10 − 5	±30

Supporting Length Bars

Length bars may be used as standards in both the horizontal and vertical positions. When vertically mounted a tiny reduction in length occurs owing to the gravitational effect on the mass of the bar, but no correction for this is applied in assessing the length of the bar and parallelism and flatness of the measuring faces is not affected.

A different situation arises when a bar is used as a horizontal beam as it is then subject to deflection over its length, because however it is supported, a deflection is always present. The solution to the problem lies in supporting the beam in such a manner that, although the beam is still subject to deflection, the slope of the beam at its ends is zero. This being the case, the end faces must be parallel. To ensure this condition the beam is simply supported at two points equidistant from each end (the Airy points), and 0·577L apart where L is the length of the beam. In Fig. 3.2 the Airy positions are indicated on all grades of length bars 250 mm in length and over. When a number of length bars are wrung together the position of the Airy points for the composite bar must be calculated. The shortening caused by the sag in the bar is negligible.

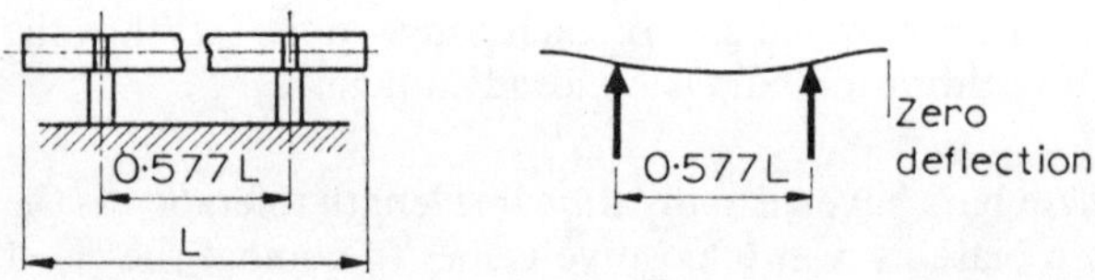

Fig. 3.2 AIRY POINT SUPPORT PROVIDING PARALLEL END FACES

Length Tolerance Accumulation

The length tolerances* on both blocks and bars increase with increase in length, e.g.

Grade	*Length* 25 mm	100 mm	1000 mm
00 block	±8	±15	—
Reference bar	±8	±20	±200

Tolerance unit 0·01 μm

* For full details of length bar tolerances see B.S. 1790:1961, *Length bars and their accessories*.

The increase in manufacturing tolerance ensures as far as possible that any group of gauges built to a given size shall carry the same total tolerance as any other group from the same set. In practice the smallest total tolerance is obtained for a given size by using the smallest possible number of gauges in combination, thus a 175 mm inspection grade end bar has a total tolerance of 1·7 μm and the same length built from a combination of three bars (100 + 50 + 25) has a tolerance of 1·89 μm, an increase of 11%.

An analysis of gauge block tolerances presents a similar picture. Selecting from a grade 0, M88/2 set, i.e. an 88 piece set of 2 mm base size, a minimum of four gauges is required to build the length 45·061 mm and a minimum of five from the remainder to build the same length. In the first case the total tolerance s 0·9 μm, and in the second 1·10 μm, an increase of 22%. The smallest measuring step for this set is 0·001 mm (1 μm) which is about twice the upper or lower accumulated tolerance limit of the gauges in question. In certain circumstances slip gauge tolerance accumulation introduces small errors in measurement, for example, in sine bar settings. When circumstances warrant it, calibrated gauges should be used.

Wringing

The ability to build up accurate lengths using gauge blocks and length bars depends on the flatness of their measuring faces. The final stages of manufacture involve lapping the faces to such a high degree of flatness that the act of rubbing two faces together causes adhesion. The faces must be cleaned with a solvent such as carbon tetrachloride, and wiped with a soft chamois leather or cloth. Placing two block gauges together to form a cross, and then rotating them under light pressure until they are parallel, provides a bond involving the least rubbing between the faces, and also tends to sweep away any tiny particles which may still adhere to the surfaces. Gauges should always be separated by sliding and never by breaking the bond. A very thin liquid film is left on surfaces cleaned with solvents, and it is intermolecular attraction between the two surfaces and the film which provides the main strength of the bond, atmospheric pressure playing a secondary part. The strength of the bond increases with time and gauges should not be left wrung together for more than a few hours as otherwise they may be damaged when separated. The wringing interval between correctly wrung gauges has a thickness of about 5 nm (5×10^{-9} m) but this may be as large as 1 μm if wringing is not fully effective; thus for maximum accuracy careful wringing is essential. The stated length of block gauges and length bars includes one wringing interval.

Damage Prevention

To prevent loss of accuracy arising from incidental and accidental damage the following precautions should be adhered to as closely as possible:

- (*a*) Protect from and remove dust, dirt, moisture.
- (*b*) Avoid magnetization.
- (*c*) Handle lapped faces as little as possible to prevent etching from finger acid. Remove all finger marks as soon as possible.
- (*d*) Wipe faces immediately before use even when use is continuous.
- (*e*) Always replace clean gauges in their box and close the lid after use. If gauges are not in frequent use they should be coated to prevent corrosion.
- (*f*) Do not handle gauges above the open box, they may cause damage to other gauges if dropped.
- (*g*) Gauges discarded for a moment during use should be laid on a soft cloth.

Building a Known Length

Gauge blocks are normally supplied in sets ranging from 32 pieces to 112 pieces at suitable intervals and in two base sizes, 1 mm and 2 mm. A 1 mm base auxiliary set comprising 27 pieces is also available. Grades 0, I and II may include two 2 mm protector blocks made of steel or tungsten carbide so that wear is confined to their faces alone. For purposes of illustration an M88/2 gauge set is used comprising the following gauges:

Size (mm)	*Increment* (mm)	*Number of Pieces*
1·005	—	1
2·001– 2·009	0·001	9
2·01 – 2·49	0·01	49
0·5 – 9·5	0·5	19
10 –100	10·0	10
	Total	88

The size to be built is 45·061 mm and two lengths are required using the minimum number of blocks in each case. Each increment of size is successively eliminated beginning with the smallest, i.e. 0·001 mm.

45·061		45·061	
−2·001	(1)	−1·005	(1)
43·060		44·056	
−2·06	(2)	−2·006	(2)
41·000		42·050	
−1	(3)	−2·05	(3)
40·000		40·000	
−40	(4)	−10	(4)
00·000		30·000	
		−30	(5)
		00·000	
4 gauges required		5 gauges required	

Finding an Unknown Length

Gauge blocks are sometimes used to find an unknown length and the following approach which involves the systematic division of an assigned tolerance is to be recommended. For the purposes of explanation it is necessary to assume the unknown length although in practice it will only be known approximately. The length to be found, 44·92 mm, is the distance between two parallel faces.

The length is estimated using rule, calipers, etc., as nearly as possible assigning a suitable zone within which the size must certainly lie, e.g. 45 ± 0·4 mm, and by trial and error converge on to the final size as follows:

Trial Length (mm)	*Result*	*Action*
45	NOT GO	Try 44·8
44·8	GO	Try 44·9
44·9	GO	Try 44·95
44·95	NOT GO	Try 44·92
44·92	GO	Try 44·93
44·93	NOT GO	Accept 44·92

The final size is seen to lie within the range 44·92 and 44·93 mm. The same process of elimination applied to the 0·001 mm value will provide a more precise measurement.

3.7 Reference Discs

Block gauges are not always suitable or convenient to use as standards when setting measuring instruments, particularly when the work is manufactured on centres and is intended to be measured using the centres as a datum. Standard

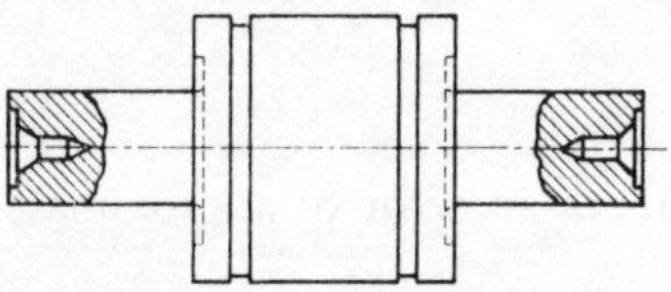

Fig. 3.3 PRECISION REFERENCE DISC

sets of reference discs are supplied with a calibration certificate which lists the differences from nominal size of each disc in the set. Thus it is a simple matter to standardize such instruments as the floating carriage diameter measuring machine. A typical set consists of 22 discs ranging in size from three to one hundred millimetres in diameter having a calibrated size determined to within a few ten millionths of a metre. For example, a disc of 30 mm nominal size has a calibrated size determined by the manufacturers of 29·9958 mm to an accuracy of ±0·0005 mm (±0·5 μm). This determination applies to that part of the disc lying between two circumferential bands marked on the disc. The purpose of this limitation is to reduce the possibility of error arising from taper on the disc (Fig. 3.3).

3.8 Sources of Error in Precision Measurement

Failure to consider the following factors may introduce errors in measurement:

(*a*) The alignment principle.
(*b*) Location of the measured part.
(*c*) Temperature.
(*d*) Parallax.

The Alignment Principle

Abbé's principle of alignment states that

"The axis or line of measurement of the measured part should coincide with the measuring scale or the axis of measurement of the measuring instrument."

The effect of simple scale alignment error is shown in Fig. 3.4.

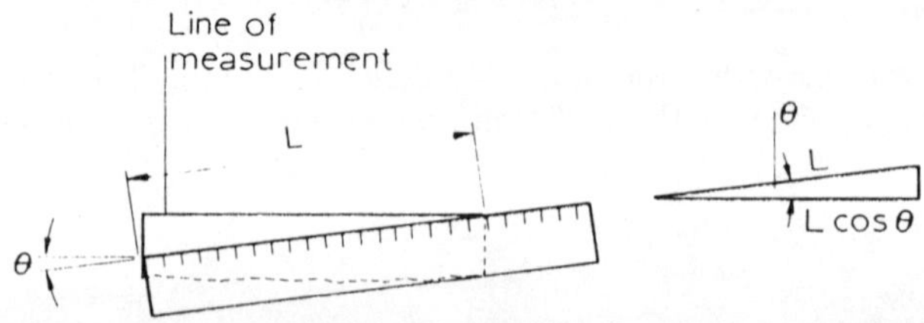

Fig. 3.4 SCALE MISALIGNMENT

If θ = angle of scale misalignment and L = apparent length,

$L \cos \theta$ = true length

If e = induced error, then

$e = L - L \cos \theta = L(1 - \cos \theta)$

An alignment error of 2° over 1 m introduces an error of approximately 0·6 mm.

Error is introduced to dial indicator readings if the plunger axis does not coincide with the axis or line of measurement (Fig. 3.5*a*).

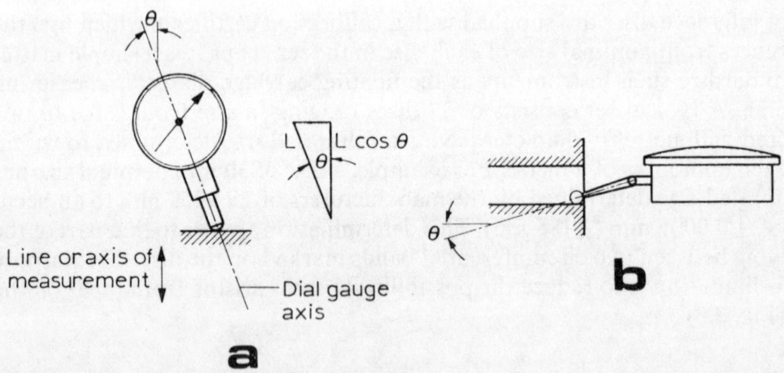

Fig. 3.5 AXIS MISALIGNMENT

If e = induced error and L = change in indicator reading,

$L \cos \theta$ = surface displacement

$e = L(1 - \cos \theta)$

To ensure correct displacement readings on the dial indicator the plunger must, of course, be normal to the surface in both mutually perpendicular planes.

Some dial indicators are fitted with a ball-ended stylus. The stylus arm is connected to the dial gauge operating mechanism in such a way that whatever the angle of the arm the gauge records the vertical displacement of the ball end (Fig. 3.5*b*).

The measuring jaws of the vernier caliper are in effect extensions of the scale markings on the instrument; thus the jaws of an instrument in good condition are parallel and remain so at any measurement within the instrument range. The length L indicated on the scale corresponding to the displacement of the jaws from each other (Fig. 3.6*a*). The effect of a bent scale beam is shown at 3.6*b*. The length L_1 between the extremes of the jaws has now become smaller

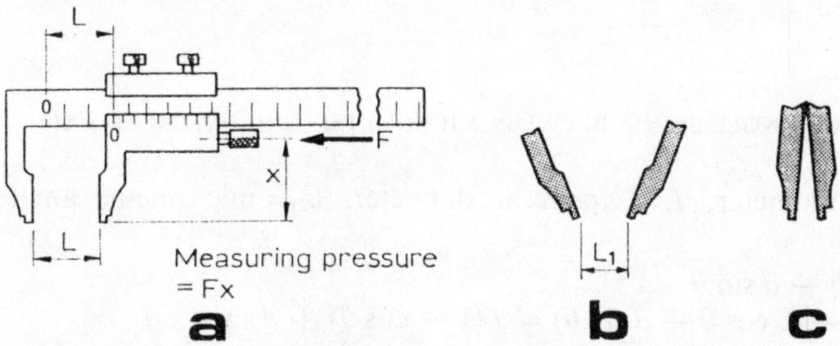

Fig. 3.6 ERRORS FROM MEASURING PRESSURE

than the length indicated on the scale. It is also evident that the error introduced is proportional to the length of the jaws and is at its minimum along the line of the scale.

A second source of error well illustrated by the vernier caliper and similar instruments or circumstances is associated with measuring pressure or "feel." The measuring pressure is applied by the adjusting screw which is adjacent and parallel to the scale. A bending moment is introduced equal to the product of the force applied by the adjusting screw and the perpendicular distance between the screw centre line and the line of measurement (Fig. 3.6*a*). Variations of force applied at the screw are augmented at the line of measurement and a not unusual form of damage to vernier calipers is permanent distortion of the measuring jaws presumably from this source (3.6*c*).

The micrometer is an instrument which fully conforms to Abbé's principle as it is clear that the scale merely records the translation of the measuring screw along its axis which coincides with the line of measurement. The measuring pressure is applied along the screw axis and the torque introduced is small, thus although variations in measuring pressure can and do occur the "feel" of the instrument is much easier to judge.

Errors in Measuring Diameter

Cylinders are often measured between centres on a diameter measuring machine (floating micrometer) and it is essential that the micrometer spindle and anvil measuring surfaces be parallel to the cylinder axis and the micrometer axis perpendicular to the cylinder axis. Fig. 3.7 illustrates the effect of angle error between the micrometer and cylinder axis.

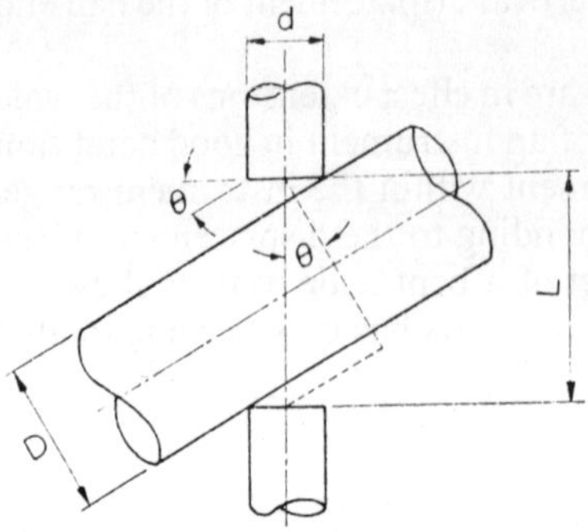

Fig. 3.7 EFFECT OF ANGLE ERROR BETWEEN MICROMETER AND CYLINDER AXIS

If D = true diameter, L = apparent diameter, d = micrometer anvil diameter,

$$D = L\cos\theta - d\sin\theta$$

$$\text{Error} = L - (L\cos\theta - d\sin\theta) = L(1 - \cos\theta) + d\sin\theta$$

When θ is small, $L \simeq L\cos\theta$ and $\sin\theta \simeq \theta$ radians, therefore

$$\text{Error} = d\theta$$

The reading error introduced by 1 minute angle error when $d = 6$ mm is approximately 1·7 μm. Floating micrometers are set by using calibrated

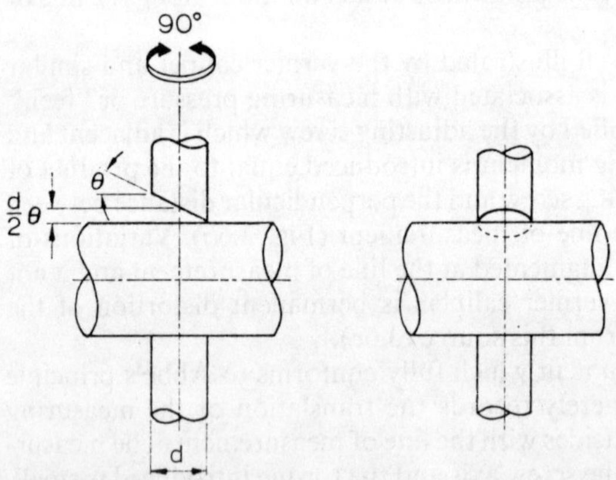

Fig. 3.8 EFFECT OF THE MICROMETER MEASURING FACE NOT BEING SQUARE TO SPINDLE AXIS

discs, thus the effect of this error source is eliminated. It is particularly important that the measuring face of a micrometer spindle be square to the spindle axis, otherwise an error of the form shown at Fig. 3.8 is introduced having a maximum value of $\frac{1}{2}d\theta$. This error source is overcome in the floating carriage micrometer by having a non-rotating spindle.

Location

When using a sensitive comparator the measured part is located on a table which forms the datum for comparison with the standard. The comparator reading is thus an indication of the displacement of the upper surface of the measured part from the datum. Faults at the location surface of the part through damage, geometrical variations from part to part, or the presence of foreign matter are also transmitted to the indicator. This provides false information regarding the true length of the part by introducing both sine and

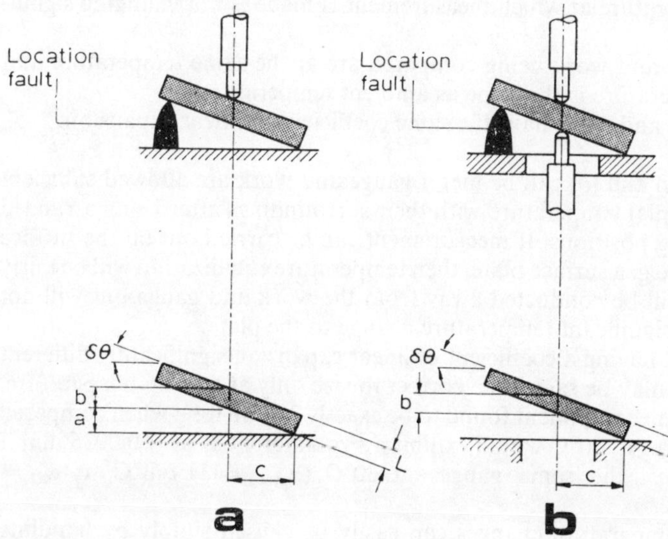

Fig. 3.9

a Surface displacement
Reading $= a + b$
$= c \tan \delta\theta + L \sec \delta\theta$
Error $= a + b - L$
$= c \tan \delta\theta + L (\sec \delta\theta - 1)$

b Differential length measurement
Reading $= b$
Error $= b - L$
$= L(\sec \delta\theta - 1)$

cosine errors (Fig. 3.9*a*). Where location conditions may not be ideal, e.g. interstage measurement during production, sensors operating on each side of the component can be used which eliminate the more serious sine-type error. A two-probe system measures length rather than surface displacement and highly sensitive electronic comparators of this type are used for slip gauge measurement (3.9*b*).

Temperature

The standard reference temperature at which line and end standards are said to be at their true length is $20 \pm \frac{1}{2}$°C and for highest accuracy in measurement this temperature should be maintained. When this is not possible and the length at reference temperature must be known, a correction is made to allow for the difference between ambient and reference temperature. The correction value required is $-0{\cdot}001\ 375$ mm ($-1{\cdot}375\ \mu$m) when a steel object exactly 25 mm long at 20°C and coefficient of linear expansion 11 μm°C/m is measured at 25°C, which is rather larger than the increment step of the M88/2 slip gauge set referred to earlier. However for less stringent measuring requirements it is not essential that correction to reference temperature is made provided that the following precautions and conditions are observed:

(*a*) the temperature at which measurement is made is not changing significantly;
(*b*) the gauge and work being compared are at the same temperature and this temperature is the same as ambient temperature;
(*c*) the gauge and work have the same coefficient of linear expansion.

Conditions (*a*) and (*b*) can be met if gauge and work are allowed sufficient time to reach equal temperature with their surroundings after being arranged in the measuring positions. If measurement can be carried out on the surface of a large mass, e.g. a surface plate, then temperature equalization will be fairly rapid as heat will be conducted away from the work and gauge but will not contribute any significant temperature change to the plate.

A component having a coefficient of linear expansion significantly different from the gauge may be said to be correct to size only at a given temperature, e.g. an aluminium component found to be exactly 30 mm long when compared with block gauges at 20°C would exhibit a size differential of $+0{\cdot}0036$ mm if measured against the same gauges at 30°C ($\alpha_{steel} = 11\ \mu$m°C/m; $\alpha_{al} = 23\ \mu$m°C/m).

Significant temperature changes can easily be caused simply by handling gauges and this source of error should not be ignored.

The Parallax Effect

On most dials the indicating finger or pointer lies in a plane parallel to the scale but displaced a small distance away to allow free movement of the pointer. It is then essential to observe the pointer along a line normal to the scale otherwise a reading error will occur. This effect is illustrated in Fig. 3.10*a* where a dial is shown observed from three positions when the pointer is set at zero on the scale. Observed from position (1), i.e. from the left, the pointer appears to indicate some value to the right of zero, and from position (2) some value slightly to the left of zero, while only at position (3) will the pointer coincide with zero on the scale. Rules and micrometer thimbles are bevelled to reduce this effect and on dials the indicator may be arranged to lie in the same plane as the scale (sigma comparator) thus completely eliminating parallax; or a silvered reflector may be incorporated on the scale so that the line between the eye and pointer is normal to the scale only when the pointer obscures its own image in the reflector (3.10*b* and *c*).

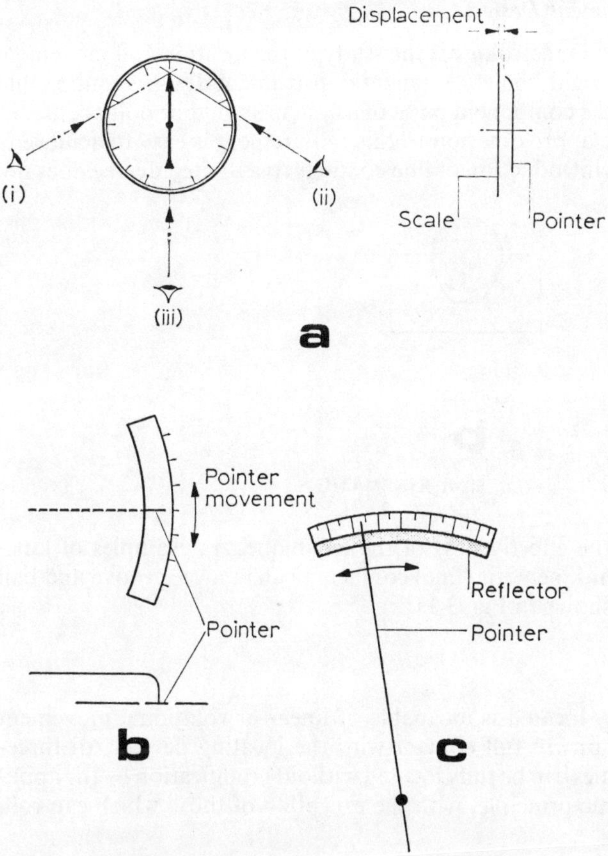

Fig. 3.10 THE PARALLAX EFFECT

3.9 Measuring Instruments

In measurement and in the design of measuring instruments it is necessary to apply principles and techniques which allow the following requirements to be fulfilled:

(1) Accurate location of one element upon another.
(2) Accurate translation (linear movement) and rotation of elements in the measuring instrument or system.
(3) Freedom from unnecessary constraints or restraints which might introduce strains between component parts.
(4) Elimination of clearances, e.g. in instrument bearings.
(5) Transmission of thrust.
(6) Freedom from unnecessary friction.
(7) Elimination of backlash.
(8) Low and constant measuring pressure.
(9) Reduction of inertia to an acceptable level.

Kinematics and Instrument Design

Kinematics as applied in metrology is the study of the location and movement of three-dimensional rigid bodies. Kinematic instrument design requires that all contacts between the component parts of a system should be point contacts. In many cases a close approximation to this requirement is easy to achieve in practice and even the introduction of line contacts to a limited degree does not

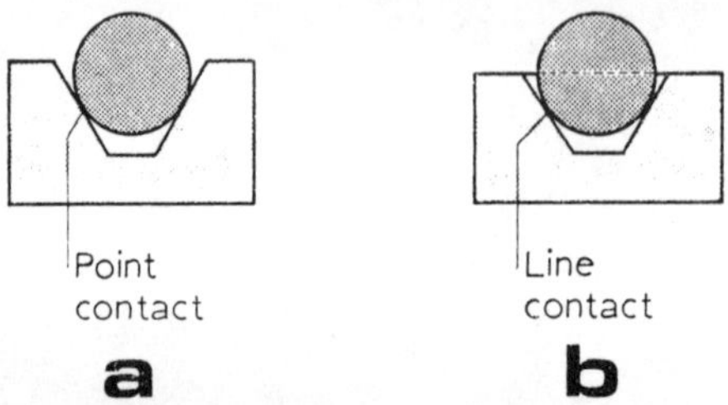

Fig. 3.11 *a* KINEMATIC *b* SEMI-KINEMATIC

significantly change the effectiveness of the technique. As examples of kinematic (point) and semi-kinematic (line) contact, a ball in a vee groove and ball in a conical hole are shown in Fig. 3.11.

Location

A body which is fully located is incapable of linear or rotational movement provided that it remains in full contact with the locating device. All three-dimensional rigid bodies can be fully located without modification by the application of the kinematic principle, with the exception of those which can roll,

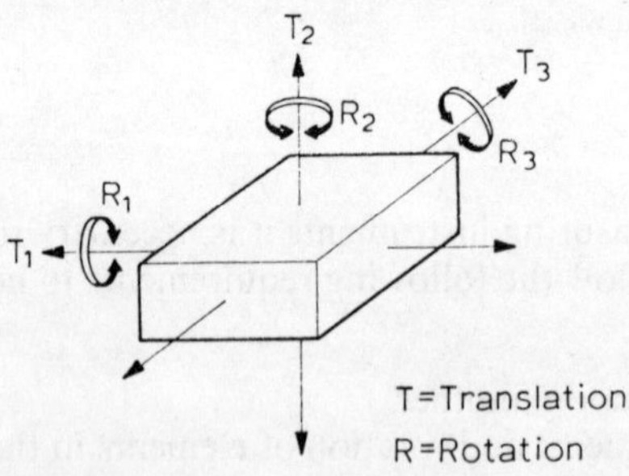

Fig. 3.12 THE SIX DEGREES OF FREEDOM OF A RIGID BODY

e.g. balls and cylinders, which can be located only to a limited extent. Further, all rigid bodies irrespective of their shape are capable of three linear or translational movements and three rotational movements either singly or in combination. Each of these possible movements is termed a degree of freedom, thus all rigid bodies have six degrees of freedom. The translational degrees of freedom correspond to three mutually perpendicular axes passing through the body and the rotational degrees of freedom correspond to rotation about those axes. The six degrees of freedom are shown in Fig. 3.12 as they apply to a simple rectangular block; the letters T and R indicate translation and rotation.

The body can only be fully located in a given position by eliminating all six degrees of freedom and to do so kinematically requires the application of no

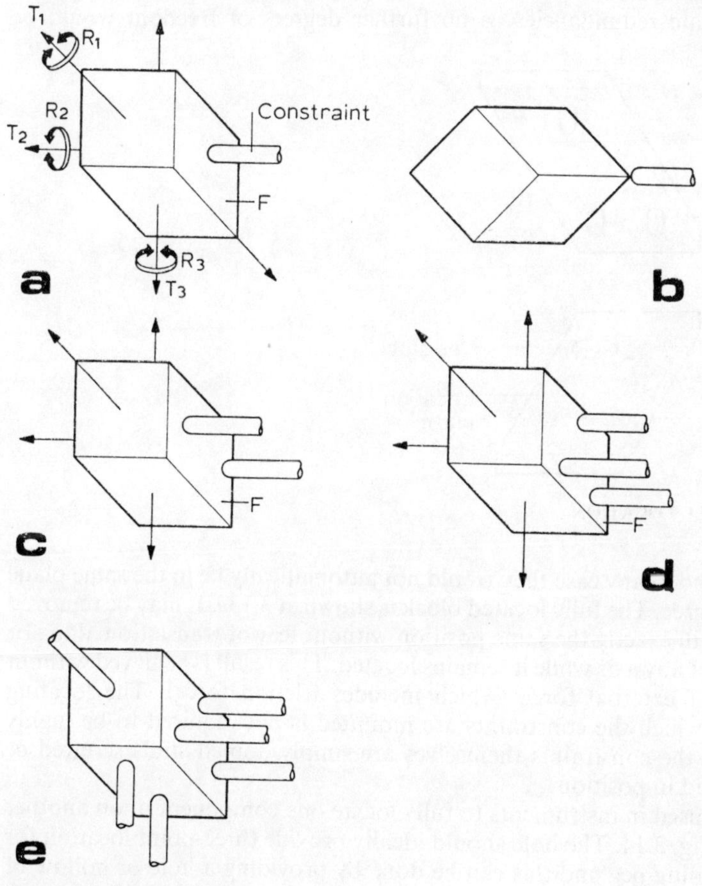

Fig. 3.13 CONSTRAINTS

more than six suitably applied point locations, each location eliminating one degree of freedom.

Constraint

Any rigid point location applied to a body is a constraint. Thus one constraint applied at face F (Fig. 3.13*a*) reduces the degrees of freedom of the block to five, two translations and three rotations. From a purely theoretical point of view the location point chosen is of no consequence, e.g. the block could be located at one corner as shown at 3.13*b*, but this is not practical as any translational movement of the body transfers the location to a new axis. Location points should be chosen so that such a transfer cannot occur.

A second constraint applied to face F eliminates rotation R_3 (3.13*c*) and a little thought indicates that a pair of constraints are necessary to eliminate a rotation. A third constraint applied to face F completes the points required to form a plane and eliminates rotation R_1 (3.13*d*). More constraints applied to face F create redundancies as no further degrees of freedom would be

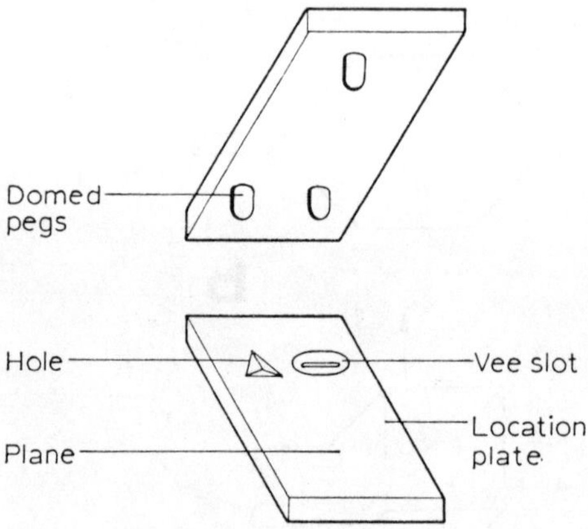

Fig. 3.14 FULL LOCATION

eliminated, and in any case they would not automatically lie in the same plane as the other three. The fully located block is shown at 3.13*e*. It may be removed and replaced in exactly the same position without fear of translation along or rotation about any axis while it remains located. This result is achieved without application of external force (which includes friction force). The locating device upon which the constraints are mounted is not required to be highly accurate and the constraints themselves are simply domed studs screwed or otherwise fixed in position.

A method used in instruments to fully locate one component upon another is shown in Fig. 3.14. The hole should ideally provide three-point location for its corresponding peg and this can be done by providing a hole or hollow of trihedral form. In practice, a conical hole is used which although it introduces line contact to the peg is much easier to manufacture and is just as effective. The vee slot can be produced by means of a simple formed cutter.

The location pegs contact the plate at six points: the slot eliminates two translations, the hole eliminates one translation and two rotations, and finally the plane eliminates the remaining rotation, a total of six.

Very often in metrology an instrument pair is required to have one degree of freedom, e.g. it may be necessary to provide a translation in one direction only. A simple method is shown in Fig. 3.15*a* where a slot and a plane are used to provide five constraints. The slot must of course be accurately machined and free from "wind" or twist, and the plane must not rise or fall in relation to the slot if straight line motion is to be achieved, otherwise rotations will be introduced.

Such a motion having very low friction characteristics can be achieved by substituting balls for the location pegs (3.15*b*). The balls are held captive in vee groves machined in the carriage so that they are able to roll; their point contacts in these grooves should not be mistaken for the locations required. Retaining pins provide a limit to the distance that the balls can roll thus preventing the system from becoming unstable. This method is used on the upper slide of the floating carriage diameter measuring machine manufactured by Coventry Gauge and Tool Co. Ltd., where it is essential that the micrometer mounted on the carriage should adjust to the measured component along a

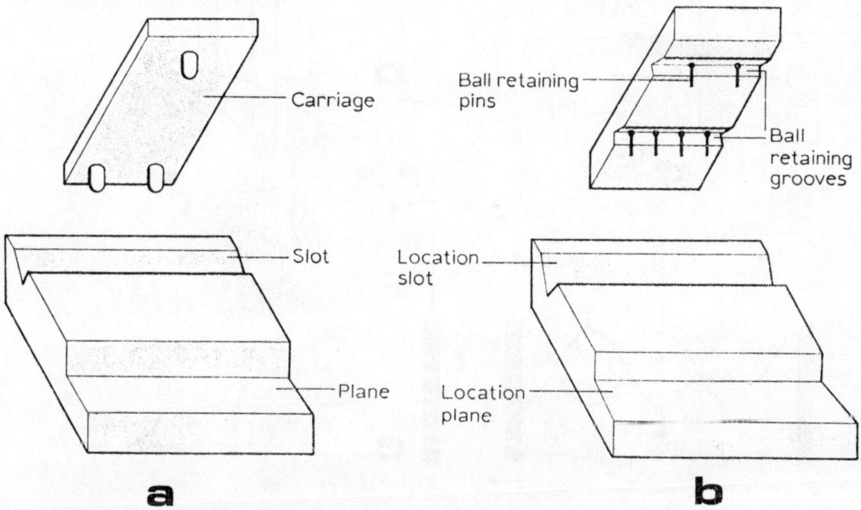

Fig. 3.15 FIVE CONSTRAINTS, ONE DEGREE OF FREEDOM

line perpendicular to the component axis and with almost complete freedom from friction.

Metal Strip Springs

Metal strip springs provide a simple and effective method of flexible connection between two elements allowing frictionless relative motion in desired directions. A number of such devices used in measuring instruments where small movements are required are shown in Fig. 3.16.

The springs range from 0·05 to 0·1 mm thick depending on the stiffness required. Reinforcing plates which prevent buckling may be used to allow the elements to be placed the desired distance apart (3.16*a*). The crossed striphinge (3.16*b*) is widely used in comparators and gauging fixtures where small rotary motions are required. At 3.16*c* rotary motion is obtained as a result of the distortion introduced to the two flexure springs by relative parallel movement

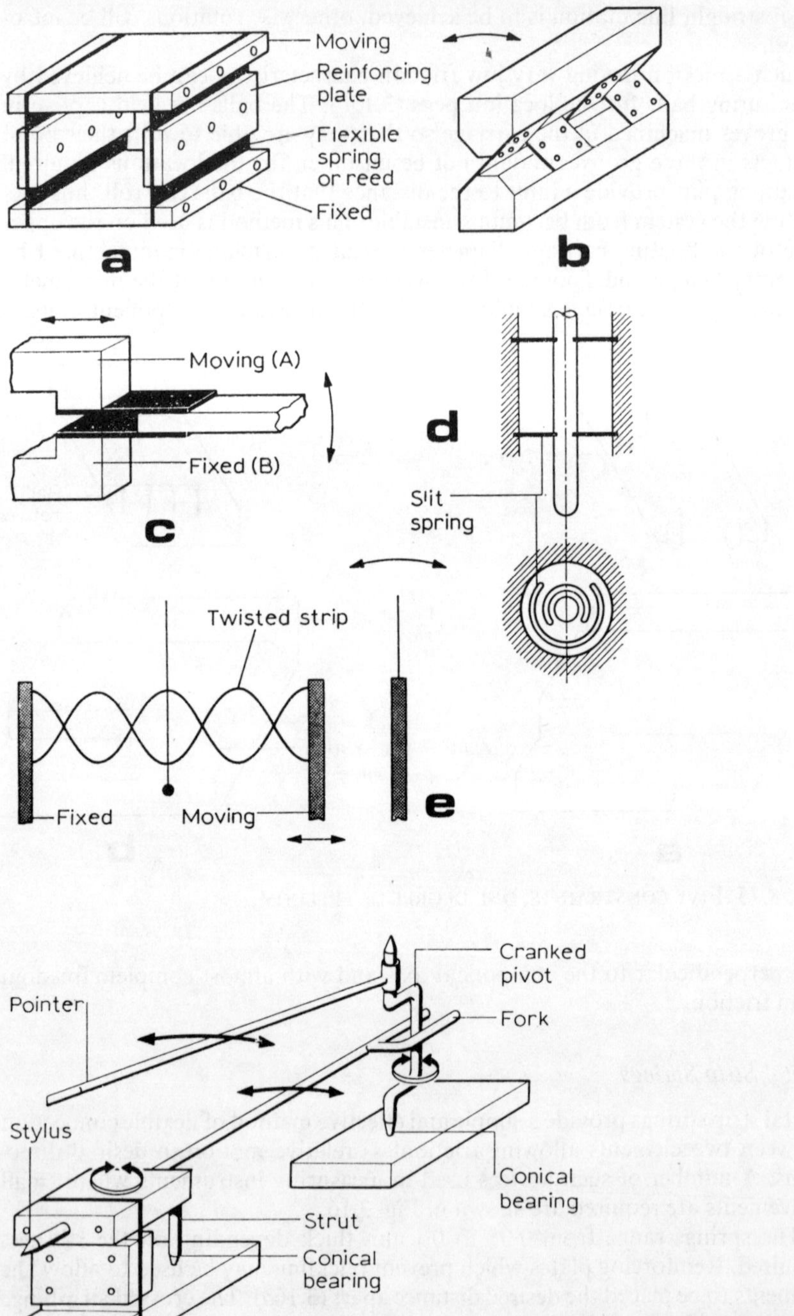

Fig. 3.16 FLEXIBLE CONNECTIONS ALLOWING SMALL MOVEMENT

of parts A and B, while at 3.16*d* linear movement of a stylus is obtained. The twisted strip shown at 3.16*e* when placed in tension causes the pointer to rotate. In all these examples sufficient stiffness, accuracy of movement, elasticity and stability, coupled with virtually zero friction and low inertia, are obtained along with the elimination of backlash. An ingenious application of the flexible metal strip is found in the design of the fiducial indicator on the N.P.L.-type pitch measuring machine manufactured by The Pitter Gauge and Precision Tool Co. Ltd. Without going into details of screw thread pitch measurement which is beyond the scope of this chapter, it is sufficient to state that the stylus which engages the thread to be measured must be moved from pitch to pitch on the thread, and an exact indication when one pitch has been moved is essential. The indicator which consists of a stylus mounted on a block supported by a flexible steel strip and a strut having conical end bearings is shown at 3.16*f*. When the stylus is located in the centre of the groove it is subject to exactly equal pressures on both sides but as soon as it is moved parallel to the screw axis it begins to climb the thread flank which causes the flexible strip to bend and twist. A fork fitted to the block transmits the twist to a cranked pivot upon which a light pointer is mounted. The pointer is immediately deflected from its index mark and will only return to the mark when the pressure is equal on each side of the stylus. This occurs first at the transition of the stylus over the crown of the thread and again when the stylus is exactly at the centre of the next groove, thus indicating one pitch movement.

Thrust Bars and Bearings

The thrust bar (Fig. 3.17) provides a simple method for transmitting thrust and eliminating errors which might arise from a more rigid connection between

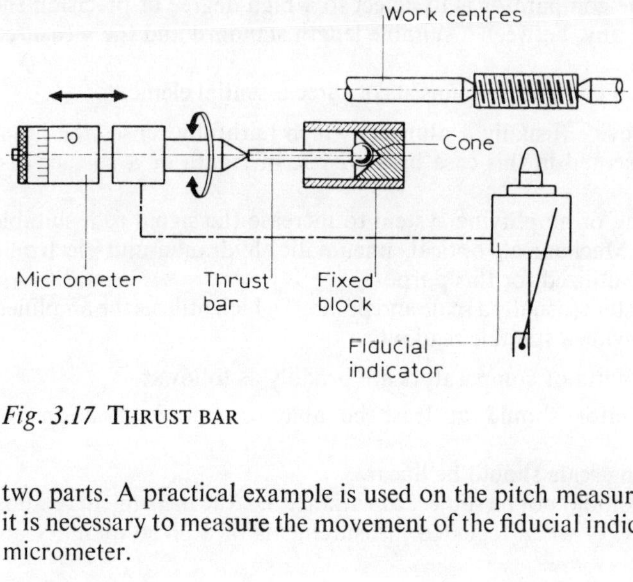

Fig. 3.17 THRUST BAR

two parts. A practical example is used on the pitch measuring machine where it is necessary to measure the movement of the fiducial indicator by means of a micrometer.

The micrometer and fiducial indicator are mounted on a kinematic carriage having one degree of freedom and the work is mounted on separate fixed centres parallel to the motion of the carriage. The thrust bar shown is a modification of the Lucas thrust bar which has two ball ends and is located in conical or trihedral seatings. The Lucas bar can also be used to provide true axial rotation (Fig. 3.18*a*). This is not easy to achieve with the normal circular bearing and spindle as the clearance and geometrical errors produce "path rotation" in which the axis of rotation does not remain in a fixed position (3.18*b*). A simple method for eliminating bearing clearances is to mount a cylindrical spindle in vee blocks instead of circular bearing, the block acting as the bearing cap being spring loaded (3.18*c*).

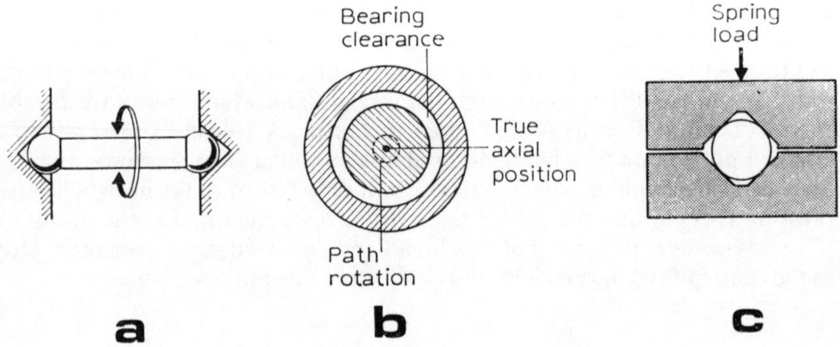

Fig. 3.18 LUCAS BAR FOR TRUE AXIAL ROTATION

3.10 Precision Comparators

The function of the comparator is to detect to a high degree of precision the small difference, if any, between a suitable length standard and the measured part.

In construction, a comparator consists of three essential elements:

(1) A sensing device (usually a plunger) which faithfully senses the input signal represented in this case by a change of length or a surface displacement.
(2) A magnifying or amplifying system to increase the signal to a suitable magnitude. Mechanical, optical, pneumatic, hydraulic and electronic methods are utilized for this purpose.
(3) A display system (usually a scale and pointer) which utilizes the amplified signal to provide a suitable readout.

The design criteria of comparators are broadly as follows:

(*a*) The comparator should at least be able to record variations of 0·0025 mm.
(*b*) The recording scale should be linear.
(*c*) The system should not be subject to variance, i.e. the instrument reading should not vary when repeated measurements of a given quantity are made.

(*d*) The scale indicator (pointer, liquid column, etc.) should be clear and free from oscillation.
(*e*) The system should be free from backlash, unnecessary friction and clearances, and inertia should be low.
(*f*) The measuring pressure should be low and constant.
(*g*) The supporting frame should be rigid and able to withstand reasonable ill usage without permanent harm.
(*h*) The instrument should be capable of a wide measuring range.

3.11 Mechanical Comparators

The essential details of two mechanical comparators are shown in Fig. 3.19 and Fig. 3.20.

The Johansson Mikrokator

This instrument shown at 3.19 makes use of the Abramson movement to provide a virtually frictionless system. The plunger is supported at its lower end by

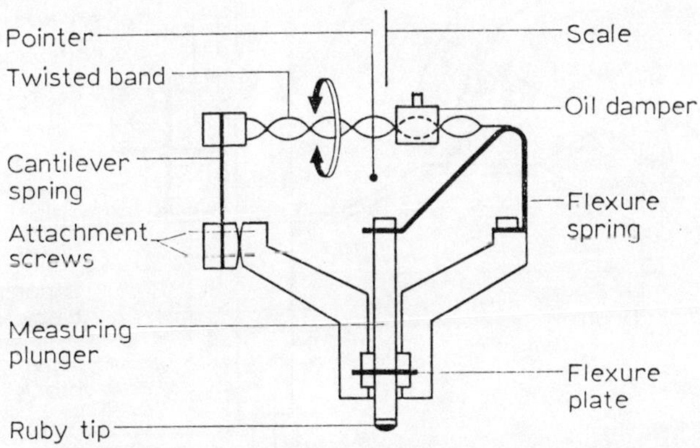

Fig. 3.19 JOHANSSON MIKROKATOR

a flexure plate and its upper end by the foot of a flexure spring. The end of the twisted metal band wraps round the knee of the spring and is firmly attached through its vertical leg. Vertical movement of the plunger is transformed into rotational movement of the vertical leg, thus stretching the twisted band which partially unwinds and rotates the attached pointer against the scale. This system provides a large angle of rotation for a small plunger movement which is further amplified by the pointer. Damping is provided by immersing a portion of the twisted band in a spot of oil. The cantilever spring fulfils two functions:

(*a*) The band can be brought to the correct tension by adjustment of the attached screws.

(*b*) The magnification can be varied by increasing or decreasing the length of the cantilever. An increased length reduces the force available to unwind the strip thus reducing magnification. The cantilever is preset in its correct position by the manufacturer.

A large range of Mikrokators are available, the most sensitive having a scale division of 0·000 01 mm, for closely controlled calibration work, while the least sensitive has a scale division of 0·1 mm. The range of instruments having magnifications up to ×5000 are available for industrial use.

The Sigma Comparator

This instrument is shown at Fig. 3.20. The plunger is attached to a rectangular bar which is supported at its upper and lower ends by flexure plates. The vertical movement of the bar is limited by stops. A knife edge is fixed to the side

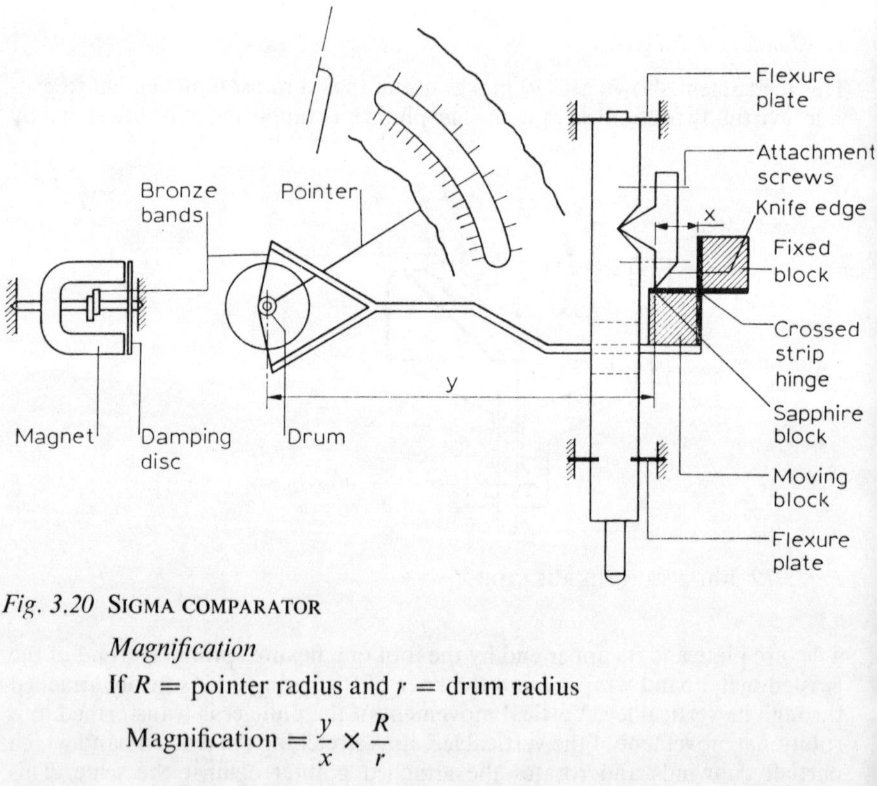

Fig. 3.20 SIGMA COMPARATOR

Magnification

If R = pointer radius and r = drum radius

$$\text{Magnification} = \frac{y}{x} \times \frac{R}{r}$$

of the bar which bears on a sapphire block attached to the moving member of a crossed strip hinge. The hinge is suitably pre-tensioned to allow it to rotate within the range of the instrument scale. A forked arm attached to the moving member of the hinge transmits rotary motion to the indicator driving drum through a bronze band wrapped round the drum. The magnification preset by the manufacturer is varied by adjustment of the knife edge attachment screws.

The instrument is damped by a horseshoe magnet fixed to the frame and a non-ferrous (aluminium) disc fixed to the pointer spindle. Rotation of the disc in the magnetic field of the magnet sets up eddy currents which are proportional to the rotational velocity and in opposition to motion.

The range of instruments available provides magnifications of ×300 to ×5000, the most sensitive models allowing scale estimations of the order of 0·0001 mm (0·1 μm) to be made.

3.12 Mechanical–Optical Comparators

Mirrors and lenses can be used to provide frictionless magnification systems.

Reflection at a Plane Surface

A ray of light falling at some angle θ to the normal of a plane (mirror) surface is reflected at the same angle from the normal as shown at Fig. 3.21*a*. If the

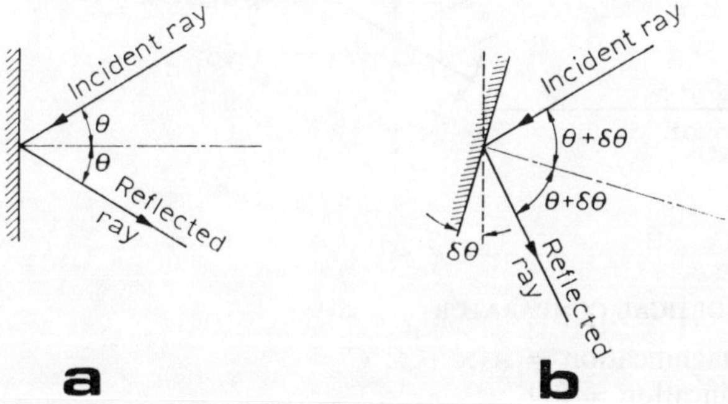

Fig. 3.21 REFLECTION AT PLANE SURFACE

mirror is rotated through a small angle $\delta\theta$ then the angle between the incident ray and the normal increases to $\theta + \delta\theta$, and the angle to the normal of the reflected ray also increases to $\theta + \delta\theta$ (3.21*b*). It is clear that the angle between the incident and the reflected ray has increased by $2\delta\theta$, that is twice the angle of rotation of the mirror. An image transmitted to a screen via a mirror has a magnified movement on the screen proportional to the distance between the pivot of the mirror and the screen and twice the angle of rotation of the mirror. Thus an optical lever has twice the magnifying power of a normal pointer. A simplified outline of a mechanical–optical comparator based on this principle is shown in Fig. 3.22.

The O.M.T. Omtimeter

This is a well-known comparator based on optical magnification. The instrument is designed to provide a moving scale against a fixed index line. The scale and index are carried side by side on a glass plate (Fig. 3.23*a*). An external beam of light is directed through the scale by a mirrored prism and transmits

its image on to a reflector which is rotated by the measuring plunger. The reflector redirects the scale image back to the plane of the index in such a manner that the scale image and the fixed index coincide, movement of the measuring plunger causing the scale image to move vertically against the index. The scale and index images now carried by the beam of light are further magnified and finally displayed on a screen at the side of the instrument. The instrument layout is shown at 3.23*b*.

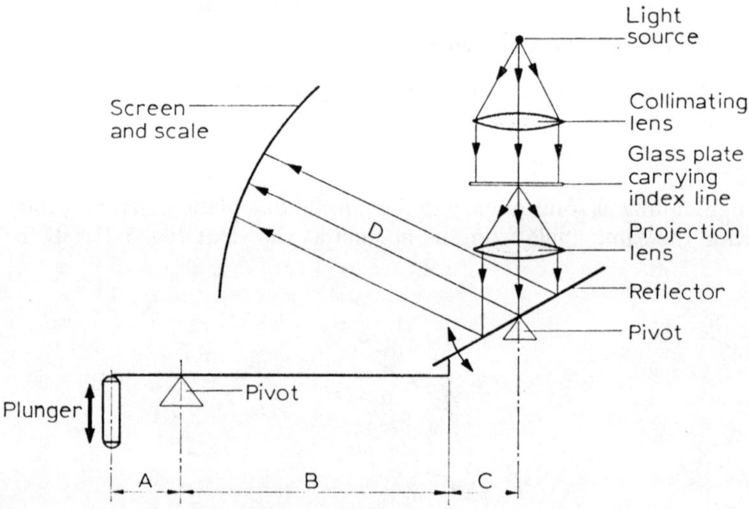

Fig. 3.22 MECHANICAL-OPTICAL COMPARATOR

Mechanical magnification = $A \times B \times C$
Optical magnification = $2D$
Total magnification = $2ABCD$

The only possible source of friction and inertia in this system is at the plunger and reflector and by kinematic design these are reduced to a very low level. The tension spring eliminates backlash, and the problem of damping does not arise. The scale range of the Omtimeter is ±0·125 mm in 0·001 25 mm graduations and the measuring force between 3 and 4 N.

The Eden–Rolt Comparator

This mechanical–optical comparator (Fig. 3.24) was originally designed for slip gauge calibration and is an excellent example of instrument design in which the accuracy of measurement is not directly dependent on the accuracy of manufacture. The mechanical part of the instrument is a combination of the flexure spring systems shown at Fig. 3.16*a* and *c*.

A long light beam terminates in a ring across which is mounted a thin strand or web of material. The plunger movement is magnified about 400 times at the web, and the web movement is magnified a further 50 times by an optical projection system, giving a total magnification of 20 000 at the scale. With this

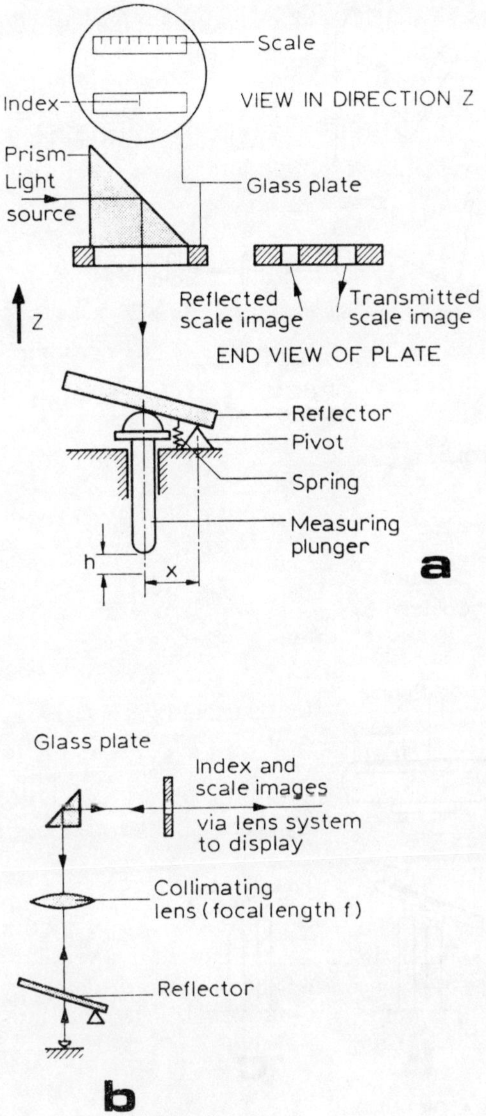

Fig. 3.23 OMT OMTIMETER

Magnification

$\delta\theta = h/x$

Scale movement $= 2f\delta\theta$

where f is focal length of lens and $\delta\theta$ the angular rotation of the reflector

$$\text{Magnification} = \frac{\text{scale movement}}{\text{plunger movement}} = \frac{2fh}{x} \times \frac{1}{h} = \frac{2f}{x}$$

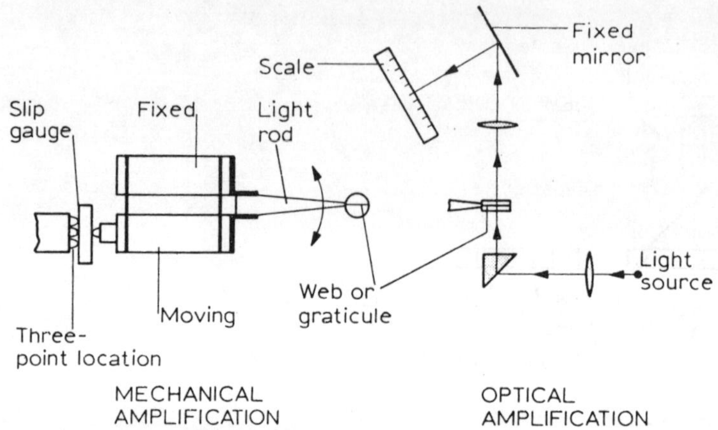

Fig. 3.24 EDEN–ROLT COMPARATOR

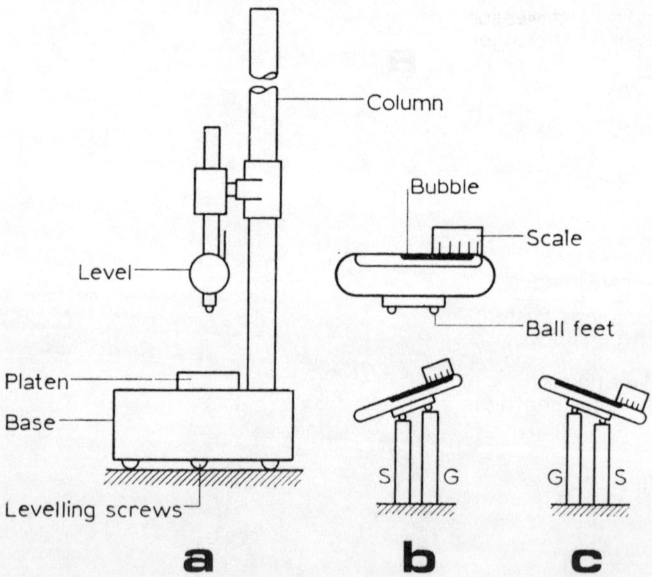

Fig. 3.25 BROOKES LEVEL COMPARATOR

instrument it is possible to estimate size differences of the order of 0·000 025 mm (0·025 μm).

The Brookes Level Comparator

This comparator, essentially a highly sensitive spirit level, provides an excellent though rather slow means for measuring long gauges, e.g. length bars. A rotatable disc or platen having mutually parallel surfaces is located on a massive base provided with three levelling screws. The upper face of the platen is a plane lapped surface while the lower face is a lapped annular ring allowing easy rotation on the base. A bracket which kinematically supports the level and allows it to be raised and lowered during measurement is mounted to the main column of the instrument. A scale is provided so that bubble displacement may be measured, see Fig. 3.25.

In operation the end of the bubble is first brought approximately to the centre of the scale by allowing the two ball feet attached to the level to rest on a surface parallel with the platen while adjusting the levelling screws on the base. A block gauge or combined block gauge and length bar may be wrung to the platen for this purpose. The standard (S) and gauge to be compared (G) are now wrung to the platen, and the ball feet which are at approximately 20 mm centres rest on the gauges forming a bridge (3.25*b*). The position of the end of the bubble against the scale is noted, the level is raised, the platen rotated through 180°, and a second reading taken (3.25*c*). The displacement of the bubble along the scale represents twice the difference in length between the gauges, thus

$$\delta h = \tfrac{1}{2}L$$

where δh = difference in length between the gauges and L = bubble displacement.

The level comparator has several advantages:

(*a*) The gauges are compared together thus making temperature equalization easier. It is usual to allow 20 min per 25 mm gauge length for this purpose.
(*b*) The rotation of the platen provides twice the bubble displacement required and at the same time eliminates the need to bring the platen exactly to the horizontal plane.
(*c*) Slight errors in scale reading are halved in the final computation of length.
(*d*) The gauges are vertically mounted and any shortening under their own weight is the same for both gauges.

Differences as small as 0·000 02 mm (0·02 μm) can be detected, and gauges up to 1 m in length can be measured. The basic design of the precision spirit level is discussed in section 3.23.

3.13 The Principle of the Pneumatic Transducer

Since pneumatic transducers were first introduced in the 1930's they have found ever widening fields of application owing to their simplicity, sensitivity, robustness and versatility. Their function is to convert changes of length or surface displacement into changes of pressure.

The principal features of the pneumatic transducer are shown in Fig. 3.26*a*. It consists of two chambers in series separated by an orifice plate or control orifice (C) of geometrical area A_c. Air passes from the first to the second chamber through the control orifice and thence to atmosphere via a second orifice, the measuring orifice or jet (M), of effective area A_m.

A continuous flow of clean dry air is supplied to the transducer through a pressure regulator which ensures that the pressure in the first chamber remains constant at all times. The pressure in the second chamber can be changed by varying the restriction applied to the measuring jet. When the measuring jet is completely closed the variable pressure p rises until it equals the operating pressure P, but if it is completely unrestricted then the variable pressure falls towards atmospheric pressure. In practice variations in p are obtained by moving a restricting surface towards or away from M.

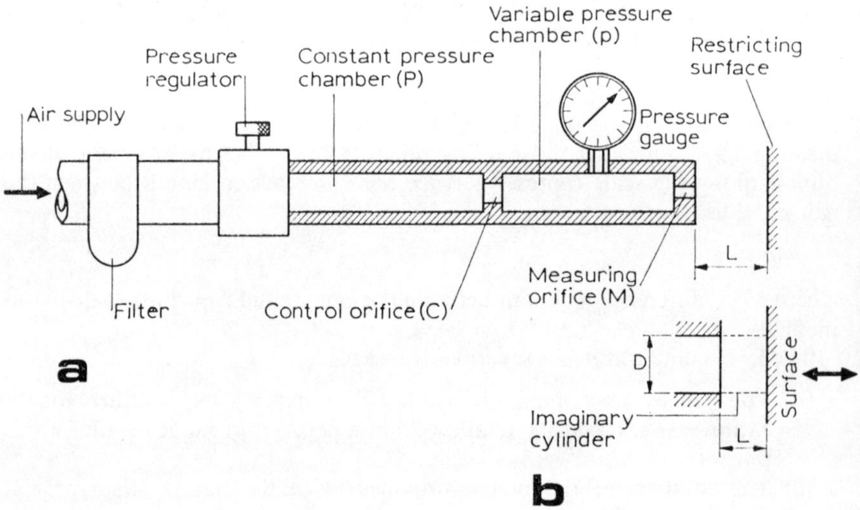

Fig. 3.26 PNEUMATIC TRANSDUCER

In a correctly designed transducer the ratio of the orifice areas is so proportioned that within a limited range of restriction the rate of change of p is uniform, i.e. dp/dL = constant, where L = displacement of the restricting surface.

The effective area of air escapement from the measuring jet is determined by the displacement of the restricting surface and the geometrical area of the orifice, i.e. the surface area of the imaginary cylinder so formed (Fig. 3.26*b*), thus,

$$\text{Effective area of measuring jet, } A_m = \pi DL \qquad \text{(i)}$$

where D = diameter of the measuring orifice.

Investigations have shown that when the pressure ratio p/P and the area ratio (A_m/A_c) are plotted over a wide range of supply pressures (15 to 500 kPa), curves having similar characteristics are obtained, each curve representing the results for a given supply pressure. The extremes of the family of curves are shown in Fig. 3.27*a*.

That portion on all the curves where p/P lies between 0·6 and 0·8 is linear to within 1% and the intercept on the p/P axis is almost exactly 1·10 in all cases.

The slope depends upon the supply pressure and ranges from 0·6 at 500 kPa to 0·4 at 15 kPa. From Fig. 3.27*b* the general linear equation may thus be stated,

$$\frac{p}{P} = E - b\frac{A_m}{A_c} \tag{ii}$$

where E = intercept on p/P axis = 1·10
b = slope ($0{\cdot}4 < b < 0{\cdot}6$)
A_c = geometrical area of control orifice.

It is evident that for a given supply pressure there must be maximum and minimum values of A_m/A_c and thus of A_m, beyond which 1% linearity does not

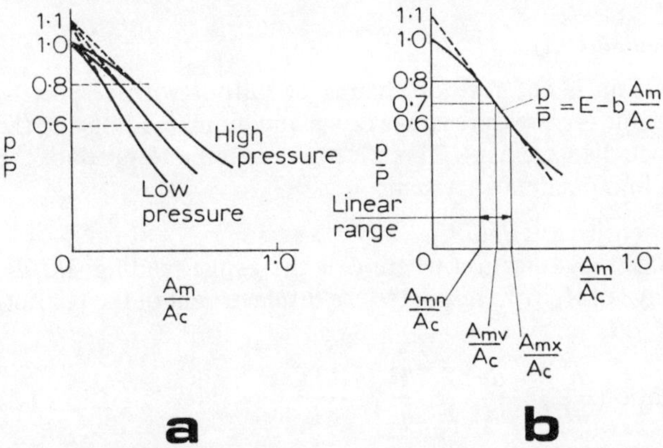

Fig. 3.27 PNEUMATIC TRANSDUCER: PLOTS OF PRESSURE RATIO AGAINST AREA RATIO

apply; this constitutes the linear range. By substituting the values of E and b into equation (ii) it can be shown that

$$A_{mx} - A_{mn} = \tfrac{1}{2}A_{mv} \tag{iii}$$

where A_{mn} is the minimum value of A_m ($p/P = 0{\cdot}8$)
A_{mx} is the maximum value of A_m ($p/P = 0{\cdot}6$)
A_{mv} is the average of A_m ($p/P = 0{\cdot}7$).

The operating pressure P does not appear in equation (iii); thus the linear range of A_m is equal to half the average value of A_m and is independent of the supply pressure.

Pneumatic Sensitivity

As p and A_m are the only variables in equation (ii), differentiating p with respect to A_m provides an expression for pneumatic sensitivity:

$$\frac{dp}{dA_m} = -\frac{b}{A_c}P \tag{iv}$$

When $A_m = A_{mv}$, $p/P = 0{\cdot}7$, i.e. the average value of p/P (1% linearity).

Substituting in equation (ii),

$$0{\cdot}7 = 1{\cdot}10 - b\frac{A_{mv}}{A_c} \qquad \text{i.e.}\ \frac{b}{A_c} = \frac{0{\cdot}40}{A_{mv}}$$

Substituting in equation (iv),

$$\frac{dp}{dA_m} = -0{\cdot}40\frac{P}{A_{mv}} \tag{v}$$

From equation (v) the pneumatic sensitivity is seen to vary directly as P and inversely as A_{mv}. The negative sign may be omitted as only the numerical value is of interest.

Magnification of a Pneumatic System

The overall magnification is the rate of change of output with respect to input. The output variable is a pressure gauge or water column reading and the input variable is surface displacement. Three factors combine to produce the overall magnification in a pneumatic system;

(*a*) The pneumatic sensitivity: dp/dA_m.
(*b*) The output gauge magnification where G is the gauge reading: dG/dp.
(*c*) The rate of change of A_m with respect to the displacement of the restricting surface: dA_m/dL.

$$\text{Overall magnification}\ \frac{dG}{dL} = \frac{dp}{dA_m} \times \frac{dG}{dp} \times \frac{dA_m}{dL} \tag{vi}$$

Mention has already been made that the effective area of M is equal to πDL, where D is the diameter of the measuring jet (Fig. 3.26*b*).

$$A_m = \pi DL \qquad \text{and} \qquad \frac{dA_m}{dL} = \pi D$$

As the linear range of L corresponds to the linear range of A_m then

$$L_x - L_m = \tfrac{1}{2}L_v$$

where L_x is the maximum value of L
L_m is the minimum value of L
L_v is the average value of L.

If the linear scale of length R provides readings of p over the range 0 to P, i.e. atmospheric to supply pressure, then

$$\frac{dG}{dp} = \frac{R}{P}$$

Substituting in equation (vi),

$$\frac{dG}{dL} = 0{\cdot}40\frac{P}{\pi DL_v} \times \frac{R}{P} \times \pi D$$

$$\frac{dG}{dL} = 0{\cdot}40\frac{R}{L_v} \tag{vii}$$

where R and L_v are expressed in the same units, e.g. millimetres. From equation (vi) the overall magnification is seen to vary directly as R and inversely as L_v.

EXAMPLE 3.1

In a pneumatic measuring system the scale length $R = 500$ mm. Find the overall magnification if the linear range is to be 0·02 mm.

$$\frac{dG}{dL} = 0{\cdot}4 \times \frac{500}{0{\cdot}04} = 5000 \qquad (Ans.)$$

Equations (v) and (vii) indicate that increased magnification could be obtained by increased operating pressure and scale length but in practice this is inconvenient if the scale is to be kept to a reasonable length. But increased magnification can be obtained by reducing the linear range and this is achieved by careful proportioning of the orifice areas. The diameters of the orifices are also a factor in determining the order of magnification; thus, for high magnifications small orifice diameters are used, e.g. 0·5 mm, while larger, more easily manufactured orifices are used when lower magnifications are required.

Speed of Response

It is difficult to measure surface displacements at frequencies greater than 2 Hz with a pneumatic transducer as time is required to establish each new value of p, but for a very wide range of measuring requirements this is not an important factor. Where speed of response must be optimized then the volume of the variable pressure chamber (which includes any volume required to operate the pressure measuring device, e.g. a bourdon tube) should be kept to a minimum.

Pressure Control

Pneumatic transducers may be operated at low, high and medium pressures, the last being of the "flow velocity" type.

LOW PRESSURE A pressure of about 7 kPa is maintained in the constant pressure chamber. Variations of pressure p are recorded on a manometer tube against a scale about 500 mm long.

Earlier versions use water in the manometer tube and excess air from the supply is allowed to bubble from a dip tube; the disadvantage of this method is the relatively high evaporation rate of the water in the system. Current models employ low viscosity oil in a manometer connected to a small chamber in which the lower end of a weight-type valve is immersed. The oil smoothes the valve movement preventing "flutter" while excess air can blow through the valve (Fig. 3.28).

A feature of interest in these systems is the interchangeable control orifice which provides the easiest means for reproportioning the orifice areas and thus changing the magnification. A variety of scales are supplied for various requirements, each scale being accompanied by its corresponding control orifice.

The "Sigma" differential pressure system operating at about 30 kPa is shown in Fig. 3.29. In this case a constant operating pressure is applied to both sides of the manometer, pressure control being maintained by a diaphragm pressure regulator. One slide of the manometer is used to zero the system by adjustment

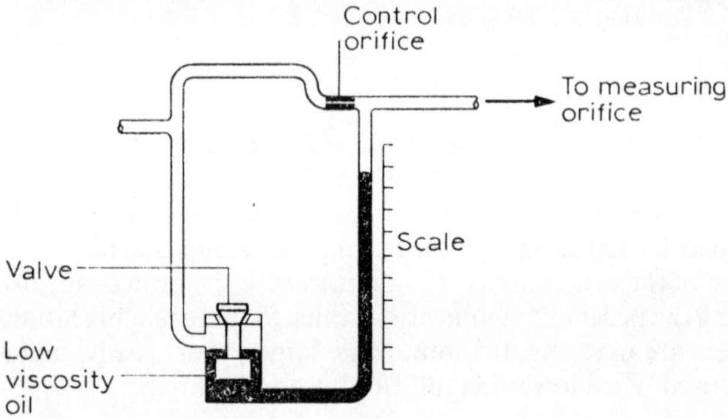

Fig. 3.28 LOW PRESSURE CONTROL FOR PNEUMATIC TRANSDUCER

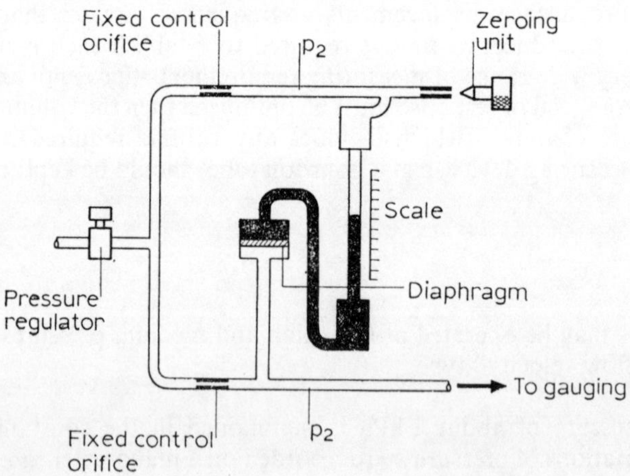

Fig. 3.29 SIGMA DIFFERENTIAL PRESSURE SYSTEM

of p_2. Because of the relatively higher operating pressure the movement of water in the tube is restricted by means of a diaphragm which maintains the scale at a reasonable length while providing high magnification. An advantage of the differential system is that the full scale reading can be made to represent the linear range of p_1, i.e. $0{\cdot}6\,P < p_1 < 0{\cdot}8P$. This can be done because the zeroing facility allows zero on the scale to represent $p_1 = 0{\cdot}6P$, and the length of the scale can be so designed that its maximum reading is $p_1 = 0{\cdot}8P$.

HIGH PRESSURE These systems use a bourdon tube pressure gauge instead of a liquid column. They have the advantage of longer range and, for measurements where swarf and coolant are present, the jet is powerful enough to break through the liquid film to the measured surface. High pressure dial-type instruments have the added advantage of being easily portable. Differential systems are also available.

MEDIUM PRESSURE The flow-velocity comparator relies on variations in the velocity of flow of air through a chamber to indicate surface displacement at the measuring jet. A constant pressure of about 70 kPa is supplied through an orifice to a glass tube which has a tapered bore and thence via a flexible pipe

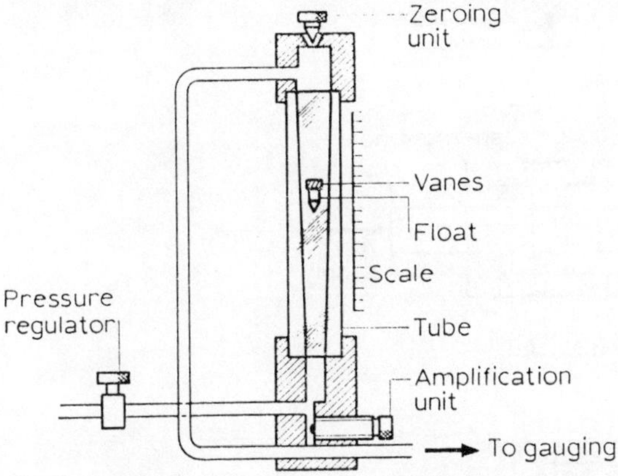

Fig. 3.30 MEDIUM PRESSURE CONTROL

to the measuring jet. The bore contains a light float equipped with small vanes causing it to rotate continuously in the air stream thus ensuring stability of the float at the centre of the tube (Fig. 3.30). When the measuring jet is unrestricted the float rises to the top of the tube but partial restriction at the jet reduces the flow velocity and the float then falls to a lower position in the tube. The position of the float against a scale beside the glass tube indicates the surface displacement at the measuring jet. The magnification of the instrument is largely determined by the tapered bore of the tube and a range of tubes allow magnification from $\times 1000$ to $\times 40\,000$. For some applications an operating pressure of 140 kPa is used, e.g. for ball-jet gauges.

This sensitive and reliable system can be used for a wide variety of pneumatic gauging applications, e.g. snap gauges, plug gauges, etc., and banks of instruments can indicate a number of dimensions simultaneously. Because increased jet constriction causes the float to fall, careful note must be taken of the sense of the float movement when changing from external to internal measurement.

Measuring Heads

Originally pneumatic gauging was developed in France to check the diameters of carburettor jets. Air at constant pressure was allowed to flow through the control jet and thence through the carburettor jet placed at the end of the second chamber. The "back" pressure measured on a manometer gave an accurate indication of jet area. This constriction technique has fairly limited application and the two types of measuring heads now in common use are the direct head and the indirect or contact head (Fig. 3.31).

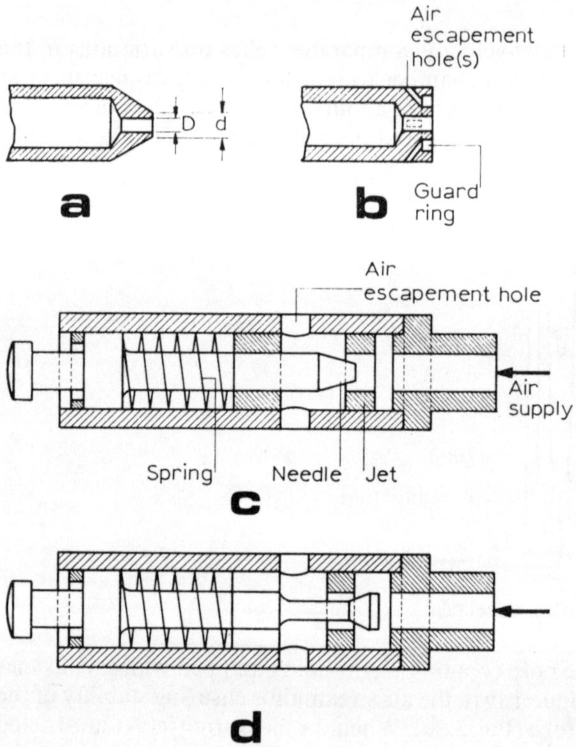

Fig. 3.31 MEASURING HEADS

a, *b* Direct heads   *c* Forward head   *d* Reverse head

DIRECT HEADS The jet shown at 3.31*a* has a tapered nose to ensure free escape of air as excessive land width would affect the characteristics of the system. The tapered nose also allows easier access for the jet in constricted measuring conditions. A satisfactory ratio between jet diameter D and land diameter d is $d = 2D$. The head shown at 3.31*b* allows considerable protection to be given to the nozzle by the provision of a guard ring and air escapement holes.

INDIRECT OR CONTACT HEADS Air escapement from the measuring jet can be controlled by a flat plate or needle-type valve, and the jet is protected from accidental damage. Forward and reverse acting valve heads are shown at 3.31*c* and *d*. The needle-type valve can be designed for short- or long-range measurement, slow taper needles providing ranges as long as 1·25 mm or more, which is not possible using direct heads. However, long range is only obtained by sacrifice of high magnification. It is interesting to note that a perfectly linear response can be obtained by the use of a parabolic needle.

3.14 Applications of Pneumatic Transducers

It is not possible or practicable to describe all the uses of pneumatic transducers, but an outline of some of the more conventional applications will provide an indication of their versatility and scope.

Pneumatic Comparator

The well known "Solex" pneumatic comparator consists of a reverse acting, indirect measuring head, mounted to a rigid vertical post upon which it may be adjusted for height.

The head is connected via a pipe to an oil or water manometer pressure control system. The scale range allows the instrument to be used above and below the linear range and thus the upper and lower portions of the scale may be non-linear. This is not a serious disadvantage as such a scale is relatively easy to produce by calibration through the pressure range p of the instrument. A typical scale for precision measurement has an effective length of about 360 mm, and scale increments of 0·001 mm. Movement of the liquid column over the full scale represents approximately 0·03 mm surface displacement. A fine adjustment device on the head allows the column to be set at any position against the scale. The advantage of the reverse acting head is that a measured part larger than the setting standard causes the liquid column to rise against the scale.

Plug Gauges

Pneumatic plug gauges are used for a wide variety of measurements. Some important examples are: single diameter, multi-diameter, average diameter, lobing, taper, straightness, squareness, ovality, and centre distance.

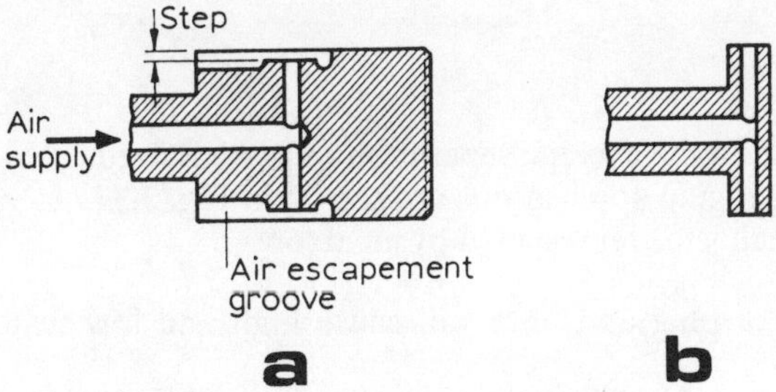

Fig. 3.32 PLUG GAUGE: SINGLE DIAMETER

SINGLE DIAMETER The gauge incorporates two equal and opposite jets in parallel as shown in Fig. 3.32*a*. The gauge can be made somewhat smaller than the bore so that it enters much more easily than the conventional plug gauge.

At first sight it might appear that the conventional plug gauge form is unnecessary and that a device simply consisting of two jets in parallel is all that is required (Fig. 3.32*b*). Unfortunately the magnitude of the "back" pressure p is not entirely independent of the relative position of the workpiece to the jets and a displacement of the workpiece from the symmetrical position changes the scale reading by about one tenth of that displacement. The cylindrical plug gauge form eliminates this effect by providing sufficiently accurate location for the jets which ensures that the total air flow is virtually constant whatever the position of the gauge in the bore. Thus within the designed range of the gauge a reduction of flow through one jet causes an equal increase of flow through the other jet, i.e. the jets are self-compensating.

One further aspect of pneumatic plug gauge design is of interest. At 3.32*a* the jets are shown slightly stepped below the surface of the plug. This gives a measure of protection to the jets but, more importantly, if the jets are flush with the skirt when holes very close to the plug diameter are to be measured, the reading becomes unstable and self-compensation is lost. The step eliminates this effect and allows holes approaching the skirt diameter to be measured. The upper limit of the measuring range is defined by a hole larger than the plug at which the reading again becomes unstable.

Twin nozzles eliminate the need to spring load the plug against one side of the bore thus simplifying gauge design but the use of twin (or multiple) nozzles in parallel also has the effect of reducing magnification. From equation (i) (p. 66),

$$A_m = \pi DL$$

Let L_1 and L_2 be the surface displacements corresponding to two opposed nozzles in parallel, then

$$A_m = \pi D(L_1 + L_2)$$

When $L_1 = L_2$

$$A_m = 2\pi DL \qquad \text{and} \qquad A_{mv} = 2\pi DL_v$$

Therefore

$$\frac{dG}{dL} = 0{\cdot}4\,\frac{R}{2L_v}$$

Thus to maintain magnification the average separation must be halved.

A number of pneumatic gauging applications are shown in Fig. 3.33*a* to *g*. For the sake of simplicity each jet is represented by an arrow.

LOBING Relative rotation of plug and bore will cause high and low scale readings (3.33*a*).

AVERAGE DIAMETER The introduction of a further three jets will provide a constant reading (3.33*b*).

MULTI-DIAMETER The progressive gauge shown at (3.33*c*) provides simultaneous indication on separate manometers for a number of "in line" bores.

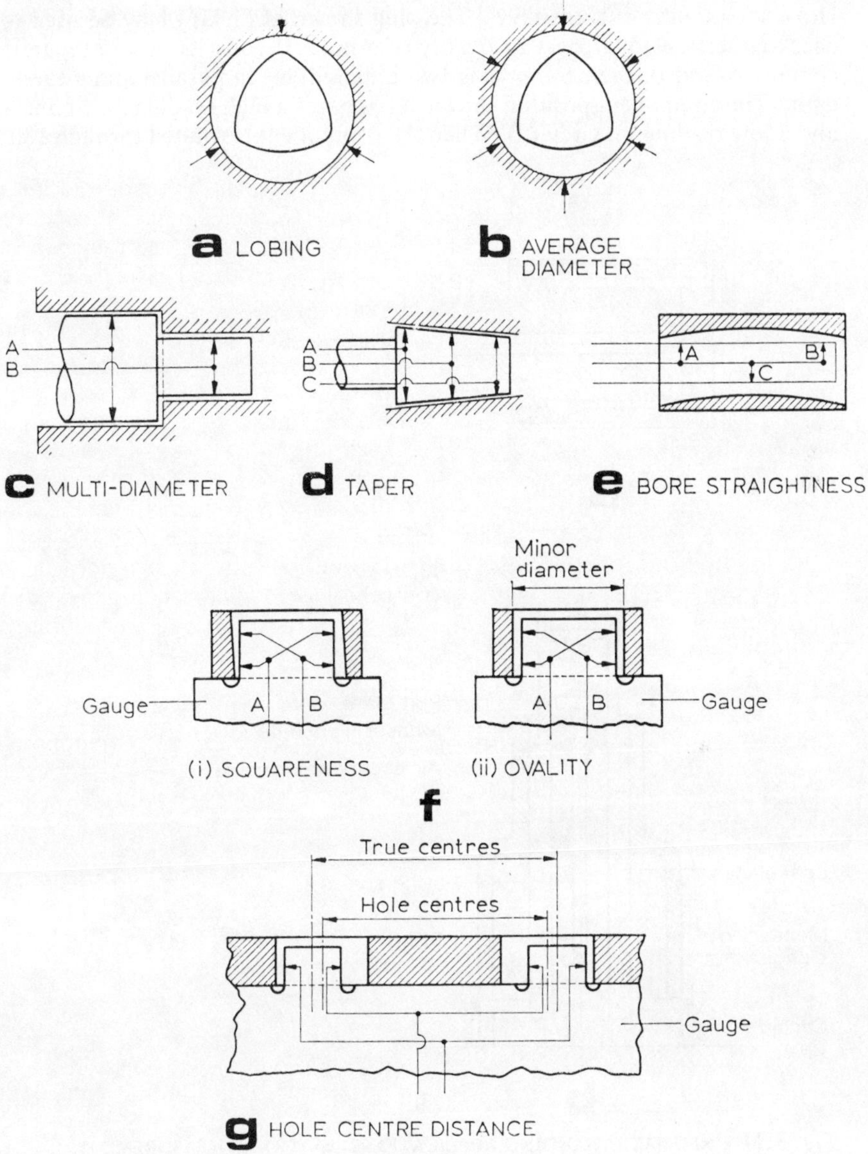

Fig. 3.33 PNEUMATIC GAUGING APPLICATIONS

TAPER Three pairs of jets will check angle, straightness of taper and diameter simultaneously (3.33*d*).

BORE STRAIGHTNESS Three jets in parallel are arranged as shown (at 3.33*e*). Jets A and B are equal and jet C has a flow equal to A + B. High and low readings will result from relative plug and bore rotation.

HOLE SQUARENESS AND OVALITY The plug shown at (3.33*f*) may be used to check either the squareness or ovality of a hole. It consists of two separate circuits, A and B, each comprising two opposed jets in parallel and an indicator. The component position shown at (i) causes a high reading on circuit A and a low reading on circuit B. When the component is rotated through 180°

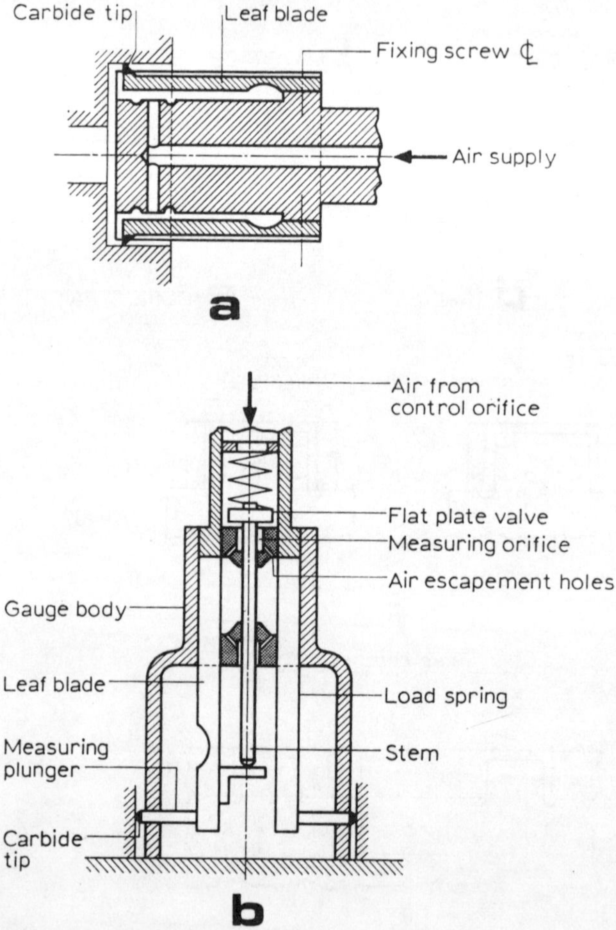

Fig. 3.34 PNEUMATIC GAUGING APPLICATIONS: EXTERNAL MEASUREMENTS

the readings are reversed. The oval condition shown at (ii) produces two low readings and on rotation successive pairs of high and low readings. Thus the behaviour of the indicators clearly differentiates between lack of squareness and ovality.

CENTRE DISTANCE Two plugs are mounted at the true hole centres. Two separate circuits are arranged in such a way that they sense variations in centre distance and yet remain insensitive to diameter variations as a change of diameter of one or both holes affects both circuits equally (3.33*g*).

Pneumatic gauging can be applied to external measurements in a wide variety of ways, such as ring gauges, snap gauges and height gauges, etc. Indirect heads are incorporated in large plug gauges for measuring cylinder bores and special direct heads are used for measuring the thickness of plastic sheet. The "leaf-jet" plug is an interesting application of the indirect or contact method which enables the diameter near the bottom of a semi-blind hole to be measured (Fig. 3.34*a*). It also has the advantage that the diameter of very shallow holes (1·5 mm) can be measured.

The comparator plug gauge (3.34*b*) is designed for use on the production line where absolutely clean conditions cannot be ensured.

Pneumatic transducers may be used to measure cutting tool deflections and thus cutting force, torque, etc. (see Chapter 7).

Two jets in parallel can be used for differential measurement.

Surface Finish

Components measured with direct jets must have a smooth surface finish as any significant surface variation gives rise to error. The source of the error is

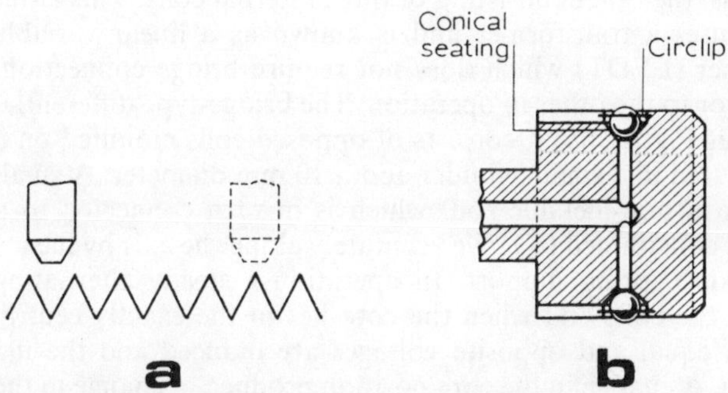

Fig. 3.35 EFFECT OF SURFACE FINISH ON GAUGING

shown in Fig. 3.35*a*, where it is evident that the effective area of the jet depends upon its position relative to the grooves, the area increasing when the jet is directly above a groove and decreasing when it is above a crest. Rough surfaces require a contact head such as the ball jet plug shown at (3.35*b*) which measures the size over the crests, which is the size usually required.

Measurement of Moving Surfaces

In general, a jet calibrated to measure a smooth stationary surface will provide the same reading when the surface is in motion, at surface speeds up to at least 30 m/s, e.g. in-process measurement during cylindrical grinding.

Calibration

When the usual slip gauge technique is not applicable, as in the case of plug gauges, standard ring gauges made to provide readings at the upper and lower end of the scale are normally adequate.

3.15 Electronic Measurement

The great advances in reliability of electronic components and circuitry brought about by the Second World War laid the foundation for modern electronic measurement. The function of electronic transducers in metrology is to convert changes of length, or surface displacement, into proportional electrical changes which can be detected, amplified and recorded or displayed on a meter. To this end, the transducer is designed to provide a linear response over a range of $\pm 0{\cdot}3$ mm. While both resistance and capacitance transducers are used for displacement measurement the inductance transducer is most suited to accurate measurement of small displacements because it suffers neither from the friction inherent in the resistance type where physical contact between a slider and resistance is used, nor from the capacitance problem caused by the low sensitivity and unreliability introduced by the large extra capacitance of the transducer leads. Differential inductance transducers of two kinds are in use, one consisting of two internal coils connected to form an inductance bridge with two external coils, the other consisting of three internal coils. This latter configuration constitutes a transformer and is known as a linear variable differential transformer (LVDT) which does not require bridge connection. Neither type is superior to the other in operation. The bridge-type differential transducer is shown at Fig. 3.36*a*; it consists of opposed coils mounted on a common axis and housed in a small cylinder about 10 mm diameter. A small iron core attached to a non-metallic rod, which is in turn connected to a measuring plunger, is constrained to move accurately along the axis by means of some form of flexure spring support. In operation a steady alternating voltage is applied to the coils and when the core lies in the exactly central position in the coils, equal and opposite voltages are induced and the net output voltage is zero. A change in the core position produces a change in the coupling between the coils which produces a differential output voltage proportional to the core displacement and thus the measuring plunger. The core position determines the direction of the meter deflection thus showing whether the plunger is above or below the null position. The block diagram shown at 3.36*b* illustrates the electronic gauging system which may be battery operated if required to provide complete freedom from the mains supply.

3.16 Advantages of Electronic Measurement

(1) Unlike mechanical systems which use part of the measuring pressure to overcome inertia, friction, etc., the electronic system is energized from an external source which allows for an extremely low measuring pressure resulting in negligible component distortion so that soft or thin walled components are easily measured.

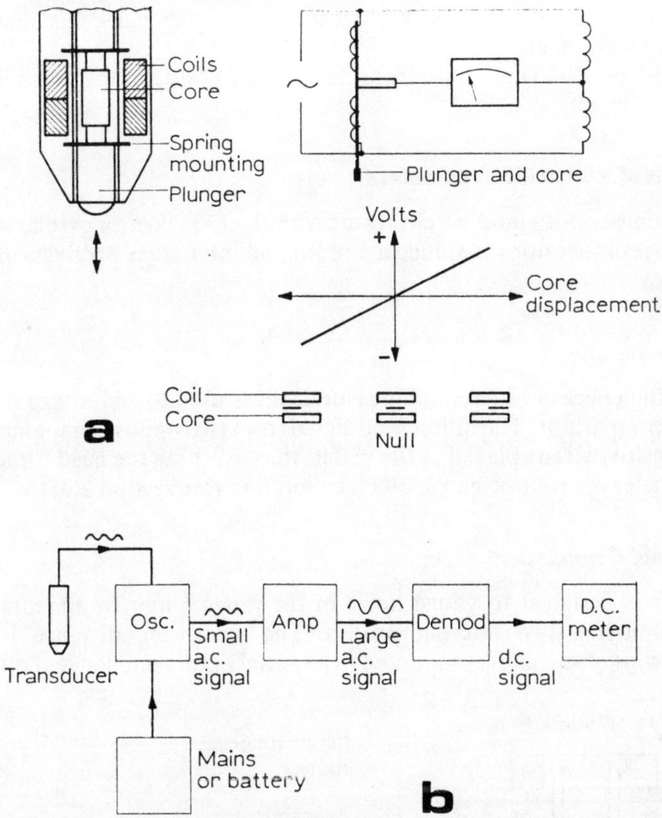

Fig. 3.36 BRIDGE-TYPE DIFFERENTIAL TRANSDUCER

(2) Friction is present at the transducer springs and at the meter springs and pivots but its magnitude is insignificant.
(3) Speed of response is of the order of 0·1 sec.
(4) The magnification is changed electrically and a number of ranges are available on one instrument at the turn of a switch. A typical instrument provides five ranges in both English and metric systems over a scale length of about 100 mm.

ENGLISH (in)	±0·0001	±0·0003	±0·001	±0·003	±0·010
METRIC (mm)	±0·003	±0·010	±0·03	±0·10	±0·30

(5) A number of transducers can be connected to one instrument and readings taken in turn by switching.
(6) Pointer zero setting is electrical which eliminates the need for fine adjustment on the transducer position.
(7) Signals from any number of transducers can easily be added or subtracted in any desired proportion, e.g. change in taper angle measured by two transducers can be expressed as one reading on the meter.

(8) The signal output can be used to operate recorders, sorting gates, machine controls, light signals, etc.
(9) Remote reading at 30 m and more from the transducer does not present a problem.

3.17 Applications of Electronic Transducers

Mounted to a comparator stand an electronic transducer makes an excellent fine measurement comparator for standards room and inter-stage production measurement.

Block Gauge Comparator

The error-reducing effect of differential measurement is used to advantage in the block gauge comparator. The differential signal from two opposed gauging heads of high sensitivity is displayed at the meter, thus avoiding the need for a gauge interferometer for routine gauge block calibration (see section 2.8).

Horizontal Internal Comparator

This comparator is designed to gauge holes in the range 6 mm to 300 mm diameter by employing two internal probes. The measuring distance is initially set by means of slip gauges mounted in a special fixture and located on

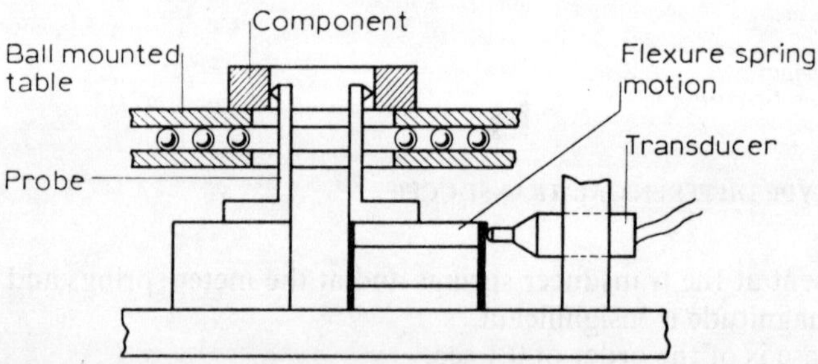

Fig. 3.37 HORIZONTAL INTERNAL COMPARATOR WITH SPRING SUSPENSION

the instrument table to ensure parallelism of the setting faces. An interesting feature is the spring suspension of the measuring probes aimed at ensuring frictionless transfer of the probe movement to a cartridge-type transducer (Fig. 3.37). Comparators using one measuring probe are also available for checking holes in the range 0·5 mm to 6·0 mm

Differential Gauging

An arrangement for taper measurement by electronic gauging is shown at Fig. 3.38*a*. Both probes have initially been set to zero from a master taper. When the part to be measured is introduced individual changes in diameters A and B may be read in turn on the meter. An equal increase or decrease in the diameters produces equal displacements of meter reading from each probe thus indicating zero change of taper. If one of the diameters changes while the other remains unchanged, the algebraic difference between the two readings, when expressed in terms of the probe separation x, represents both the change in taper and its sense, i.e. an increase or a decrease. This result holds true if both diameters change. The evaluation of the algebraic difference (or sum), of two separate probe readings is time consuming and open to error and it is much easier to perform these operations electronically, displaying the result as a single meter reading. Such gauges incorporate a switch position which allows single probe readings to be taken for initial setting purposes, etc., and a second switch position for differential gauging.

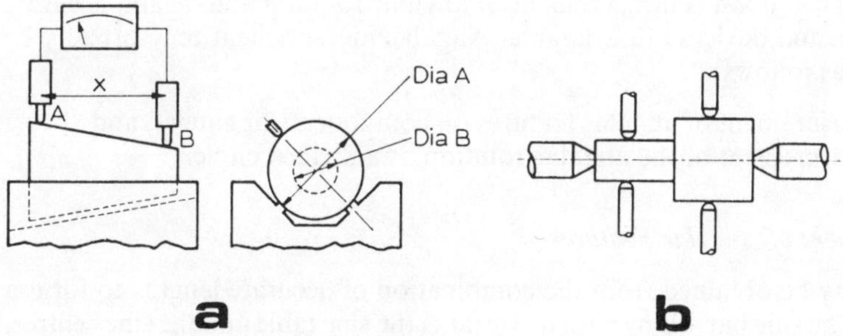

Fig. 3.38 DIFFERENTIAL GAUGING

a Taper measurement by electronic gauging
Let master gauge reading for A and B equal zero,
i.e. $A = B = 0$
Increase in dia. $A = \delta_1$ Increase in dia. $B = \delta_2$
Differential reading: $(A + \delta_1) - (B - \delta_2) = \delta_1 + \delta_2$

b Eccentricity error

The arrangement shown at Fig. 3.38*a* can, of course, be used for checking cylinders. Taper is evaluated as before, and by electronic addition of the probe readings, twice the diameter error midway between the probes is produced. One probe is insufficient when checking the diameters of work mounted on centres as half the diameter error plus twice the eccentricity error is sensed when the work is rotated. Eccentricity error is eliminated by the use of two opposed probes (Fig. 3.38*b*), as the algebraic addition of the probe outputs when responding to eccentricity is always zero, and changes in diameter only are recorded.

The applications of electronic transducers to fine measurement is by no means limited to the examples described above. They have been and are currently used for grading diameters, checking squareness of end face to axis, roundness measurement both internal and external either by means of a static probe and rotating workpiece or by transferring signals from a rotating probe through a slip ring, contour tracing, and surface texture measurement. In some of these applications side acting transducers are used and the usual precautions against "sine" and "cosine" error must be observed.

3.18 Measurement of Angle

Angle measurement has already been mentioned in the discussion on precision measurement applications of transducers. Such methods require the manufacture of an accurate master gauge or gauges from which the initial transducer settings are made and against which the setting may be checked from time to time. There are also many occasions when gauges are required for directly checking tapered holes and in addition components are manufactured incorporating tapered features. Accurate measuring devices and techniques have been developed for these purposes. A second aspect of angle measurement not so far touched upon is measurement of angular rotation and, again, accurate techniques and devices are available. Angular measurement may broadly be classified as follows:

(1) Measurement of angular features on components or gauges, and
(2) Measurement of the angular rotation of a divided circle.

Measurement of Angular Features

Angles may be obtained from the combination of accurate lengths to form a triangle. The sine bar, along with its variants the sine table and the sine centres, is designed with this in mind. The bar is essentially a hardened steel beam mounted on two hardened cylinders and conforming to dimensional and geometrical tolerance requirements essential to the accurate establishment of a desired angle (see Table 3.3). The cylinder centre distance (L) provides the hypotenuse of the triangle and block gauges (H), the opposite side (Fig. 3.39*a*).

TABLE 3.3 *Sine Bar Tolerances*

Nominal cylinder	*Tolerances* (mm)	
Centre distance (mm)	100	200 and 250
Dimension		
Cylinder centre distance	±0·0025	±0·005
Cylinder diameter equality	0·0025	0·0025
Geometry		
Upper surface flatness	0·0013	0·0025
Upper surface parallel to datum surface	0·0013	0·0025
Cylinder axes parallel and coplanar	0·0025	0·0025

B.S. 3064:1959, *Sine bars and sine tables (excluding compound tables).*

Sine bars are most frequently used as shown at (3.39*b* and *c*), but for angles greater than 45°, the bar should be set at the complement of the angle as shown at (3.39*d*), as the effect of manufacturing errors in the bar and also those arising from probable accumulated error in the gauge blocks begin to increase rapidly

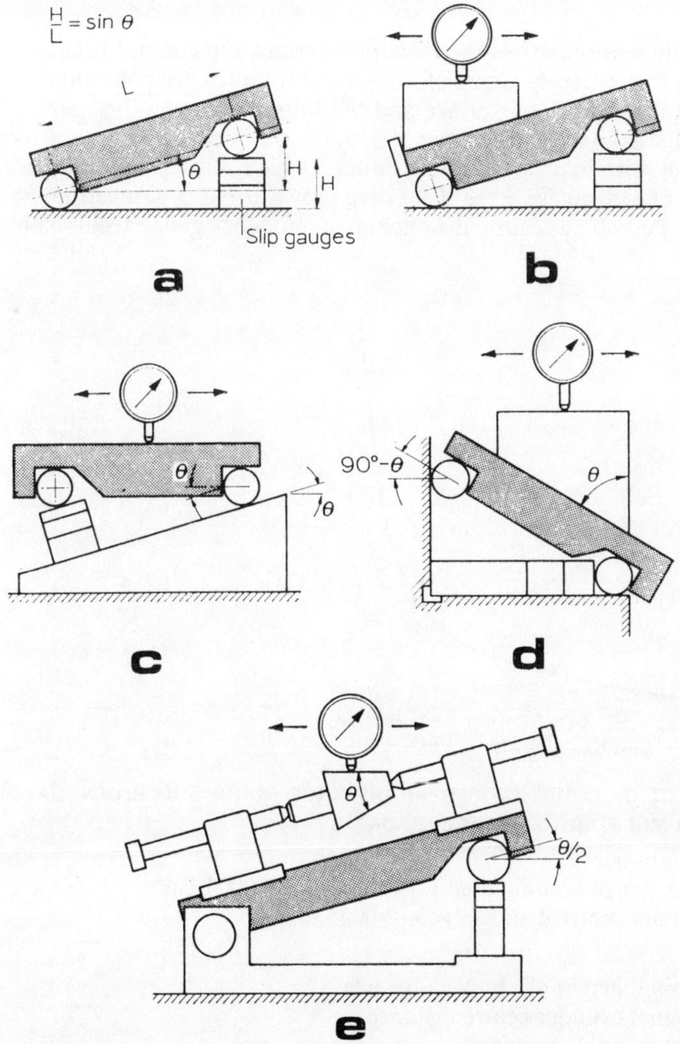

Fig. 3.39 Measurement of angular features

beyond this point. Sine tables have one hinged cylinder and tee slots to facilitate work holding. Compound tables are also available with which the required compound angle is set in two stages. The sine centres shown at (3.39*e*) facilitate measurement of taper on circular work manufactured on centres, or work which may be mounted to a mandrel. It is evident that when the bar and work angles are equal the work surface, or bar surface as the case may be, lies parallel to the datum surface, with the exception of the sine centres when the bar angle must equal one half of the included angle of the work. This condition may be checked with a dial test indicator or other instrument of suitable sensitivity mounted on the datum surface and near to the bar.

Sources of Error

Variations of dimension or geometry within the manufacturers' tolerances introduces small angle errors. Some of these are constant, e.g. a constant angle error is caused if the working surface and the cylinder axes are not parallel, while others, such as an error in cylinder centre distance, introduce a progressive increase of error with increase of angle. Block gauge tolerance accumulation is also a source of progressive error. The curve shown at Fig. 3.40 illustrates the possible effect of combined centre distance error and block gauge accumulated tolerance error.

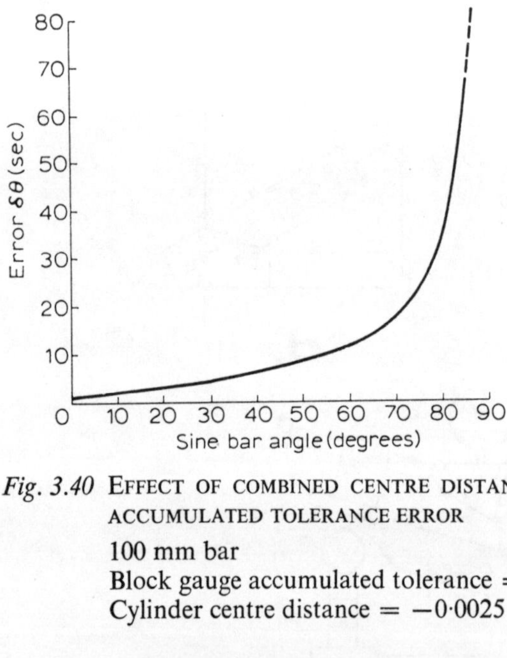

Fig. 3.40 EFFECT OF COMBINED CENTRE DISTANCE ERROR AND BLOCK GAUGE ACCUMULATED TOLERANCE ERROR

100 mm bar
Block gauge accumulated tolerance = +0·0005 mm
Cylinder centre distance = −0·0025 mm

If H = nominal height of block gauge pile
L = nominal cylinder centre distance
θ = nominal angle required

$$\sin \theta = H/L$$

and by partial differentiation,

$$\delta\theta = 1/L\,(\sec \theta \,.\, \delta H - \tan \theta \,.\, \delta L) \quad \text{approx.}$$

where $\delta\theta$ = angle error
δH = accumulated block gauge tolerance error
δL = centre distance error.

Maximum error occurs when H and L are of opposite sign.

Block gauges can provide lengths only in small steps and it may not be possible to build a gauge pile to the exact length required for a given angle.

The error introduced is small but its magnitude depends on the step change of the gauge set (0·0005, 0·001 or 0·005 mm). This and many other error sources are reduced by using a long rather than a short bar.

To ensure that compound angle error is not introduced the axis of the work must always be parallel to the axis of the bar.

3.19 Combination Angle Gauges

An angle gauge is a hardened steel block approximately 75 mm long and 16 mm wide which has two lapped flat working faces lying at a very precise angle to each other (Fig. 3.41*a*). They are supplied in sets and can be wrung together to form angles up to 90° when used in conjunction with a precision square block. The angles of the gauges in a thirteen piece set, which also includes a square block, are shown in Table 3.4. This arrangement allows angles in 3 second steps to be obtained; thus no required angle expressed in whole seconds can be more than 1·5 seconds in nominal error.

TABLE 3.4 *Nominal Angles of Combination Angle Gauges*

Degrees	1	3	9	27	41
Minutes	1	3	9	27	
Fractions of minutes	0·05	0·1	0·3	0·5	

Each angle gauge is a wedge, thus two gauges wrung with their narrow ends together provide an angle which is the sum of the angles of the individual gauges. Subtraction of angles is obtained when the narrow ends are opposed (3.41*b*).

The following example illustrates how an angle is obtained systematically.

Required angle = 33°16′42″

	Gauges required	
Degrees	27 + 9 − 3	33°
Minutes	27 − 9 − 3 + 1	16′
Fractions of Minutes	0·5 + 0·3 − 0·1	42″
A total of 10 gauges		

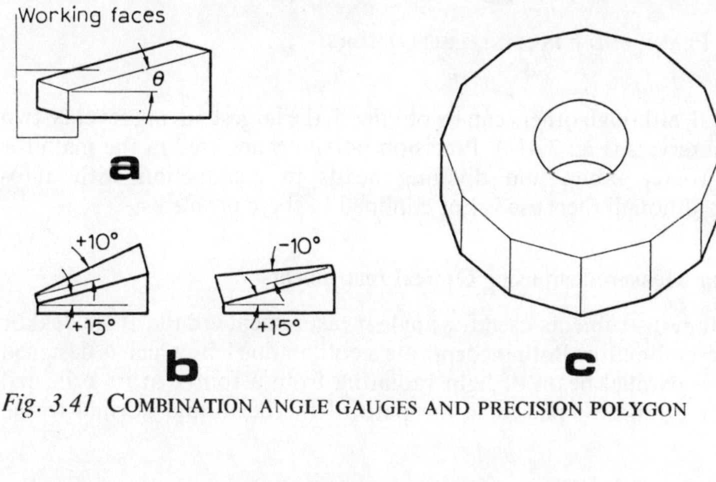

Fig. 3.41 COMBINATION ANGLE GAUGES AND PRECISION POLYGON

A manufacturer's calibration certificate is supplied with the gauges which enables the user to assess the effect of angle tolerance build up. This is, of course, very small but precise angles in terms of seconds of arc must be treated with caution, as tolerance build-up can amount to a few seconds of arc. Because of their high accuracy, angle gauges find wide application in the comparative measurement of angles in conjunction with angle dekkors, auto-collimators, etc.

Precision Polygons

These are hardened and stabilized steel blocks having lapped working faces normal to equal divisions of a circle. Normally a polygon has twelve faces each

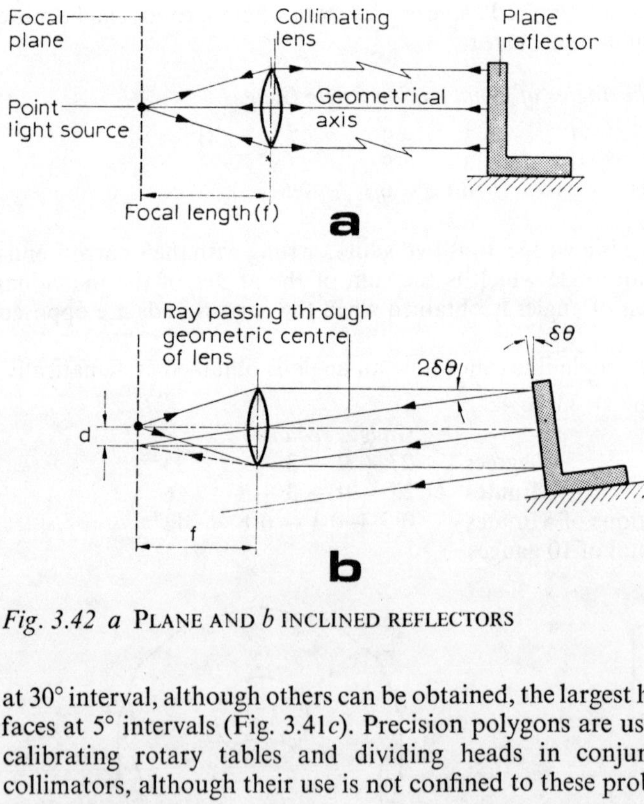

Fig. 3.42 a PLANE AND *b* INCLINED REFLECTORS

at 30° interval, although others can be obtained, the largest having seventy-two faces at 5° intervals (Fig. 3.41*c*). Precision polygons are used in the main for calibrating rotary tables and dividing heads in conjunction with auto-collimators, although their use is not confined to these problems.

3.20 Angular Measurement using Optical Instruments

Two important instruments used for angle measurement are the angle dekkor and the auto-collimator. Both incorporate a collimating lens which is designed to transmit a parallel beam of light radiating from a source at its principal focus. A plane reflector placed in the path of the beam and normal to the geometric axis of the lens will reflect the light along the transmission path to be refocused at the source (Fig. 3.42*a*). If the reflector is inclined at a small angle $\delta\theta$ to the normal, the beam is reflected at an angle equal to $2\delta\theta$ from its transmission path (see section 3.12). Any portion of the reflected beam passing through the lens will be refocused at the focal plane a distance d from the point of principal focus (3.42*b*). An examination of the triangle formed by the ray passing through the geometric centre of the lens and the focal length f, shows that

$$d = 2f\delta\theta \qquad \text{when } \delta\theta \text{ is small}$$

Four important points should be noted:

(1) The point at which the reflected beam is refocused depends on the angle of the reflector to the normal ($\delta\theta$), and is independent of the distance of the reflector from the lens.
(2) The angle $\delta\theta$ must not be so large that for a given instrument to reflector distance the beam of reflected light bypasses the lens altogether.
(3) A long focal length is required for high magnification.
(4) Angular rotation of the reflector in the horizontal plane merely displaces the reflected image to the left or to the right of the source.

Auto-collimator

In the microptic auto-collimator, a target wire is mounted at the focal plane of the lens and illuminated from behind. Both the target wire and its reflected image are seen through a microscope eyepiece with which the instrument is equipped (Fig. 3.43). The eyepiece incorporates a scale graduated in 0·5 minute

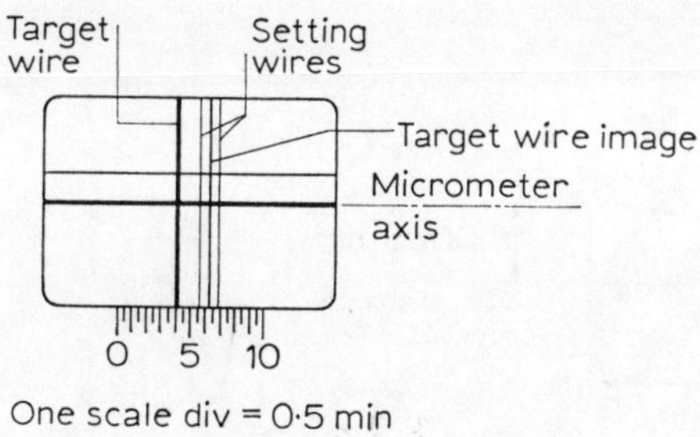

Fig. 3.43 AUTO-COLLIMATOR EYEPIECE

intervals and a pair of parallel setting wires which can be adjusted to straddle the reflected image at each change of angle. Movement of the wires is effected through a micrometer graduated in 0·5 second intervals, one rotation of the drum equalling one scale division movement of the wires, i.e. 0·5 minutes. The instrument is designed to be rotated through 90° about its longitudinal axes so that angles in both the horizontal and vertical planes can be measured. The instrument can be used at a considerable distance away from the reflector although the further the instrument is away the smaller the range of angle measurement. At a distance of 10 m, an angle range $\delta\theta$ is of the order of 10 minutes of arc.

3.21 Applications of Auto-collimators

Auto-collimators are used to solve a wide variety of measuring problems, and the following applications are by no means exhaustive.

Slideway Straightness

The instrument is used in conjunction with a reflector mounted to a base of such a length as to provide 0·5 μm change of height for each 1 second of arc

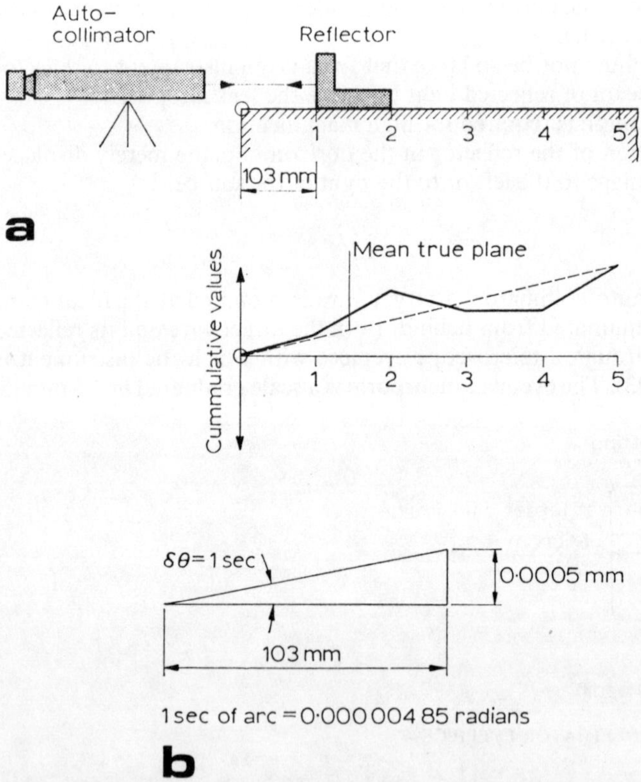

Fig. 3.44 SLIDEWAY STRAIGHTNESS USING AUTO-COLLIMATOR

In *b*, Linear rise or fall *h* (1 second of arc)
$= 103 \times 0{\cdot}000\,004\,85 = 0{\cdot}0005$ mm (0·5 μm)

tilt. The slideway is divided along its length at intervals equal to the basic length and may be numbered as shown in Fig. 3.44*a*. The reading obtained from the first position (0, 1) of the reflector is taken to be zero and the reflector moved to the second position (1, 2). Any change of instrument reading at this position represents an incline or decline of the surface relative to point 1 on the slideway. At the third position (2, 3) the incline or decline is relative to point 2 and so on. The change of angle at each position is easily converted to linear rise and fall and the cumulative values (from zero) plotted to form a profile of the surface (Fig. 3.44*b*). By joining the end points of the plotted profile the mean true plane of the surface is defined and deviations from this line represent the straightness errors of the slideway.

Slideway Squareness

An optical square may be used in conjunction with an auto-collimator to determine the squareness and straightness of machine slideways relative to each other (Fig. 3.45). The square is a prism designed to turn a beam of light through 90° irrespective of the angle of incidence.

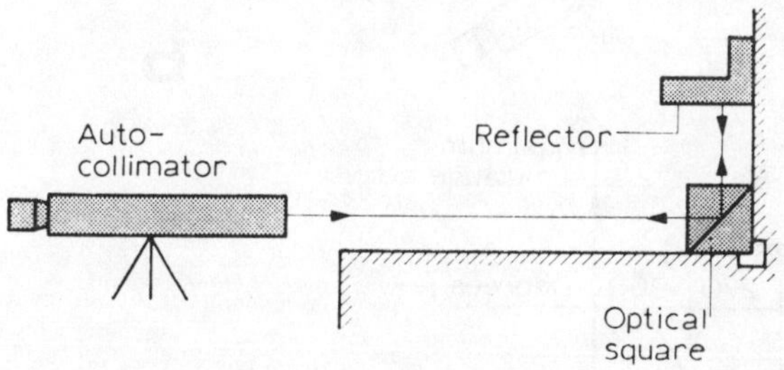

Fig. 3.45 SLIDEWAY SQUARENESS

Axial Slip or Float

It is particularly important that a machine tool lead screw should rotate accurately about its axis. Any longitudinal translation arising for example from lack of squareness of the end thrust bearings should be of insignificant proportions otherwise periodic errors are transmitted to the thread being cut. The arrangement shown at Fig. 3.46 may be used to measure the longitudinal float of a shaft. A float of 0·25 μm is easily detectable by this method.

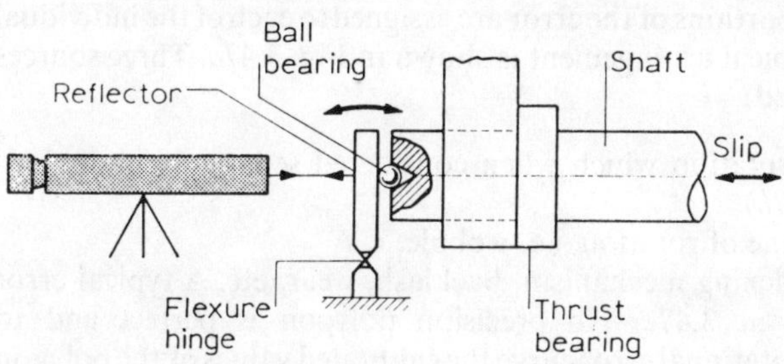

Fig. 3.46 AXIAL SLIP

Angular Rotation

This is most easily done by mounting a precision polygon to the indexing device, e.g. a rotary table or dividing head. The auto-collimator is aligned on a face of the polygon and a full check made by rotating the device to ensure that a reading is obtainable from all faces. The table is set at 0° and the instrument reading at this position taken to be zero. A second reading is taken by indexing

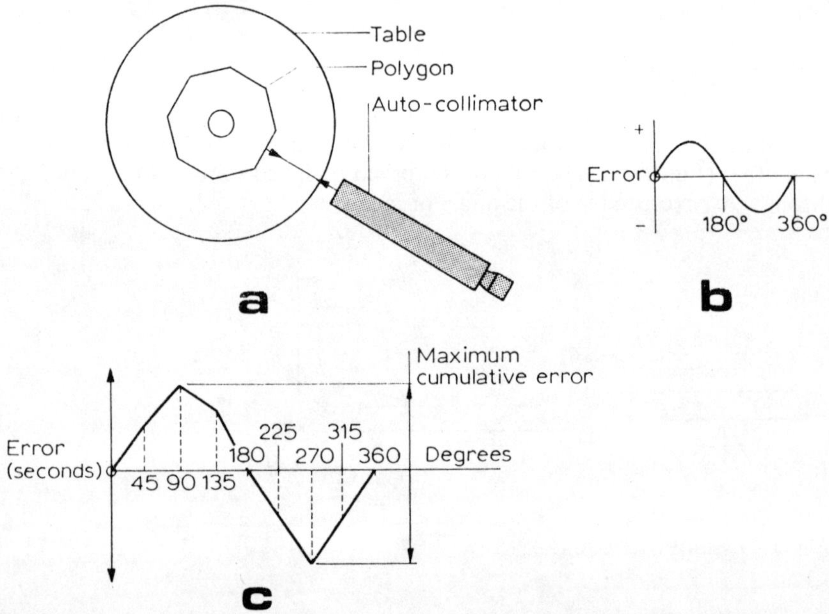

Fig. 3.47 ANGULAR ROTATION

through the nominal angle of the polygon and the difference between this and the first reading equals the error in indexing, assuming a perfect polygon. Successive readings are taken until the device has indexed through 360°. The last reading is, of course, a repeat of the first and should be zero. In a case when it is not zero the error represents the indexing error of the total angle turned through and equal proportions of the error are assigned to each of the individual angles measured. A typical arrangement is shown in Fig. 3.47*a*. Three sources of error can be identified:

(1) Eccentricity of rotation which when considered separately is of sinusoidal form (3.47*b*).
(2) Errors in the plane of rotation, i.e. wobble.
(3) Errors in the indexing mechanism, backlash, wear, etc. A typical error curve is shown at 3.47*c*. No precision polygon is perfect and to establish a true rotational error curve the calibrated values of the polygon must be known and the instrument readings corrected accordingly.

3.22 Angle Dekkor

The angle dekkor is an angle comparator designed for use in conjunction with a setting master which may be a sine bar or a combination of angle gauges. A microscope eyepiece, a ground glass screen and a collimating lens are mounted

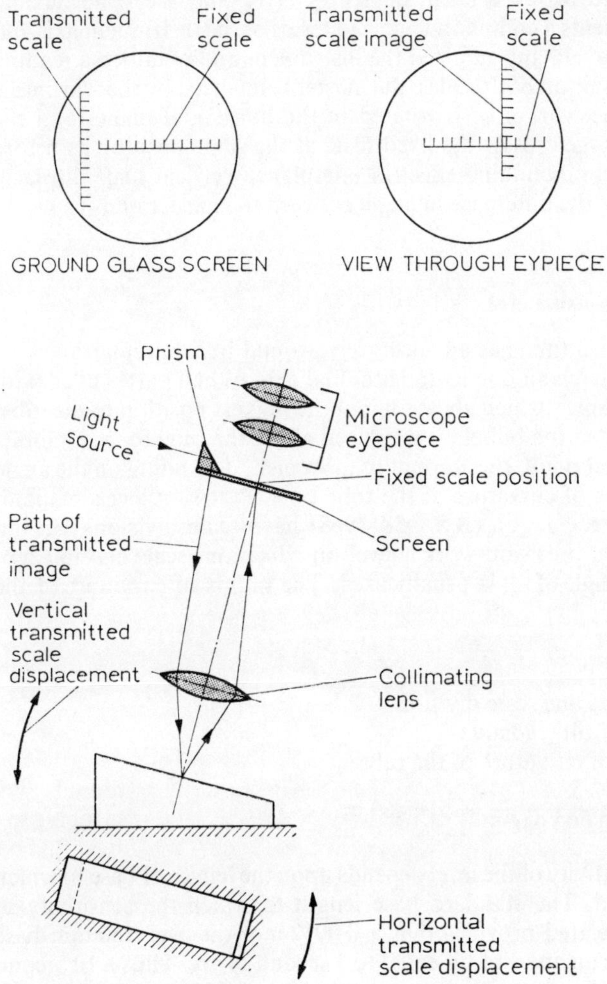

Fig. 3.48 ANGLE DEKKOR

in a tube to form the instrument (Fig. 3.48). The screen, which lies at the focal plane of the lens, has two mutually perpendicular scales marked on it which are graduated in minutes of arc enabling estimations to about 0·2 minutes to be made. The essential function of the instrument is to transmit an image of one of the scales to a reflecting surface which returns the image through the lens to be superimposed on to the other scale which is fixed. Both scales are then viewed simultaneously through the eyepiece. The effect of rotating the reflecting surface through an angle is to displace the transmitted scale image against the fixed scale horizontally or vertically.

The instrument is mounted by means of an adjustable arm to a small surface table. Both scales are seen when the optical axis of the instrument is normal to the reflecting surface within a small degree of error and the scale reading obtained thus represents a compound angle in terms of the instrument axis and the reflecting surface. Having adjusted the instrument and obtained a reading from the setting master on both scales, the master is replaced by the workpiece to be measured. The workpiece is rotated in the horizontal plane until the transmitted scale image crosses the fixed scale at the same position as before thus repeating the compounding effect in one plane. Vertical scale displacement now represents the difference in angle between the master and the workpiece.

3.23 The Precision Spirit Level

A precision spirit level tube has an accurately ground bore, circular in cross-section and barrel-shaped along its longitudinal axis. When partly filled with liquid a bubble is formed which always lies at the highest position in the tube. When the tube is tilted the bubble is displaced along the tube to maintain its position at the highest point, the amount of movement depending on the angle of tilt and the radius of curvature of the tube bore. Precision levels made to British Standard Specification (B.S. 958:1968) have scale divisions 2·5 mm apart and for general precision work a level on which one scale division represents 10 seconds angle of tilt is usually used. The radius of curvature of the tube is given by

$$R = l/\delta\theta$$

where l = length of one scale division
$\delta\theta$ = angle of tilt (radians)
R = radius of curvature of the tube

$$= \frac{2{\cdot}5}{10^3 \times 0{\cdot}000\,048\,5} = 51{\cdot}5460 \text{ m}$$

The ultimate sensitivity of the level depends upon the length of base to which the tube is mounted. The standard base length to which the sensitivity of precision levels is related provides that a 0·0012 mm tilt over 250 mm base length shall produce an angle of tilt equal to 1 second of arc. Thus a 10 second bubble on a 250 mm base requires a tilt equal to 10 × 0·0012 mm for one scale division of bubble movement. The sensitivity of a level increases in inverse proportion to its base length, e.g. a 10 second bubble mounted to a 125 mm base requires only (125/250) × 0·012 mm tilt per division, thus its sensitivity has doubled. One type of block level has a lapped flat base which allows slip gauges to be wrung to the surface so that sensitivity may be varied when required.

When the base of the level lies exactly in the horizontal plane the bubble should lie at the centre of the scale markings and this condition should be repeated when the level is rotated end for end in the same position. However, to eliminate the small error in vial setting relative to the base which must be assumed to exist, two readings are taken at any one position by reversing the level, the mean value indicating the true error in level of the surface.

Levels are used in the main for checking the straightness or flatness of surfaces but one special application is the level comparator (see section 3.12).

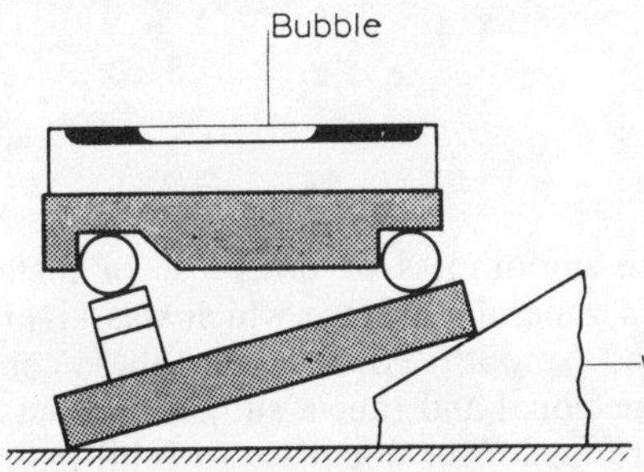

Fig. 3.49 USE OF SPIRIT LEVEL AS AN ANGLE COMPARATOR

A level can also be used as an angle comparator in conjunction with a sine bar or master setting gauge as shown in Fig. 3.49. The wedge provides stepless adjustment of the surface table so that the bubble can be set approximately at the centre of the level scale; this reading then represents the angle of the sine bar. If the component is placed on the table parallel to the sine bar with the level mounted to it, any difference of reading represents the difference in angle between the bar and the component.

4 Assessment of Surface Finish

The earliest method used to assess the smoothness or otherwise of metal surfaces was by sight and touch, and in general a surface which was highly reflective was considered to have a good (smooth) finish. Under these conditions surface quality depended on a personal and thus a subjective assessment. By the 1930's such methods were no longer adequate to fulfil many design requirements and particular attention began to be focussed on methods whereby a quantitative, i.e. numerical, assessment of surface texture could be made which would be independent of personal opinion. Subsequent experience has shown that no single parameter, e.g. average peak to valley height, is capable of providing a fully adequate description of surface texture characteristics. Present practice is to specify one parameter. This, together with the production process to be used and the direction in which the measurement is to be made, is considered sufficient to control the surface finish required for many purposes although it must be understood that this by no means provides all the information about a surface.

Before any attempt can be made to provide a measure of the characteristics of a surface it is necessary to define what is generally meant by the terms "surface finish" or "surface texture", as developed by machining processes and finishing techniques, such as grinding, lapping, turning, milling, die casting, forging, etc.

4.1 Surface Texture

Those regular or irregular surface spacings which tend to form a pattern on the surface. The general aspect of the surface produced by the manufacturing process.

The component parts of the surface texture may be defined as follows.

ROUGHNESS The short wavelength irregularities arising from the production process which comprises individual scratch or tool marks such as that produced by a single traverse of a planing tool across the surface. Such marks contain within them further small irregularities which are also included in the definition.

THE LAY The tool or scratch marks taken collectively which characterize the particular process. Where these show definite directional characteristics as, for example, in planing, then this is called the "lay" of the surface. Surface finish measurements are usually, but not always, taken across the lay as the surface is at its roughest in this direction. The effect of inclining the direction of measurement from the direction normal to the lay is to cause an apparent increase in the wavelength of the lay grooves, thus giving a false impression of the shape of the surface (Fig. 4.1).

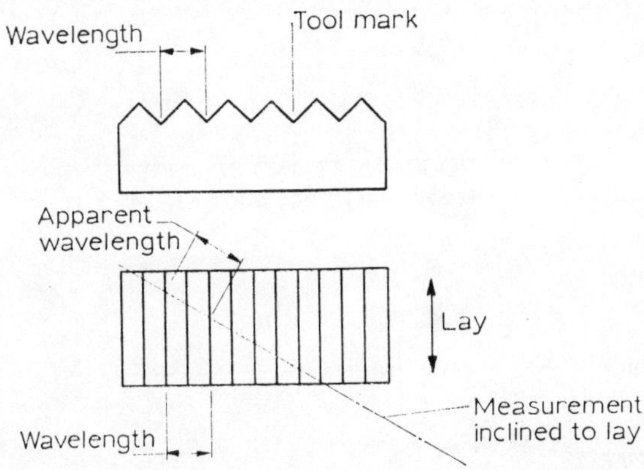

Fig. 4.1 ERROR FROM MEASURING AT AN ANGLE TO THE LAY

WAVINESS The longer wavelength irregularities upon which roughness is superimposed. Waviness may be induced by vibration, hard spots, imperfect truing of a grinding wheel, chatter, heat treatment, etc.

General Errors of Form

These may arise from general kinematic faults in machine tools, such as slideway wear, or from induced kinematic errors such as tool above or below centre when taper turning. Again, there may be general flexure of the workpiece during machining giving rise to errors of form. Such errors are not generally included in surface finish measurement because of their long wavelength. The diagrams shown in Fig. 4.2 illustrate schematically the ways in which a surface may deviate from perfect smoothness and geometrical form.

From Fig. 4.2 it is clear that the characteristics of a surface are closely associated with the wavelength and amplitude of the surface irregularities and, in particular, wavelengths from different sources may be superimposed one upon another to form a complex wave pattern. The signals from a modern electronic stylus instrument may be used to form a graphical record of a portion of the profile of the surface under investigation or to provide a numerical parameter. When a parametric assessment of a surface is required the presence of long wavelengths will affect the value obtained usually by increasing it; thus the influence of longer wavelengths must be suppressed. The form that this suppression takes depends on the basic design of the stylus instrument.

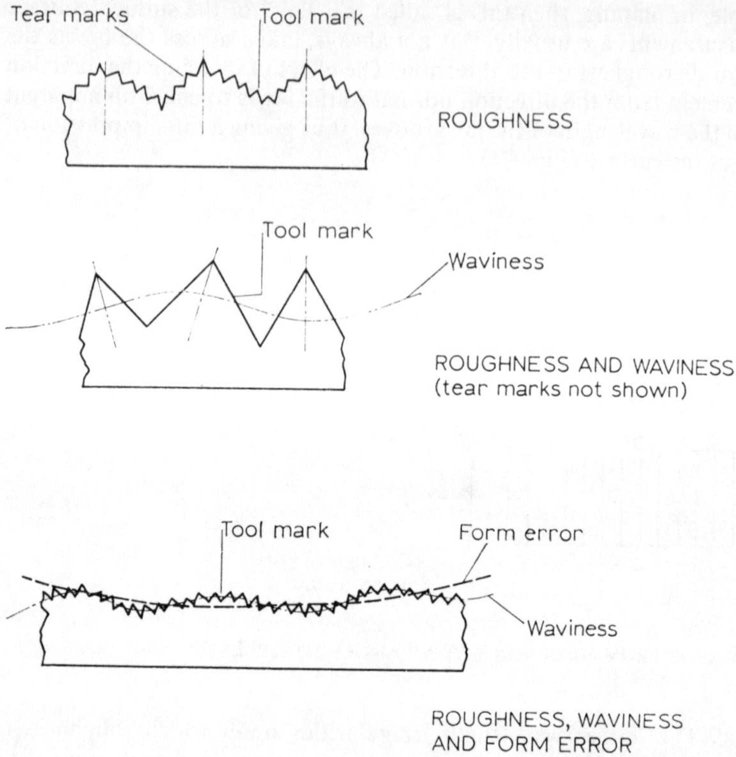

Fig. 4.2 ERRORS OF FORM

4.2 Parameters used in Surface Finish Measurement

A number of parameters have been specified and used in various countries throughout the world and the following names and their abbreviations are probably the most well known:

Centre line average: R_a, h_{cla}, R_u, *a.a.*

Root mean square: r.m.s., R_q.

Maximum peak to valley height in the sample: R_t, R_{max}.

The average of the R_t values found in five consecutive samples: R_{tm}.

The average of the five highest peaks and five deepest valleys in the sample: R_z.

The average or levelling depth of the profile: R_p.

The parameter which has become the most widely accepted over the years is the centre line average height h_{cla} which is a mean line or M-parameter. Arising from the current and projected change to SI units, international agreement has been reached on the terminology, standards for instruments and roughness comparison specimens. The new British Standard has now been published (1972), incorporating the necessary revisions.

Two parameters are described in the standard:

(*a*) Arithmetical mean deviation (R_a). This is the primary parameter and is identical to the centre line average height of the former Standard B.S. 1134:1961. The advantage of this parameter is that it lends itself to electrical integrating stylus instruments which provide a direct meter reading.

(*b*) Ten point height of irregularities (R_z). Parameters involving peak to valley assessment have been more popular on the Continent because development there has tended to the production of purely graphical records from which it is a rather tedious process to obtain an integrated average value. In cases where the surface is too short for integrated values to be obtained, peak to valley assessments are of use and they are also used where the parameter has special significance for the function of the surface, for example its frictional characteristics.

The Arithmetical Mean Deviation (R_a)

When obtaining the roughness average (R_a) from a graphical trace of the surface profile, it is necessary first to establish the mean line. This may be generated on the trace by a simple two-stage *CR* filter network whose characteristics are now agreed and standardized by those countries subscribing to I.S.O. recommendations.

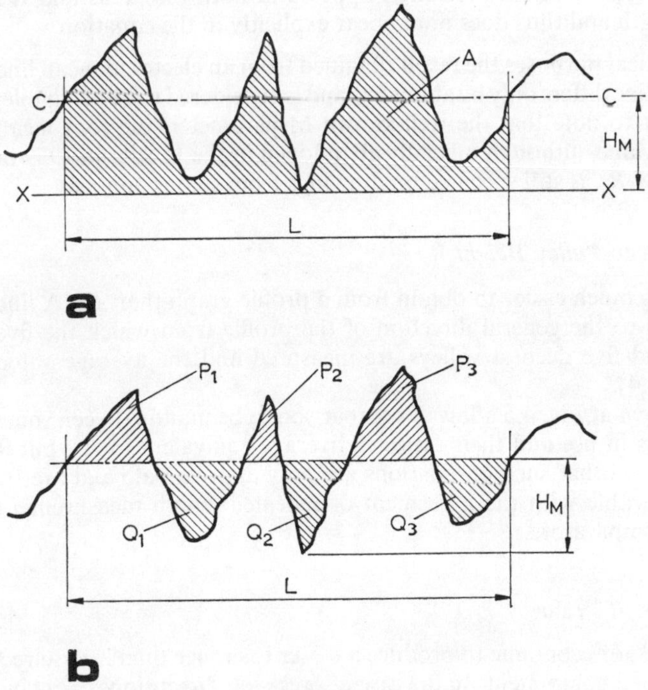

Fig. 4.3 OBTAINING THE ROUGHNESS AVERAGE

The arithmetical mean deviation (R_a) of a surface may be established graphically as follows:

(1) Draw a straight line X–X in the general direction of the surface and for convenience touching the deepest valley (Fig. 4.3*a*).
(2) Select a suitable sampling length L such that if the surface has a distinguishable waveform a whole number of waveforms are enclosed.
(3) Find the area A under the curve using a planimeter or ordinate method. The height of the centre line C–C may be found from

$$H_m = \text{area } A/L$$

where H_m is the perpendicular height from X–X to C–C.
(4) The trace is now divided into two halves so that the sum of the enclosed areas above the line ($P_1 + P_2 +$ etc.) equates the sum of the enclosed areas below the line ($Q_1 + Q_2 +$ etc.) (see Fig. 4.3*b*). R_a may now be determined:

$$R_a\ \mu\text{m} = \frac{\text{areas } P + \text{areas } Q}{L} \times \frac{1000}{V_v}$$

where V_v = vertical magnification and the areas P and Q are expressed in square millimetres and the length L in millimetres.

The average of a number of samples is taken as the R_a for the measuring traversing length.

N.B. The horizontal magnification appears in both the areas and the sampling length and thus does not appear explicitly in the equation.

For most practical purposes the result obtained from an electrical mean line or a drawn mean line differs only by about 5% and is considered to be negligible.

It is of interest to note that the well-known M-parameter, the root mean square or r.m.s. value, although officially abandoned in the U.S.A. in 1959 in favour of h_{cla} now R_a, is still specified on drawings from some countries.

The Average Peak to Valley Height R_z

This parameter is much easier to obtain from a profile graph than R_a. A line is drawn parallel to the general direction of the profile from which the five highest peaks and five deepest valleys are measured and the average value computed (Fig. 4.4).

The graph shown at Fig. 4.5 allows a comparison to be made between some of the parameters in use and their expected average equivalent in R_a, but it must be emphasized that such correlations are only approximate and are by no means comparable with the agreement of repeated length measurement using different comparators.

4.3 Tolerances for R_a Values

It is not desirable nor economic to produce a closer tolerance than is required for the function of a component. In the case of precision dimensions in engineering the tolerance in 95% of all cases is less than 1% of the nominal dimen-

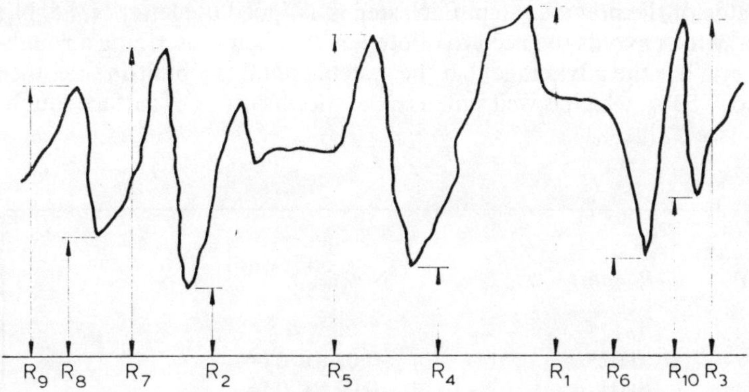

Fig. 4.4 OBTAINING AVERAGE PEAK TO VALLEY HEIGHT

$$R_z = \tfrac{1}{5}[(R_1 + R_3 + R_5 + R_7 + R_9) - (R_2 + R_4 + R_6 + R_8 + R_{10})] \times \frac{1000}{V_v}$$

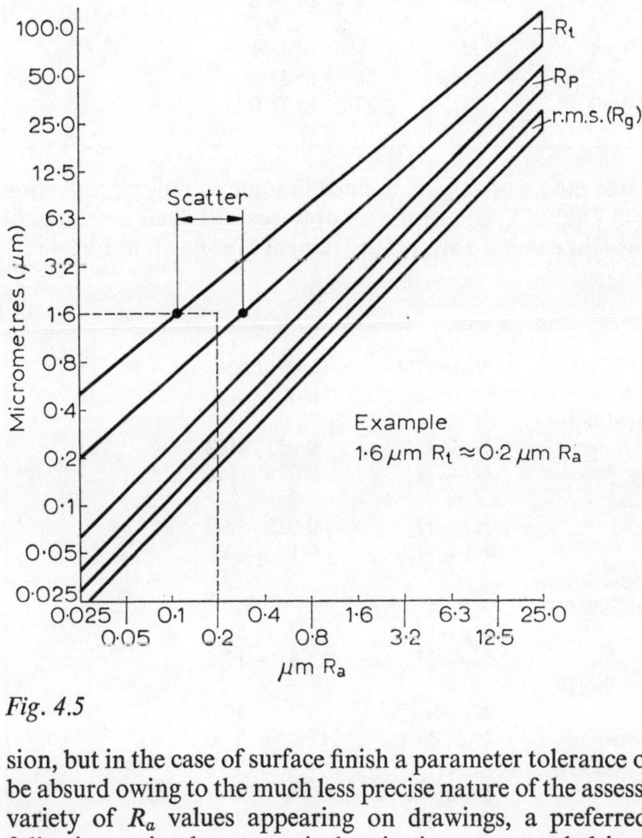

Fig. 4.5

sion, but in the case of surface finish a parameter tolerance of this order would be absurd owing to the much less precise nature of the assessment. To limit the variety of R_a values appearing on drawings, a preferred range of values following a simple geometrical series is recommended in which each step is twice the value of the previous step. Each step is assigned the letter N, i.e. N_1, N_2, N_3, etc., which avoids the need to quote μm R_a when prescribing a roughness value, and has the advantage that the system contains a built-in tolerance of −25% to +50%, which is well suited to the specifications of surface roughness values (see Table 4.1).

TABLE 4.1

Central values c.l.a. (μm)	R_a (μm)	N	N $^{+50\%}_{-25\%}$ (μm)
0·5	0·125	—	— —
1	0·025	N_1	0·02 to 0·04
2	0·05	N_2	0·04 to 0·08
4	0·1	N_3	0·08 to 0·15
8	0·2	N_4	0·15 to 0·3
16	0·4	N_5	0·3 to 0·6
32	0·8	N_6	0·6 to 1·2
63	1·6	N_7	1·2 to 2·4
125	3·2	N_8	2·4 to 4·8
250	6.3	N_9	4·8 to 9·6
500	12·5	N_{10}	9·6 to 18·75
1000	25·0	N_{11}	18·75 to 37·5
2000	50·0	N_{12}	37·5 to 75·0

A guide to probable ranges of values obtained in some common production processes is shown in Table 4.2, and a number of traces with their corresponding average R_a values taken on a Talysurf instrument are shown in Fig. 4.6.

TABLE 4.2

Process	*Range* (N)	*Central values* (R_a)
Flat and cylindrical lapping, superfinishing	$N_1 - N_4$	0·025 – 0·2
Diamond turning	$N_1 - N_6$	0·025 – 0·8
Flat and cylindrical grinding	$N_1 - N_8$	0·025 – 3·2
Linishing	$N_4 - N_8$	0·1 – 3·2
Face and cylindrical turning, and milling, reaming	$N_5 - N_{12}$	0·4 – 50·0
Drilling	$N_7 - N_{10}$	1·6 – 12·5
Shaping, planning, horizontal milling	$N_6 - N_{12}$	0·8 – 50·0
Sand casting and forging	$N_{10} - N_{11}$	12·5 – 25·0
Extruding, cold rolling, drawing	$N_6 - N_8$	0·8 – 3·2
Die casting	$N_6 - N_7$	0·8 – 1·6

The use of unequal magnifications to produce a graphical trace is generally unavoidable because the variations in height of the surface irregularities are much smaller than their wavelength. The resulting distortion gives a false impression of the shape of the surface profile and it is important to remember that they are in fact much smoother than traces of this type would suggest.

4.4 Instruments and Techniques

A number of instruments and techniques are in use for the purpose of surface finish assessment but the following are probably the most important from the point of view of the engineer:

(1) Electrical integrating stylus instruments.
(2) Tactile (finger-nail) surface assessment and its development in the form of mechanical instruments.
(3) Light interference microscopes of various types.
(4) The production of plastic replicas of the surface.

Although (4) is not an independent method it is useful, for instance, in cases where the surface is inaccessible to instruments.

Light interference and replica methods are area sampling methods, while stylus and tactile methods sample a number of profiles or points on the surface.

Electrical Integrating Stylus Instruments

These instruments provide the fundamental information required for surface measurement, i.e. a numerical assessment displayed on a meter or a graphical trace. Other methods of assessment can be and are referred to the results obtained by these instruments when required. The overall calibration of electrical stylus instruments is made against standard etched or electroformed surfaces having, for example, a uniform zig-zag profile of known dimensions. To check the minimum surface variation obtainable by the instrument a scratch-free optical flat may be used as a test piece.

Electrical integrating instruments employ an arm on which a diamond stylus is mounted. The stylus is drawn across the sample surface and its vertical movement transduced to electrical signals. The principal features of electronic stylus instruments are shown in Fig. 4.7.

At Fig. 4.7*a* the movement of the arm which constitutes an armature varies the air gap at the outer limbs of the E-shaped core; the coil-wound limbs form two branches of a bridge circuit. A carrier wave is supplied to the bridge and the effect of varying the air gaps is to modulate or vary the amplitude of the waveform. The modulated wave is then amplified, demodulated (or smoothed), and either filtered to provide a meter readout or in the unfiltered condition made to operate a rectilinear recorder thus providing a graphical trace.

A moving coil type instrument is shown at 4.7*b*. The coil is mounted above the stylus and moves with the stylus relative to a radial magnetic field provided by a magnet. The current generated by the vertical stylus movement is proportional to the amplitude and the wavelength or frequency of the stylus movement; the frequency depends on the spacing of the irregularities and the speed

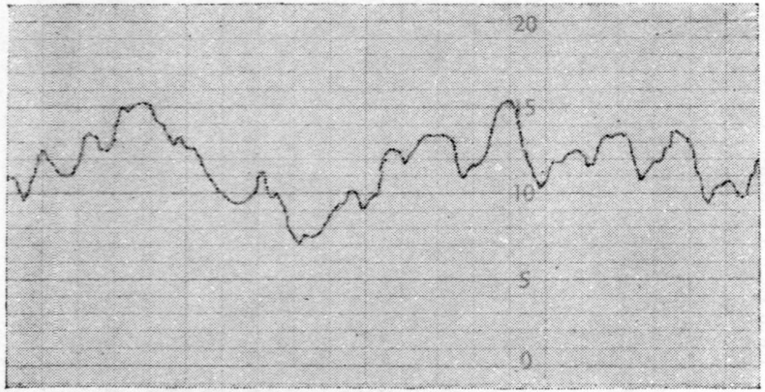

(*a*) *Shaped surface*
Feed 0·25 mm; depth of cut 0·25 mm
Talysurf settings: mag. ratio V/H = 1000/100
meter cut-off = 0·75 mm

Results: Average of 10 measuring traversing lengths = 3·52 $\mu m R_a$
Maximum = 4·5 $\mu m R_a$ Minimum = 3·0 $\mu m R_a$
R_a of graphical sample = 3·75 $\mu m R_a$

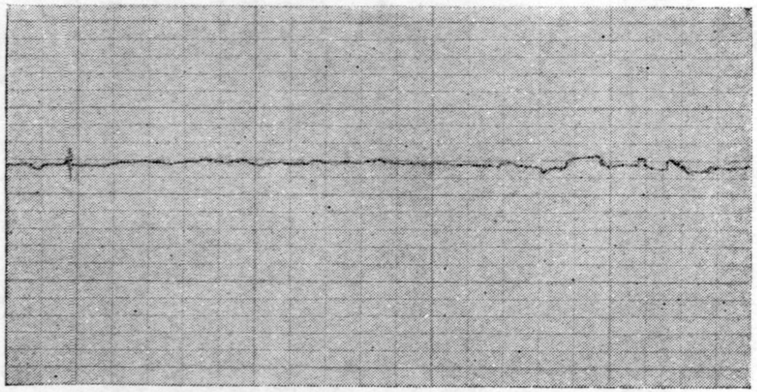

(*b*) *Lapped gauge block*
Talysurf settings: V/H = 50 000/100
meter cut-off = 0·75 mm

Results: Average of 5 measuring traversing lengths = 0·008 $\mu m R_a$

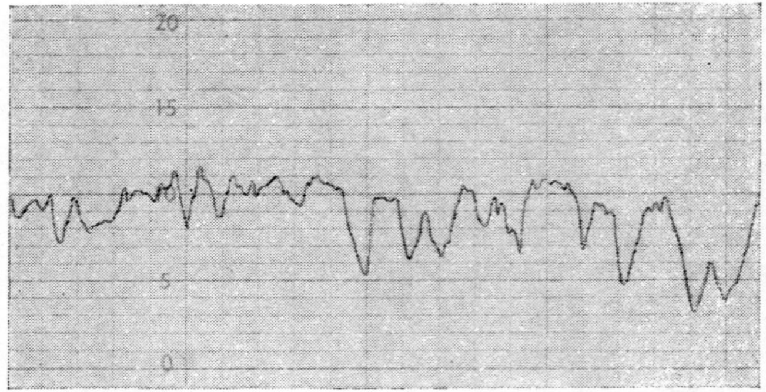

(*c*) (i) *Face milled surface*
Feed = 0·3 mm per tooth; depth of cut 0·2 mm
Talysurf settings: V/H = 2000/100
meter cut-off = 0·75 mm

Results: R_a of graphical sample = 2·22 $\mu m R_a$

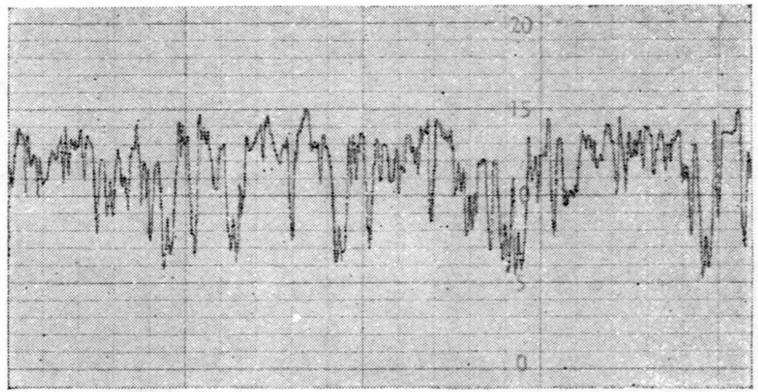

(*c*) (ii) *Face milled surface* As (*c*) (i), V/H = 2000/20

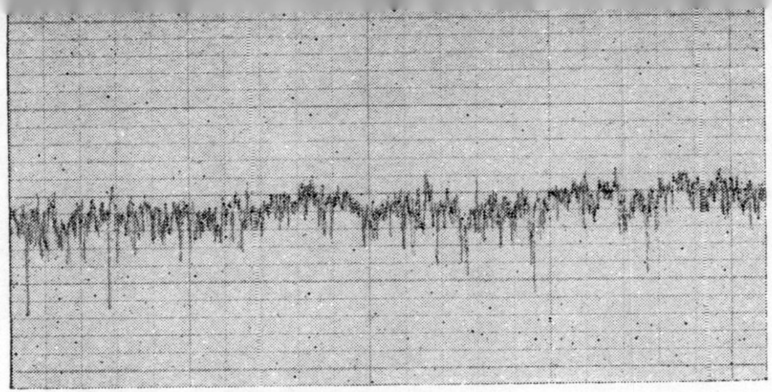

(*d*) (i) *Ground surface (horizontal)*
Talysurf settings: V/H = 20 000/20
meter cut-off = 0·75 mm

Results: R_a of graphical sample = 3·8 μmR_a

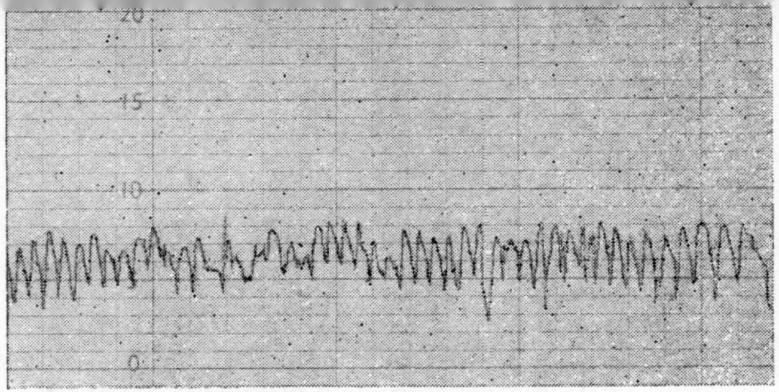

(*e*) (i) *Diamond turned aluminium alloy piston (skirt)*
Talysurf settings: V/H = 5000/20
meter cut-off = 0·75 mm

Results: Average of 5 measuring traversing lengths = 0·51 μmR_a
Maximum = 0·55 μmR_a Minimum = 0·48 μmR_a
R_a of graphical sample = 0·5 μmR_a

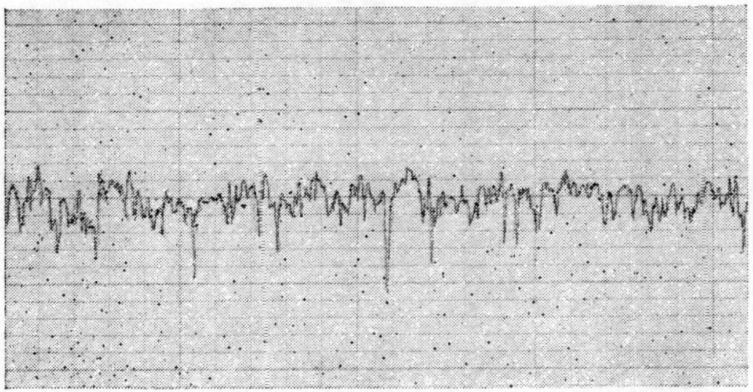

(*d*) (ii) *Ground surface (horizontal)*
As (*d*) (i), V/H = 20 000/100

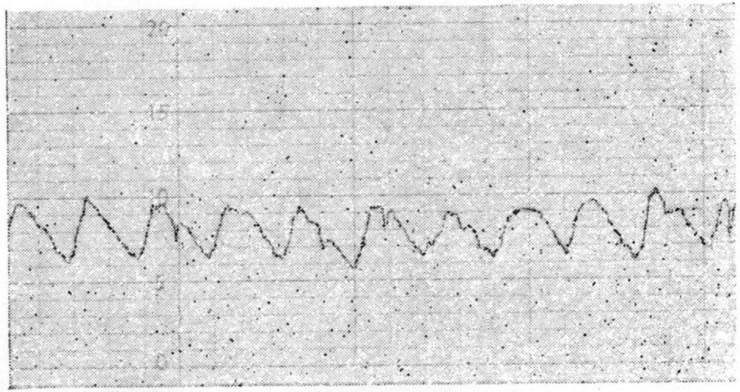

(*e*) (ii) *Diamond turned aluminium alloy piston (skirt)*
As (*e*) (i), V/H = 5000/100

Fig. 4.6 PORTIONS OF TRACES TAKEN ON A TALYSURF INSTRUMENT, SHOWING AVERAGE R_a VALUES

at which the stylus is traversed across the surface. Because the instrument must be insensitive to frequency within its working range, an *RC* correcting circuit is inserted between the receiving amplifier and the amplifier which transmits the signal to the meter. This eliminates unwanted frequencies from the signal before it arrives at the second amplifier. Such instruments are suited only to numerical assessments. They may be drawn across the surface by hand, but are usually motorized.

A piezo-electric (voltage generating) instrument is shown at 4.7*c*. The stylus arm is connected to a crystal which is pivoted to the body of the instrument.

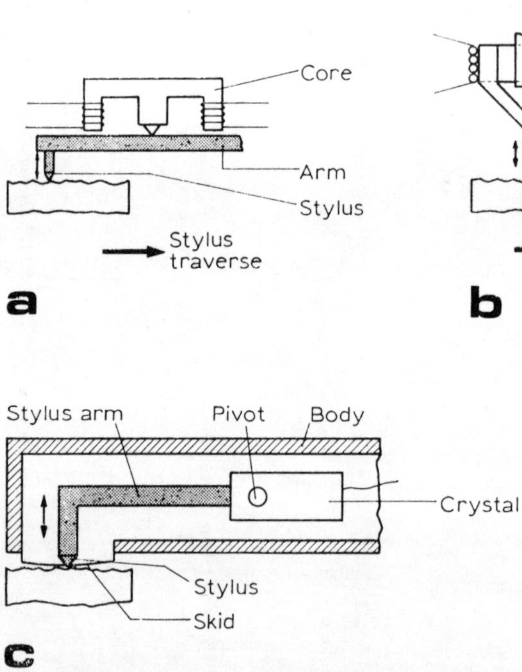

Fig. 4.7 PRINCIPAL FEATURES OF ELECTRONIC STYLUS INSTRUMENTS

The head incorporates a skid which slides over the surface. At low frequencies the crystal and stylus rotate together but at higher frequencies the inertia of the crystal opposes rotation and the stylus then moves relative to the crystal. The effect of this movement is transmitted through the stylus arm causing mechanical vibrations at the crystal which are transformed into proportional voltages. The amplified signal can be used to provide a meter reading or a graphical trace.

THE STYLUS Styluses are normally conical or pyramidal in form and the tip is either flat or radiused. A rounded tip will fully explore any scratch mark provided that it is sufficiently shallow and its radius is equal to or larger than the tip radius. Tip radii vary in size depending on requirements, but for general profile recording 2 μm is suitable, while for instruments designed to give average readings only, 10 μm is acceptable. A flat-tipped stylus is largely independent of depth and will explore any scratch which is wider than the flat if the slope of the sides of the scratch are not sheer. There may of course be deep narrow scratches which a tip cannot explore, but experience has shown that these are usually of little significance.

THE DATUM A datum is required in all instruments which mechanically explore surface irregularities and should conform to the nominal geometry of

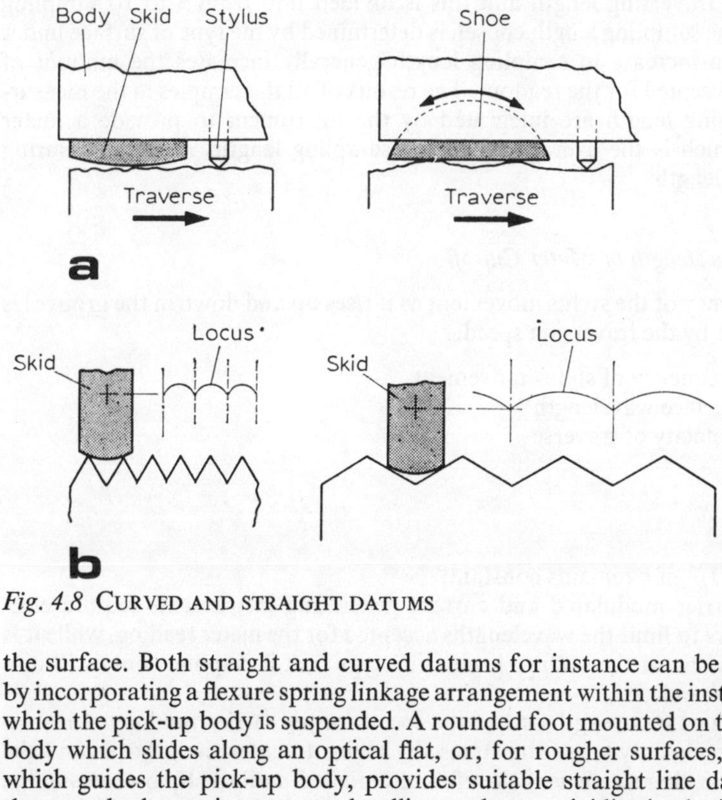

Fig. 4.8 CURVED AND STRAIGHT DATUMS

the surface. Both straight and curved datums for instance can be developed by incorporating a flexure spring linkage arrangement within the instrument on which the pick-up body is suspended. A rounded foot mounted on the pick-up body which slides along an optical flat, or, for rougher surfaces, a slidway which guides the pick-up body, provides suitable straight line datums. All these methods require accurate levelling and great rigidity in the instrument framework and it is more usual to use a single rounded skid or a flat swivelling pad or shoe as approximate datum forming devices. Skids or shoes are mounted to the pick-up body, which is hinged to the driving mechanism of the instrument (Fig. 4.8*a*).

The datum formed by a skid is the locus of the centre of curvature of its radius which usually ranges from about 6 mm to 50 mm. The greater the spacing of the crests of the surface, the more the skid movement departs from a straight line (Fig. 4.8*b*). This increases the relative movement between the skid and the stylus and introduces phase effects which may give a false impression of the surface. However, provided attempts are not made to measure surfaces beyond the scope of the instrument, e.g. rough milled surfaces, such phase effects do not seriously affect the results. The shoe-type pick-up is used when the surface crests are wide apart, i.e. up to half the length of the shoe. There is no relative position of the stylus to the skid or shoe where phase effects are eliminated.

4.5 Measuring Traversing Length and Sampling Length

One complete traverse of the stylus of an electronic instrument is called the measuring traversing length and this is divided into from 3 to 10 sampling lengths. The sampling length chosen is determined by the type of surface under test and an increase in sampling length generally increases the amount of waviness accepted for the readout. The results of all the samples in the measuring traversing length are integrated by the instrument to provide a meter reading which is the average of all the sampling lengths in one measuring traversing length.

Cut-off Wavelength or Meter Cut-off

The frequency of the stylus movement as it rises up and down in the grooves is determined by the traversing speed.

If f = frequency of stylus movement
λ = surface wavelength
v = velocity of traverse

then

$$f = v/\lambda$$

Thus, $f \propto 1/\lambda$, if v remains constant.

Both carrier modulated and current generating instruments require electronic filters to limit the wavelengths accepted for the meter reading, while it is a characteristic of the voltage generating type that it responds only within a given waveband.

At the short wavelength end of the spectrum the carrier modulated instrument responds only when the stylus tip begins to enter the grooves and the transmission characteristic is therefore attenuated until the tip bottoms in the grooves. An electronic filter is used to reject those frequencies which result from very short wavelengths as these signals are attenuated and also to reject those of greater wavelength than required; thus the meter reading is the product of a given waveband known as the pass band. The transmission characteristic is shown in Fig. 4.9*a*. The response of the current generating instrument is attenuated at high frequencies by the stylus effect and because the magnitude of the current generated depends on the relative velocity of the coil to the magnet, it is also attenuated at low frequencies (long wavelength), as the output then tends to zero. The transmission characteristic is shown in Fig. 4.9*b*. The upper and lower limits of wavelength cut-off are 75 ± 5% of the transmission characteristic. The wavelength or meter cut-off is made equal to the sampling length and can be varied by switching to suit the type of surface under test.

For surfaces produced by single point tools, a useful guide to determining meter cut-off is that the cut-off should exceed one feed spacing, but for many

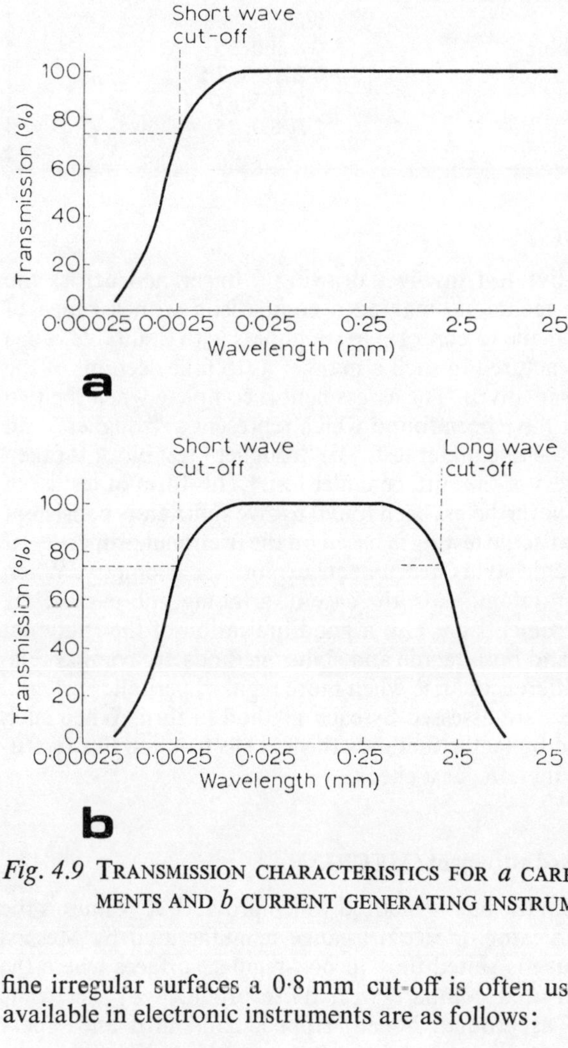

Fig. 4.9 TRANSMISSION CHARACTERISTICS FOR *a* CARRIER MODULATED INSTRUMENTS AND *b* CURRENT GENERATING INSTRUMENTS

fine irregular surfaces a 0·8 mm cut-off is often used. The cut-offs usually available in electronic instruments are as follows:

Meter cut-off (mm)	0·08	0·25	0·8	2·5	8·0	25·0

The 0·08 mm cut-off is generally used for fine surfaces too short for longer cut-offs.

Finishing processes and suitable sampling lengths (meter cut-offs) are shown in Table 4.3.

TABLE 4.3

Finishing processes	*Meter cut-offs* (mm)
Superfinishing, lapping, honing, diamond boring and turning, spark machining, polishing and buffing	0·25 and 0·8
Turning, reaming, broaching	0·8 and 2·5
Grinding	0·25, 0·8, 2·5
Boring, milling, shaping	0·8, 2·5, 8·0
Planing	2·5, 8·0, 25

Tactile Surface Assessment

This essentially comparative test involves drawing a finger nail across the surface and by the same technique making a comparison with a range of electroformed test blocks made to conform in roughness with a suitable range in the N-series and manufactured in such a manner as to take account of the actual production process involved. The assessment is complete when the two standard blocks in the set have been found which represent a "rougher" and "smoother" surface than the one under test. The "rougher" test block is taken to be the grade of roughness of the surface under test.* This form of test is, of course, subjective but has nevertheless been found to give sufficiently consistent results in many instances. Tactile testing is based on the frictional properties of the specimen while electronic stylus instruments explore the geometry. When surface irregularities are random, as in the case of grinding, the parameters obtained from stylus instruments are also a good indication of the frictional properties of the surface and both tactile and stylus methods are comparable. However, considerable differences arise when more regular, periodic surfaces, e.g. rough turned surfaces, are assessed by each method in turn. When such surfaces are to be assessed by tactile methods then an electronic stylus instrument must be used to evaluate R_a as a check.

4.6 Mechanical Roughness Instrument (MECRIN)

A recent development from the tactile method which provides R_a values is the mechanical roughness indicator or MECRIN gauge manufactured by Messrs. Rubert and Co. This gauge is suited only to non-regular surfaces where the frictional properties measured compare with those provided by electronic stylus instruments, such as ground, honed, lapped, hand filed and emery clothed surfaces, etc., to approximately 4·5 μm R_a (grades N_8 to N_9) and on the finer grades of turning and milling to 0·8 μm R_a (grade N_6). On regular surfaces of rougher values the instrument can be used as a comparator against surface roughness blocks but the readings then bear no relation to R_a values.

* B.S. 2634, *Roughness comparison specimens*

The Principle of the MECRIN Instrument

The MECRIN instrument assesses the surface through its frictional properties and the average slope of the irregularities.

A thin metallic blade when pushed against a surface at a certain angle will either slide or buckle depending on two factors:

(*a*) the surface roughness,
(*b*) the angle of attack, i.e. the angle of the blade to the surface.

The essential behaviour of the blade is shown in Fig. 4.10. At angles below the critical angle the blade tip will simply slide along the surface (4.10*a*), but

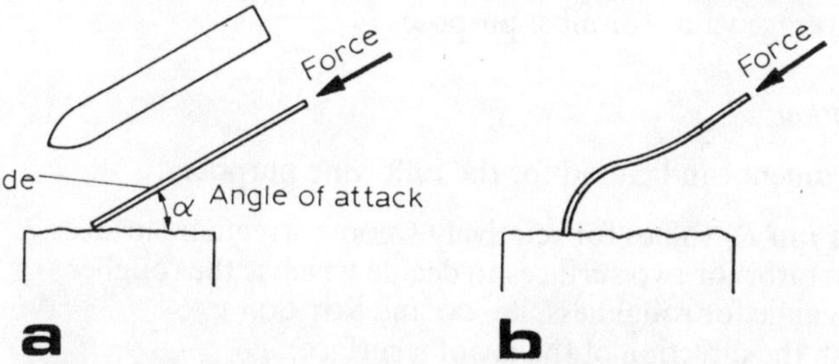

Fig. 4.10 BEHAVIOUR OF THE BLADE IN THE MECRIN INSTRUMENT, SHOWING *a* SLIDING BLADE AND *b* BUCKLING BLADE

if the angle is successively increased, the angle will be found at which the blade just begins to buckle (4.10*b*). This angle is the critical angle of attack. The smoother the surface the greater the critical angle and thus this angle is a function of the degree of roughness encountered. The essential form of the instrument is shown at Fig. 4.11 and the angle of attack is measured with a

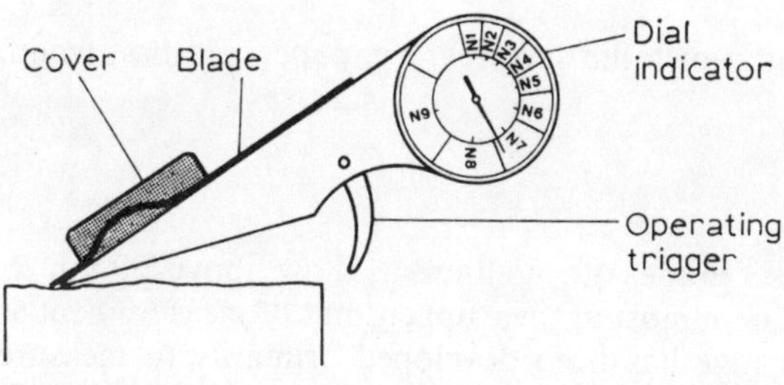

Fig. 4.11 BASIC FORM OF THE MECRIN INSTRUMENT

gravity dial indicator which incorporates a pendulum and gear mechanism and the buckling blade is observed through the transparent cover. The dial is calibrated in terms of the N grades and the corresponding central R_a values.

Calibration of the instrument is against hardened carbon steel specimens having nominal values of 0·1 and 0·4 μm R_a which are accurate to within $\pm 12\%$.

Accuracy of Determination

It is good practice to take the average of at least five readings when using an electronic stylus instrument, because the R_a value of each measuring traversing length on machined surfaces scatters by anything from 10% to 100% depending on the machining method. Each measuring traversing length, however, consists of from three to ten sampling lengths, thus the scatter from one sampling length to another may be greater than 100%. Because the MECRIN is a purely mechanical instrument it evaluates each individual sampling length and thus its scatter of results is generally larger than those obtained from electrical stylus instruments. However, the manufacturers claim that for general workshop purposes three to five readings on different parts of the surface produces a good enough average value for most purposes.

Range of Application

The MECRIN instrument can be used for the following purposes:

(*a*) To provide μm R_a values for relatively smooth irregular surfaces.
(*b*) As a comparator for two surfaces to decide which is the rougher.
(*c*) As a limit gauge for roughness, i.e. GO and NOT GO.
(*d*) To establish the direction of the lay of a surface.
(*e*) In the case of a surface having a definite lay to determine the direction of maximum roughness, e.g. a saw tooth profile of unequal angle. Electronic stylus instrument meter readings do not differentiate this condition.
(*f*) To establish a numerical scale of the frictional qualities of rough, regular surfaces. No correlation exists between such values and μm R_a.
(*g*) For use in conjunction with electrical stylus instruments to establish whether R_a values are also descriptive of frictional qualities.
(*h*) To measure short lengths down to 1 mm which may be inaccessible to stylus instruments.
(*i*) As a comparator for non-metallic surfaces, e.g. paper, plastics, wood, painted surfaces, etc.

The Dial Depth Gauge

It is useful to define the coarser grades of roughness as those above 5·0 μm R_a and most stylus instruments are almost at their upper limit of measurement at this value. The dial depth gauge has been developed primarily to measure rougher surfaces such as rough turned, shaped, milled, grit and shot blasted, spark eroded or cast surfaces, etc. The R_z and similar parameters are those most usually preferred for this type of work as an assessment of the total depth of the irregularities is usually of more interest than the R_a parameter. An advantage of the instrument is that a quick measurement can be made without the need for producing a graphical trace.

The gauge is essentially a dial test indicator in which the plunger has been replaced by a hardened steel or diamond stylus and to which two-point or three-point datum attachments can be fitted depending on whether the surface tested is flat or cylindrical (Fig. 4.12). The gauge is first set to zero on a setting standard and the surface measured by pressing the gauge on to it ensuring that the stylus engages a groove in the surface. The feet of the datum attachment are then pressed against the surface and a reading taken. This procedure is repeated at at least four other points on the surface and the average of the readings almost exactly equals the R_z value; the highest reading is the R_t value or maximum depth. When for any reason the R_a value is required an almost exact conversion can be made by dividing the R_z value by four as for most coarser roughness grades $R_z = 4R_a$, to a tolerance of $\pm 15\%$. It must be emphasized,

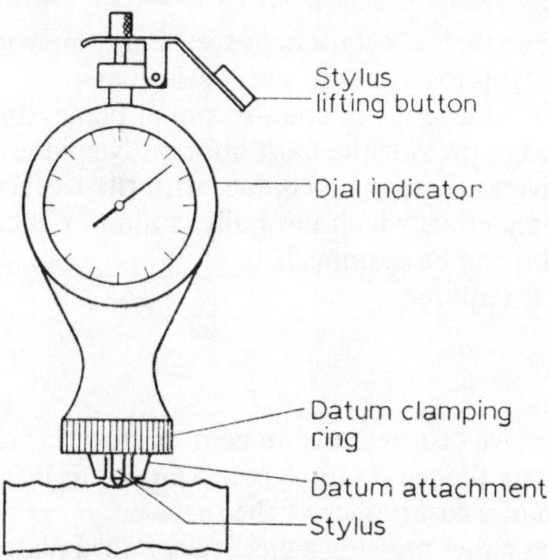

Fig. 4.12 THE DIAL DEPTH GAUGE

however, that excessive reliance should not be placed on such empirical relationships, as under other circumstances R_z may be as much as seven times R_a.

4.7 Light Interference Microscopes

The technique used to develop interference fringes on nominally flat or curved surfaces is explained in Chapter 2, and the basic optical system used in a suitably modified form can be applied to microscopes for surface finish examination.

The advantage of the interference microscope is that it allows an *area* of the surface to be examined and the more expensive versions provide a range of magnifications and provision for a camera, so that a permanent record of the fringe pattern may be obtained if desired. Very high resolution is possible with these instruments and scratch spacings 0·5 μm wide are fairly easily seen. This order of resolution compares favourably with the normal electronic stylus instrument where response begins to fall off at scratch widths below 2·5 μm. High resolution of scratch depth can also be obtained and scratches as shallow as 0·05 μm or even less can be resolved under good conditions. Multiple beam interference microscopes in which the interfering light beams are made to traverse the interference space several times permit scratch depth measurement of the order of 0·002 μm. Comparison microscopes allow the fringes appearing on a master simultaneously to be compared with the fringes on the component. The surface finish and sphericity of ball bearings are checked by this method.

Because of their high resolving power micro-interferometers are limited to the examination of smoother surfaces (peak to valley heights of less than 2 μm) and by virtue of their construction can only be used for direct observation of external surfaces on components within the size capacity of the instrument. Assessment of surface faults cannot directly be related to the R_a parameter but a good correlation with peak to valley parameters is possible. The main advantages are as follows:

(*a*) The area sample obtained allows surface details to be seen as a composite in the field of view, whereas stylus exploration is very selective.
(*b*) The fringes can be thought of as a series of cross-sectional planes and their direction can be adjusted to provide the most informative pattern.
(*c*) The method is non-destructive as there is no contact with the surface. The surfaces of soft or thin materials which normally could not effectively be explored with a stylus can be examined.
(*d*) No instrument recalibration is required.

4.8 Surface Replicas

A simple but effective method of surface texture measurement where a surface is inaccessible to instruments or where the part is too large to mount or bring to an instrument is the preparation of a cast replica of the surface.

Essentially, the technique involves either pressing a piece of softened plastic on to the surface or placing sticine wall or metal ring round the specimen area and pouring in a plastic fluid. In either case, after a suitable setting time the replica may be removed and an instrument reading taken using a lightly loaded stylus.

Very high fidelity approaching 100% is obtainable using the fluid method and softened plastic sheet offers about 80% fidelity. Special kits with full instructions for use are available from manufacturers.

Where circumstances permit, the fidelity of the replica may be checked by comparing the profile graph of the sample with that of the replica. To ensure that both readings are taken along the same path, a 45° cross may be scribed on the sample and stylus readings taken through the intersection. If the stylus passes to one side of the cross two disruptions will appear on the trace indicating that the setting must be corrected.

5 Strain Measurement

Until recent times one of the most difficult problems facing engineers had been to find the stresses acting upon a body by the direct application of measuring devices. Before the invention of the electrical resistance strain gauge the most important means available for this purpose was the extensometer in its various forms (mechanical, optical, etc.) but these have a number of disadvantages, the most important of which are their relative bulk, which makes application impossible in conditions of limited space, and the gauge length required for application, usually not less than 12 mm. Thus many engineering stress problems formerly were solved on a purely theoretical basis coupled with trial and error tests and large factors of safety which were introduced to allow for the lack of precise knowledge of stress conditions at a particular point or points in the structure.

The principle on which electrical resistance strain gauges are based was established by Lord Kelvin in 1856 when he showed that a stress applied to a metal wire, besides changing its length and diameter, also changed its electrical resistance. In the 1930's this effect was used by American workers to develop independently the bonded wire electrical resistance strain gauge which is attached by means of a suitable cement to the surface under test and subjected to the same strain. Since this type of gauge was developed a number of gauges have become available based on effects other than change of electrical resistance but the resistance gauge is still by far the most widely used. The most important gauges used for strain determination are as follows:

Gauge types	*Effect of strain*
Bonded wire Bonded foil Semiconductor	Change of electrical resistance.
Photoelastic	Fringe displacement.
Extensometers	These include a variety of methods for measuring extension, i.e. mechanical, optical, pneumatic, inductance, capacitance, and acoustic responses.

5.1 Strain

A body subjected to external forces is in a condition of both stress and strain. Stress cannot be directly measured but its effect, i.e. change of shape of the body, involves change in the fundamental quantity, length, which can be measured; thus, provided there is a known relationship between stress and strain, the stresses occurring in a body can be computed if sufficient strain information is available.

Consider the simple case of a uniform bar loaded first in tension and then in compression (Fig. 5.1). It is seen that strain appears along the axis of the applied load and, in all directions normal to that axis, the bar grows longer and thinner (tension) or shorter and thicker (compression).

The change in length in the direction of the direct stress caused by the load P is the linear extension δL but for a given load this extension is dependent upon the length of the bar: the greater the length the greater the extension. Strain

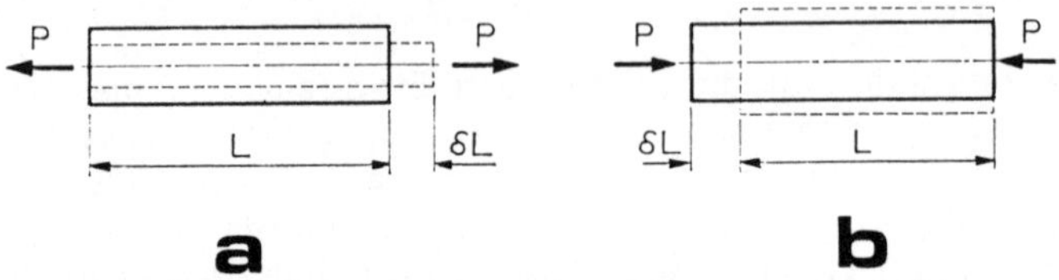

Fig. 5.1 UNIFORM BAR LOADED *a* IN TENSION AND *b* IN COMPRESSION

may be defined as extension per unit length irrespective of whether the bar is lengthened or shortened; thus in each case direct or longitudinal strain is $\delta L/L$, where L is the length prior to the application of the load.

Lateral or Transverse Strain

Lateral strain arises from the increase or decrease in the dimension normal to the applied load, and is computed in the same way as direct strain. The ratio of lateral strain to direct strain is called Poisson's ratio, i.e.

$$\text{Poisson's ratio} = \frac{\text{Lateral strain}}{\text{Direct strain}}$$

and is taken to be positive.

This ratio is particular to any given metal or alloy; it is as low as 0·21 for zinc and as high as 0·35 for brass. The value for steel is 0·29.

The constant connecting stress and strain in elastic materials under direct stresses is the modulus of elasticity E, which is the slope of the stress/strain curve, thus

$$E = \frac{\text{Stress}}{\text{Strain}} \qquad \text{and} \qquad \text{Stress} = E \times \text{Strain}$$

Static, Transient and Dynamic Strain

STATIC AND TRANSIENT STRAIN Strain which remains at a constant value for an appreciable length of time is called static strain and can occur in stationary or moving structures. A rotating engine shaft under constant torque is subject to steady or only slightly varying strain, a condition equivalent to static strain. Likewise a cantilever under a constant bending load or a rod or spring under a constant tensile or compressive stress are in a state of static strain (Fig. 5.2*a* to *d*).

A strain/time relationship is shown graphically at 5.2*e*. Notice that the curve is divided into separate parts, transient and static. The transient portions indicate an increase or decrease in strain with time as load is steadily increased

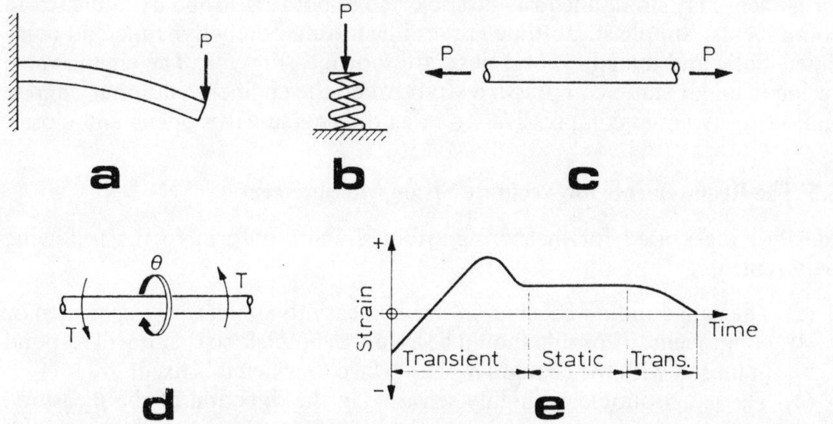

Fig. 5.2 STATES OF STATIC STRAIN: *a* CANTILEVER, *b* SPRING, *c* ROD, *d* ROTATING SHAFT UNDER CONSTANT TORQUE

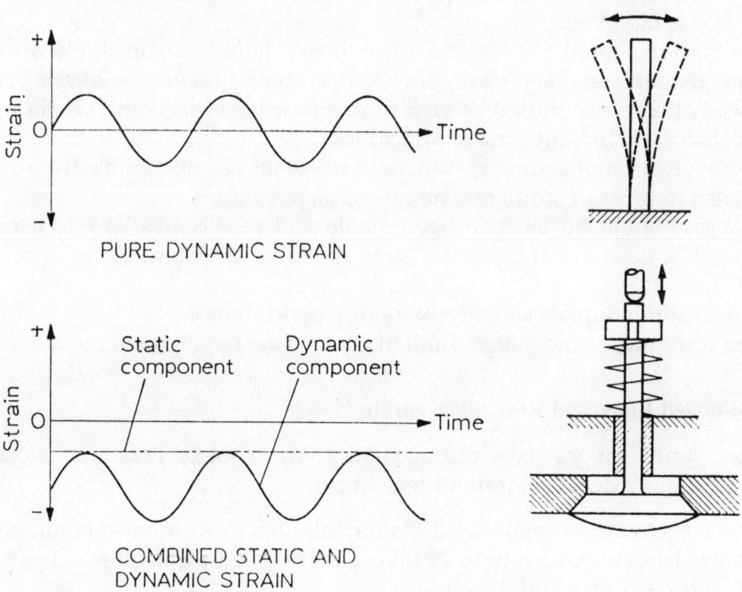

Fig. 5.3 STRAIN/TIME CURVES

or decreased to the required value when it remains constant. Transient strain is a form of dynamic strain because it changes with time.

DYNAMIC STRAIN Strain which is changing with time is dynamic strain and the problems associated with recording such a strain depend on the rate at which it is changing, as there is obviously a limit to the ability of an observer to follow and plot, on a time base, strain variations indicated by a scale pointer or even from an instrument displaying a digital readout. Furthermore, an element may simultaneously be subjected to both static and dynamic strain components. Simple strain/time curves illustrating purely dynamic and combined static and dynamic conditions are shown at Fig. 5.3. The engine valve spring is under static compressive strain when the engine is stationary and is subject to dynamic compressive strain as the engine valve opens and closes.

5.2 The Requirements for Accurate Strain Measurement

Ideally gauges used for measuring strain should conform to the following requirements:

(*a*) The gauge should be small in size and easy to attach to the specimen or component; its profile should be as low as possible so that it will respond in unison with the changes in the surface to which it is fixed.
(*b*) The gauge should be highly sensitive in the direction of the measured strain but of low sensitivity in the transverse plane (cross-sensitivity). The strain/resistance characteristics should be linear.
(*c*) Stiffness in all directions should not be such that the stiffness of the tested surface is modified.
(*d*) Calibration should be easy and, once made, should remain stable with time, dynamic loading, changes of temperature, pressure or humidity.
(*e*) Speed of response should be high so that time lag is negligible. Remote indication should not present difficulties.
(*f*) The evaluation of complex strain patterns should be obtainable from as small a number of strain measurements as possible.
(*g*) Gauges should be inexpensive, reliable and readily available. There should be a variety of types and sizes suited to a wide range of applications.
(*h*) Immersion in liquids should not modify performance.

Bonded resistance strain gauges fulfil many of these requirements.

5.3 The Bonded Electrical Resistance Strain Gauge

Resistance strain gauges have characteristics which make them the most versatile existing devices for strain measurement.

(*a*) Gauge lengths are as small as 0·127 mm although those in most common use range from about 5 mm to 25 mm. The backing to which the gauge is fixed increases the overall length.
(*b*) Weight is negligible and attachment to the tested body is fairly easy.
(*c*) Strain sensitivity is quite low, e.g. a constantan wire gauge of 100 Ω resistance would show an increase of approximately 0·25 Ω on low carbon steel strained to the elastic limit.
(*d*) Temperature, pressure and humidity, behaviour in liquid environments, etc., all affect performance but protection or compensation can usually be provided.

(*e*) Strain gauges can indicate static, transient and dynamic strain (50 000 Hz) or more, and remote indication and recording presents little difficulty.
(*f*) Wire and foil gauges are similar in price, costing from about 80p upwards depending upon their complexity.

Metal resistance strain gauges are made in two basic forms, wire and foil. The foil types, which are etched or pressed from thin sheet foil, have the advantage that the gauge can take almost any shape that can be drawn; they also have a lower profile, very good linear cross-sensitivity, and, because the grid is rectangular in form and very thin, heat is dissipated more readily than from wire gauges. Because of their generally superior qualities foil gauges have largely superseded wire gauges except for post-yield applications and where grids in excess of 150 mm long are required (Fig. 5.4).

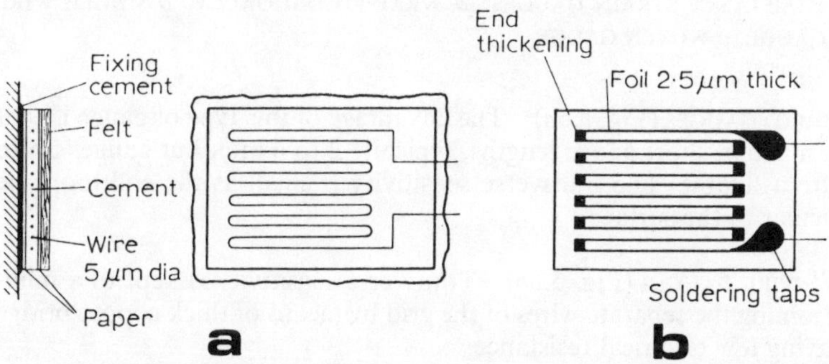

Fig. 5.4 METAL RESISTANCE STRAIN GAUGES: *a* FLAT GRID TYPE WIRE GAUGE, *b* ETCHED FOIL GAUGE

The resistance R of a metal depends on its electrical resistivity ρ, its area a and its length L according to the equation

$$R = \rho L/a$$

Thus, to obtain a high resistance gauge occupying a small area, the metal chosen has a high resistivity, a large number of grid loops, and a very small cross-sectional area. Wire gauges are embedded in plastic cements and sandwiched between thin paper or plastic insulators, and a felt cover is sometimes provided to give added protection. The cement holds the wire rigidly on to the backing so that it will not buckle under strain. Foil gauges are usually cemented on to a backing but may also, by means of a special process, be transferred directly to an insulated surface which eliminates the operating temperature limitations of paper or plastic backings. Two important features of foil gauges not present in wire gauges are the thickenings at the end of each loop which, by introducing a low transverse electrical resistance, reduce cross-sensitivity to almost negligible proportions, and the provision of relatively large tabs tapered into the grid provides easy soldering or welding of the leads and reduces the danger of fatigue failure.

Besides the conventional wire and foil gauges there are a number of other types much less widely used (Fig. 5.5).

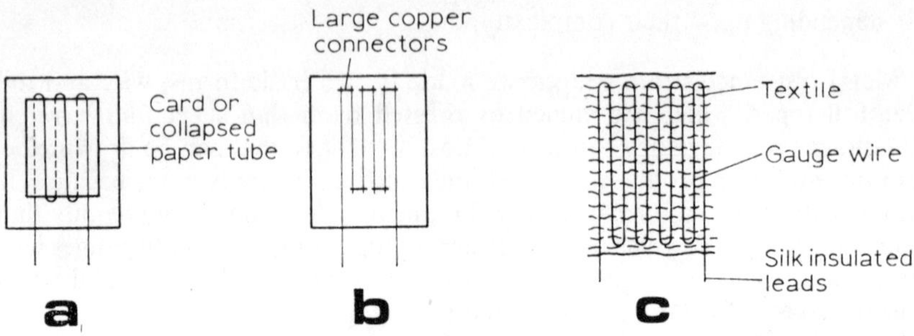

Fig. 5.5 RESISTANCE STRAIN GAUGES: *a* WRAP-ROUND GAUGE, *b* SINGLE WIRE GAUGE, *c* WOVEN GAUGE

WRAP-ROUND GAUGES (Fig. 5.5*a*) The advantage of this type of gauge is that it can be made in short gauge lengths, typically 2 to 6 mm, but gauges up to 12 mm are available. The transverse sensitivity is small as the end loops lie perpendicular to the surface.

"SINGLE" WIRE GAUGES (Fig. 5.5*b*) Transverse sensitivity is kept to a minimum by joining the separate wires of the grid by means of thick copper bridge pieces having low electrical resistance.

WOVEN-TYPE GAUGES (Fig. 5.5*c*) These gauges are used to measure large strains in fabrics, etc.

It is highly desirable that the change of resistance $\delta R/R$ of a strain gauge should bear a linear relationship to the measured strain $\delta L/L$ and metallic strain gauges fulfil this requirement. This relationship is called the *gauge factor* λ.

5.4 The Gauge Factor or Strain Sensitivity

The electrical resistance of a metallic conductor varies directly as its resistivity and length and inversely as its cross-sectional area; this follows from $R = \rho L/a$. The metallic crystal lattice forms a regular atomic pattern and the particular form of bonding between the atoms—metallic bonding—involves mutual sharing of all the valency electrons by all the atoms in the metal. Thus current passes along a conductor in the form of directional motion of the electrons, called electron flow. The increased length and decreased area of a gauge under tensile strain accounts for part of the increase in resistance as the metallic lattice will suffer distortion but these changes alone do not fully account for the total resistance change; thus other changes in the metal lattice must also occur bringing about a change in the resistivity of the metal. This latter effect is an important consideration as it is well known that the resistivity of metals also changes with temperature.

The gauge factor λ is the ratio of change of resistance $\delta R/R$ to the change of gauge length $\delta L/L$, thus

$$\lambda = \frac{\delta R/R}{\delta L/L}$$

where R is the nominal resistance of the gauge.

The gauge factor is supplied by the manufacturer and may range from 1·7 to 4 depending largely on the length of the gauge, although most wire and foil gauges have a gauge factor of about 2. Thus, provided that a means is available for measuring the change of resistance, strain can be determined and hence stress. The following example illustrates this point.

EXAMPLE 5.1

A strain gauge (gauge factor 2) measures the strain on the surface of a shaft subjected to a tensile stress. The nominal resistance of the gauge R is 120 Ω and the change of resistance $\delta R = 0{\cdot}012$ Ω. Strain $e = \delta L/L$ and from the above equation $\delta L/L = \delta R/R\lambda = e$. Therefore

$$e = 0{\cdot}012/120 \times 2 = 0{\cdot}000\,05$$

E for steel = 206 GPa and stress = $E \times$ strain. Therefore

$$\text{Stress} = 206 \times 10^9 \times 0{\cdot}000\,05 = 10{\cdot}3 \text{ MPa}$$

The example illustrates that δR is very small and to measure this accurately a Wheatstone bridge circuit is used incorporating a sensitive galvanometer. A rough guide to the sensitivity of the gauge under discussion is that a change of stress intensity of 0·5 MPa will bring about a change in δR of approximately 0·0006 Ω. It follows that an increase in gauge factor increases sensitivity, but this is limited in metal gauges, largely because of the relatively low resistivity of metals, a limitation which is overcome in semiconductor strain gauges which have gauge factors of the order of ± 100 or more.

5.5 Strain Gauge Alloys

The alloys most widely used for static strain measurement are the variously named 55% copper, 45% nickel alloys known as "Advance", "Copel" and "Constantan". For dynamic strain measurement the 52% iron, 36% nickel, 8% chromium, 0·5% molybdenum alloy "Iso-elastic" is usually used. To avoid confusion the name "Advance" will be used to identify the copper/nickel alloys.

Advance has a gauge factor of about 2 which remains constant well into its plastic range and its temperature coefficient of resistance is ± 2 $\mu\Omega$°C/Ω which is very low compared with say nickel (6000) or iron (5000). Corrosion resistance as would be expected of this type of alloy is also good, up to 400°C. The resistance of a 100 mm length of 25 μm diameter wire is approximately 95 Ω compared with nickel and iron (approximately 23 Ω). Iso-elastic alloy has the following characteristics: temperature coefficient of resistance $+175$ $\mu\Omega$°C/Ω, resistance approximately 232 Ω, gauge factor 3·5. The higher temperature coefficient of resistance of iso-elastic is the limiting factor which makes it suitable only for dynamic strain measurement as in these circumstances the strain is changing rapidly, the measurements are short term and the effect of temperature variation is usually insignificant.

5.6 The Wheatstone Bridge

The highly sensitive Wheatstone bridge is used to detect changes in strain gauge resistance. An outline of a simple direct current strain instrument and gauges is shown at Fig. 5.6. It consists of a galvanometer, four resistors (R_1, R_2, R_3, R_4), and a battery. Resistance R_1 is the gauge used for strain measurement. Resistance R_4 is a second strain gauge which for the present purpose is assumed to remain at constant resistance. Resistances R_2 and R_3 are variable resistors which by adjustment are used to balance and rebalance the bridge, i.e. to reduce the current across the galvanometer arm to zero. When the bridge is balanced the ratio of the gauges and variable resistances are equal, thus:

$$\frac{R_1}{R_4} = \frac{R_2}{R_3} \qquad \text{i.e. } R_1 = R_4 \times \frac{R_2}{R_3}$$

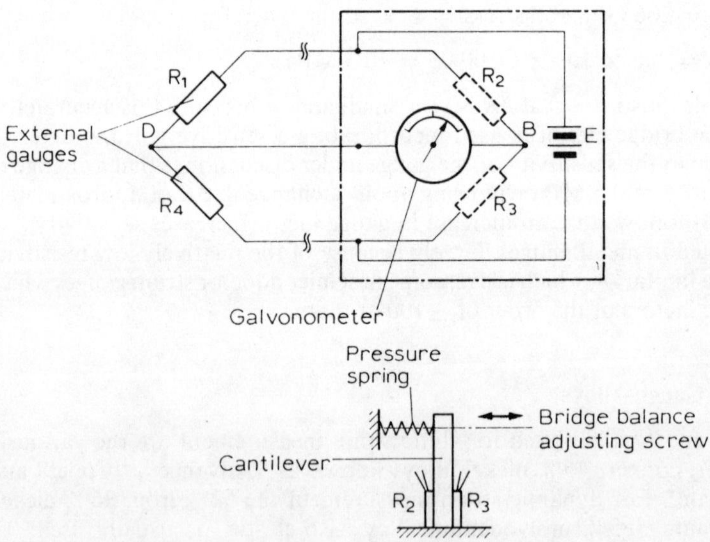

Fig. 5.6 SIMPLE STRAIN GAUGE INSTRUMENT

An important characteristic of the Wheatstone bridge is that its output voltage is a non-linear function of resistance change in the bridge; thus the above ratio is strictly true only when the bridge is balanced, i.e. when the voltages at D and B are equal and the current through the galvanometer arm is zero. However, provided that the resistance change in the bridge is small the change in bridge output voltage is virtually linear, e.g. for a resistance change of 1% the departure from linearity will be of the order of 0·005% of this change.

It is evident that two methods are available for static strain measurement: continuous reading from one change of strain to another without rebalancing, which is convenient but less accurate as the bridge output tends to non-linearity if the strain is successively increasing or decreasing, or the null balance method in which, after each successive change in strain, the bridge is rebalanced. As the strain indicator will be calibrated under null balance conditions this method offers the highest accuracy while also maintaining linear bridge output.

The strain indicator is calibrated for gauges of a given gauge factor, thus it provides accurate readings only when gauges having the same gauge factor are used. Most instruments incorporate a variable resistor which allows the strain indicator to be matched to gauges having somewhat different gauge factors; however, in the event that no such adjustment exists or that the gauge factor required is beyond the range of the adjustment, the strain can be computed if the indicated strain is multiplied by the ratio of the instrument gauge factor value to the gauge factor of the gauges in use. For example,

Instrument gauge factor value = 2
Strain gauge factor = 3·5

$$\text{True strain} = \text{Indicated strain} \times \frac{2}{3{\cdot}5}$$

Overall Measuring Accuracy

Overall measuring accuracy must be apportioned between the strain gauges and the instrumentation. It will depend on the gauge factor tolerance, temperature coefficient of resistance, gauge resistance tolerance, bonding quality, number of gauges used and how well they are matched, etc. The total error (gauges plus instrumentation) should be within ±5% for static or very slowly changing strains.

5.7 Simple Strain Gauge Applications

It is usual to incorporate more than one strain gauge in the bridge for the following reasons:

(*a*) to increase output for a given strain, thus increasing sensitivity,
(*b*) to provide temperature compensation, and
(*c*) to eliminate unwanted effects, e.g. a rod subjected to a tensile force will almost certainly suffer some bending; thus if the strain due only to tension is required the gauges must be so arranged on the rod and in the bridge that the bridge output corresponds to strain produced by tension alone.

Wheatstone Bridge Rules

The following rules offer a useful guide to the effects of the positions of the gauges in the bridge arms:

(*a*) Bridge unbalance increases in proportion to the *algebraic difference* of changes of resistance in two adjacent arms.

(*b*) Bridge unbalance increases in proportion to the *algebraic sum* of changes of resistance in two opposite arms.

Fig. 5.7 illustrates that there are four pairs of adjacent arms (R_1, R_2; R_2, R_3; etc.) and two pairs of opposite arms.

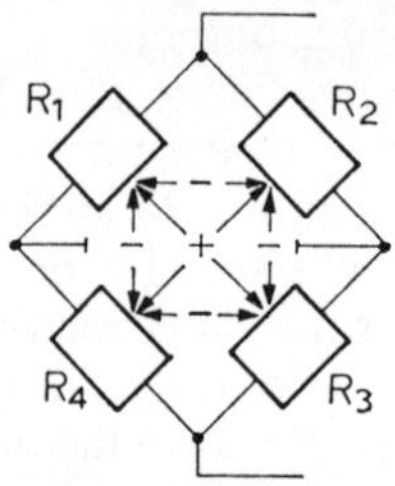

Fig. 5.7 BRIDGE ARM RELATIONSHIPS

BRIDGE RULE 1 A simple cantilever beam is under pure bending stress and loaded, as shown, Fig. 5.8. Two active gauges, each of nominal resistance R ohms, are applied as near to the position of maximum bending moment as

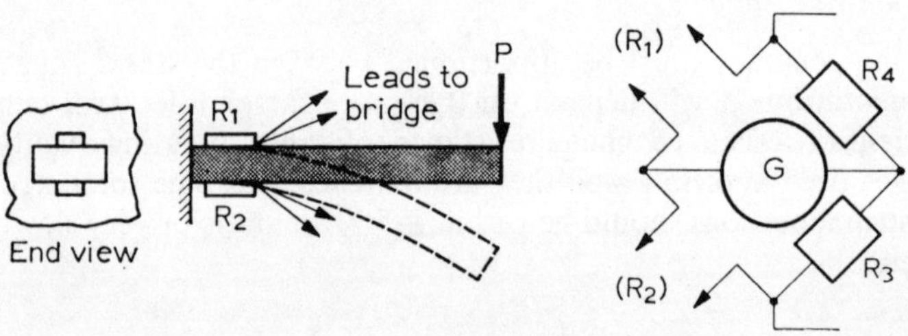

Fig. 5.8 GAUGES APPLIED TO A SIMPLE CANTILEVER

possible and with the gauge longitudinal axis parallel to that of the cantilever. The gauge on the upper surface of the beam is subjected to tensile strain, the lower gauge to an equal compressive strain; thus the resistance changes in each case (δR) are virtually equal, δR being positive in tension and negative in compression. Applying rule 1,

$$(\delta R_1) - (-\delta R_2) = 2\delta R \qquad \text{where } \delta R_1 = \delta R_2 \tag{i}$$

Thus the bridge output is doubled. Further consideration indicates that temperature compensation is also achieved in this case.

The electrical resistivity of most alloys changes with temperature, increasing as temperature rises and decreasing as it falls. This effect is independent of any strain applied to the gauge, thus temperature changes during measurement will be a source of false strain readings. In addition, unless the coefficient of linear expansion of the gauge is the same as that of the material to which it is bonded, then temperature change during measurement will also be a source of false strain readings due to differential expansion. The temperature coefficient of resistance of Advance alloy is much smaller than those of many other alloys (see section 5.5), but for static strain measurement temperature compensation is always applied whenever possible as this improves accuracy. Gauges R_1 and R_2 (Fig. 5.8) are close together and any temperature change occurring should apply equally to both. This being the case, any change of resistance due to temperature change will be equal and of the same sign for both gauges. Applying rule 1,

$$(\delta R_1 \pm \delta R_t) - (-\delta R_2 \pm \delta R_t) = 2\delta R \pm \delta R_t \mp \delta R_t = 2\delta R + 0 \qquad \text{(ii)}$$

where $\pm\delta R_t$ is the resistance change due to temperature. Thus temperature compensation is achieved. The above result can of course be obtained using the arms R_1, R_4; R_2, R_3; etc.

BRIDGE RULE 2 A rod is subjected to a tensile load as shown in Fig. 5.9 but is also subject to some bending owing to difficulties in applying a purely tensile load.

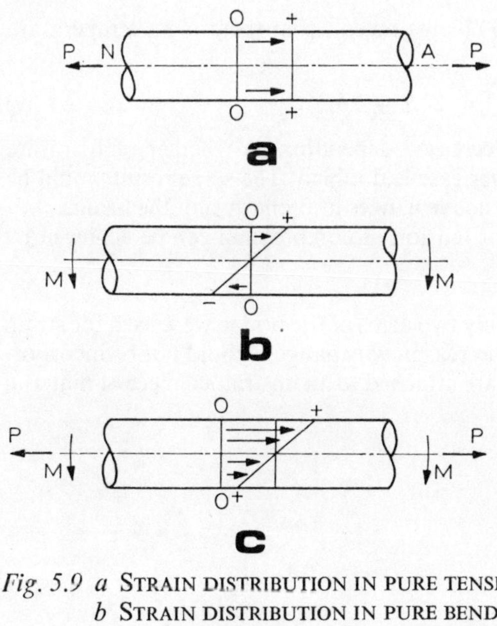

Fig. 5.9 *a* STRAIN DISTRIBUTION IN PURE TENSION
b STRAIN DISTRIBUTION IN PURE BENDING
c STRAIN DISTRIBUTION IN COMBINED TENSION AND BENDING

Pure tensile strain is uniform across the rod cross-section as shown at (5.9*a*), while the effect of bending is to produce maximum strains at the surfaces of the rod which are equal and opposite, i.e. tensile and compressive (5.9*b*). The combined effect (5.9*c*) indicates an increase in tensile strain on the "humped" or convex side of the rod and an equal and opposite decrease in tensile strain on the concave side. Two active gauges are bonded to the rod under tension as shown in Fig. 5.10.

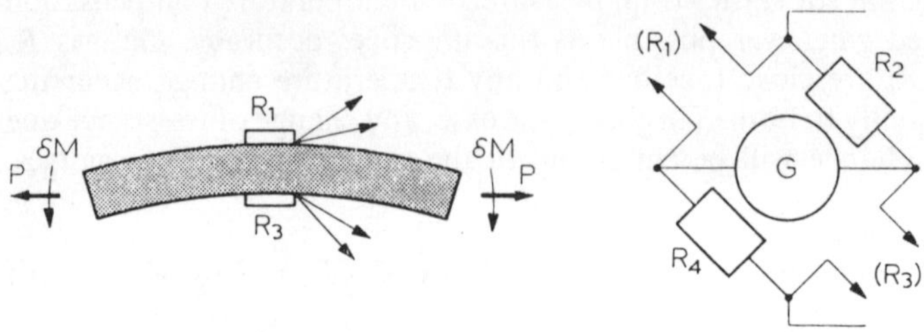

Fig. 5.10 ACTIVE GAUGES FOR SIMPLE DIRECT STRESS

Applying rule 2,

$$(\delta R_1 + \delta R_B) + (\delta R_3 - \delta R_B) = 2\delta R + 0 \qquad \text{(iii)}$$

where $\delta R_1 = \delta R_3 =$ resistance change due to tensile strain,
and $\delta R_B =$ resistance change due to bending strain.

Consideration of temperature change reveals that there is no temperature compensation in this case, thus

$$(\delta R_1 \pm \delta R_t) + (\delta R_3 \pm \delta R_t) = 2\delta R \pm 2\delta R_t \qquad \text{(iv)}$$

Bridge output is increased or decreased depending on whether temperature rises or falls and its effect in either case is doubled. The same result would be obtained by introducing the two active gauges into one arm of the bridge.

Temperature compensation for tension or compression can be achieved by means of "dummy" gauges.

DUMMY GAUGES In Fig. 5.10, only two arms of the bridge were used for strain gauges (R_1 and R_3) but there is no reason why gauges should not be incorporated at R_2 and R_4. Strain gauges are attached to an unstrained piece of material

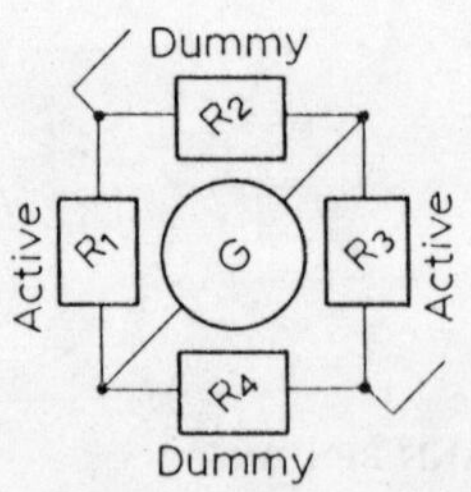

Fig. 5.11 DUMMY GAUGES FOR TEMPERATURE COMPENSATION

which is similar to the material under test. Provided that the unstrained material experiences the same temperature changes as the strained material, then the dummy gauges will compensate only for temperature changes when introduced to adjacent bridge arms (Fig. 5.11). Using the result obtained at

(iv) and subtracting the effect of the two dummy gauges (R_2 and R_4), by rule 1,

$$(2\delta R \pm 2\delta R_t) - (\pm\delta R_t \pm \delta R_t) = 2\delta R + 0$$

Thus temperature compensation is achieved.

THE POISSON METHOD The dummy gauge is bonded at right angles to the active gauge on a member in simple tension (Fig. 5.12). Gauge R_2 is reduced in length

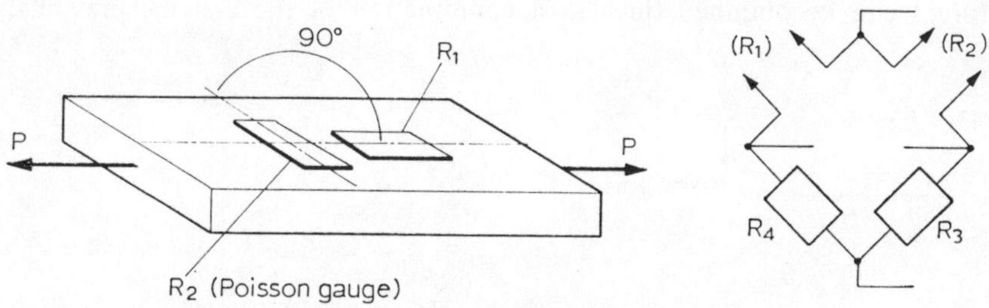

Fig. 5.12 THE POISSON METHOD FOR TEMPERATURE COMPENSATION

depending on Poisson's ratio for the material, approximately 0·3 for steel, thus

$$(\delta R_1 \pm \delta R_t \pm \delta R_B) - (-0{\cdot}3\delta R_2 \pm \delta R_t) = 1{\cdot}3\delta R \pm \delta R_B \qquad \text{(v)}$$

This arrangement provides increased sensitivity, full temperature compensation, but no compensation for bending, and is used when it is not convenient or possible to apply gauges in corresponding positions to the underside of the member. It is also necessary to find the precise output ratio by applying known loads to the member. A further increase in sensitivity and compensation for both bending and temperature can be achieved by applying gauges to both sides of the member and in the same relative positions, R_3 becoming the active gauge on the underside.

Separate Measurement of Combined Forces and Bending Moments

Further consideration of the simple cantilever suggests that gauges may be so arranged in the bridge that they provide an output from forces applied in one direction but not in another. The cantilever shown at Fig. 5.13*a* is subjected to both an axial and a bending load. With the bridge arrangement shown the gauges provide an output only from the axial load. Gauges R_1 and R_3 are in opposite bridge arms, thus any resistance changes are additive but such changes, if they are of equal magnitude, must be of the same sign to provide an output. When P_B is applied the resistance changes are of equal but opposite sign, thus output from the bridge is zero. Consideration of the effect of the axial force P_A indicates that both gauges are subjected to equal resistance changes of the same sign; thus output from the bridge is proportional to the sum of these resistance changes. The same result is obtained for any combination of directions of P_A and P_B. Gauges R_2 and R_4 are mounted either as dummy or Poisson gauges to provide temperature compensation.

An arrangement which provides an output from P_B only is shown at 5.13*b*. Gauges R_1 and R_3 are in opposite arms and are of the same sign and likewise R_2 and R_4, thus their output is proportional to the sum of their resistances. When the axial force P_A is considered all gauge resistance changes are of the same sign and by their bridge positions mutually cancel each other. Temperature change has the same effect on the gauges as axial force, thus the system is fully temperature compensated.

It is evident that simultaneous measurement of both bending and axial forces can be obtained through a combination of the systems previously

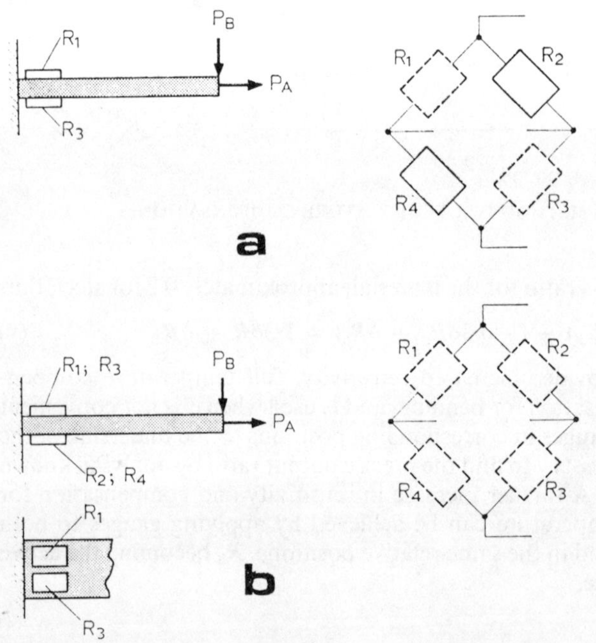

Fig. 5.13 SEPARATE MEASUREMENT OF COMBINED DIRECT AND BENDING STRAINS

described. Two separate bridges are required and gauges R_2 and R_4 are used as dummy or Poisson as before.

A further development of this approach is the use of a cantilever cylinder which senses strains in two mutually perpendicular directions (Fig. 5.14).

If the effect of the horizontal applied force P_H is considered it is clear that gauges R_H will respond to bending in the horizontal plane, e.g. R_{1H} and R_{2H} are in adjacent arms and as their signs are opposite they are additive and similarly R_{3H} and R_{4H}. If the gauges mounted to sense the vertical force P_V are now examined it is seen that the horizontal force P_H causes them to bend slightly, those to the left of the vertical axis, R_{1V} and R_{2V} being in compression and those to the right, R_{3V} and R_{4V} being in tension. An assessment of their bridge positions suggests that the bending effect of P_H is virtually zero. An examination of any slight compressive or tensile effects arising from P_H, i.e. P_A, indicates that the resultant strain indication will be zero. A consideration of the effect of the vertical applied force P_V produces a similar result, i.e. P_B. The system is also fully temperature compensated and output is augmented.

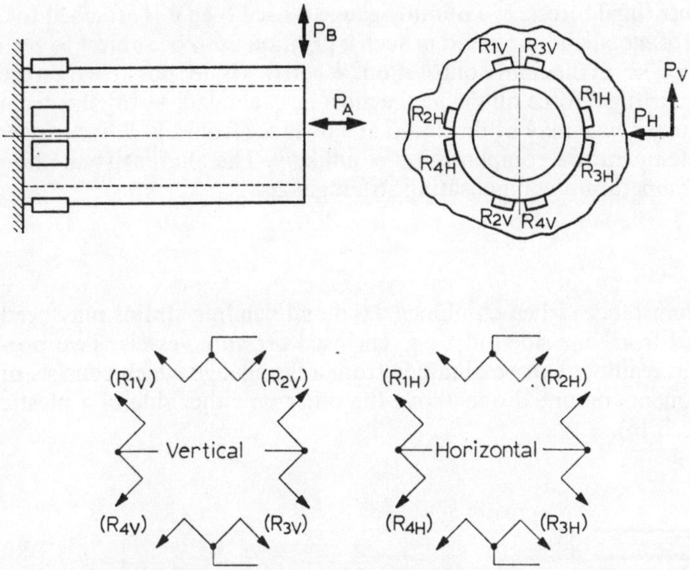

Fig. 5.14 SIMULTANEOUS MEASUREMENT OF DIRECT AND BENDING STRAINS

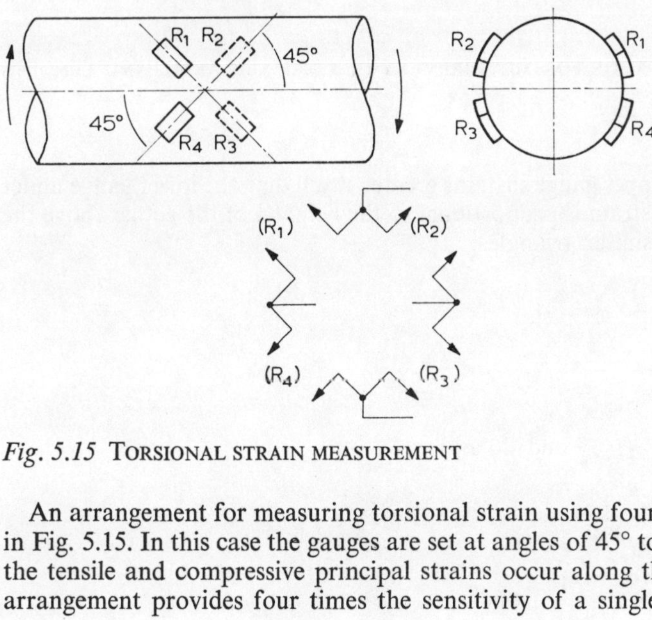

Fig. 5.15 TORSIONAL STRAIN MEASUREMENT

An arrangement for measuring torsional strain using four gauges is shown in Fig. 5.15. In this case the gauges are set at angles of 45° to the shaft axis as the tensile and compressive principal strains occur along these planes. This arrangement provides four times the sensitivity of a single gauge with full temperature compensation and eliminates bending and axial tensile or compressive effects

When the static or steady strain on a rotating body is non-torsional, then the problem of temperature compensation arises because of the difficulty in placing the temperature compensating gauge in a position where it will not be affected by centrifugal forces. If a dummy gauge is used then it is attached to a separate piece of metal and mounted in such a position as to be subject to zero centrifugal force, i.e. at the centre of rotation. When this is not possible then the the effect of centrifugal force on the gauge must be evaluated. Using this technique the dummy gauge may be displaced at a distance from the active gauges and thus full temperature compensation is unlikely. The alternative method is to use self-temperature-compensating strain gauges.

Flexagauges

There are circumstances when combined axial and bending strains may need to be measured from one side only, e.g. enclosed pressure vessels. Two proportional strain readings can be obtained from a flexagauge which consists of two gauge filaments mounted one above the other on either side of a plastic separator (Fig. 5.16).

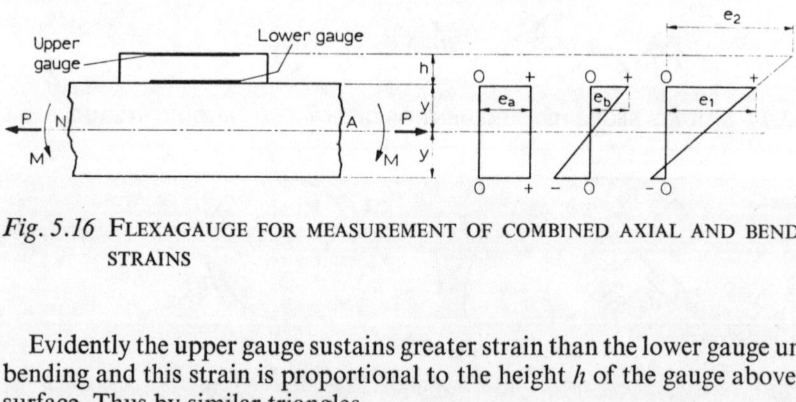

Fig. 5.16 FLEXAGAUGE FOR MEASUREMENT OF COMBINED AXIAL AND BENDING STRAINS

Evidently the upper gauge sustains greater strain than the lower gauge under bending and this strain is proportional to the height h of the gauge above the surface. Thus by similar triangles,

$$\frac{e_b}{h} = \frac{(e_2 - e_1)}{y}$$

Therefore

$$e_b = \frac{h}{y}(e_2 - e_1) \qquad \text{and} \qquad e_a = e_1 - e_b$$

Therefore

$$e_a = e_1 - \frac{h}{y}(e_2 - e_1)$$

where e_a = axial strain
e_b = bending strain at surface
e_1 = indicated strain from lower gauge
e_2 = indicated strain from upper gauge
y = half material thickness
h = thickness of flexagauge.

5.8 Self-temperature-compensated Strain Gauges

A completely self-compensated gauge should exhibit zero change of resistance with change of temperature when bonded to a specified material. Consider a conventional advance gauge bonded to a steel base. The coefficient of linear expansion of this alloy is 15 μm°C/m, while the coefficient of steel is 12 μm°C/m;

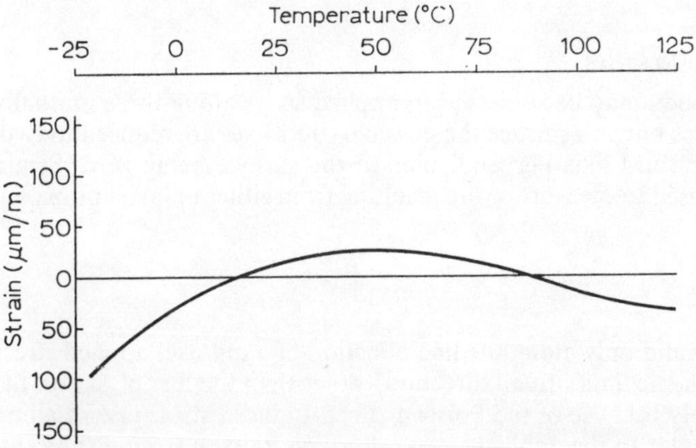

Fig. 5.17 ERROR CURVE FOR A TEMPERATURE COMPENSATED STRAIN GAUGE

thus under rising temperature the free expansion of the gauge is prevented and as a result the gauge is in compression and $\delta R/R$ from this source is negative. Conversely, the thermal coefficient of resistance of the gauge itself is positive, thus $\delta R/R$ from this source is positive. If these resistance changes are equal and opposite then change of resistance due to temperature is zero from these sources. It is of interest to note that a similar gauge bonded to aluminium, coefficient 23 μm°C/m, would be in tension. The most widely used self-compensated gauges are made from specially prepared "selected melt" alloys for use on specified materials.

The temperature coefficient of resistance of the gauge is, by suitable heat treatment during manufacture, matched to the coefficient of linear expansion of the material on which it is to be used. For physical reasons such matching can only be ensured over a limited range of temperature and to overcome this limitation the manufacturers calibrate selected gauges from each batch produced and supply an apparent strain/temperature graph for use with the gauges. A typical graph is shown at Fig. 5.17.

The following are some of the materials for which self-temperature-compensated gauges may be purchased.

Material	*Coefficient of Linear Expansion* (μm°C/m)
Steel	12
Austenitic steel	16
Aluminium	23
Plastic	65
Titanium	9

Another type is made by connecting two different wires in series to form the grid, one having a positive and the other a negative gauge factor. Increase in resistance in one wire due to temperature is almost equally compensated by a similar decrease in the other. Such gauges must still be matched for use on specified materials.

5.9 Strain Gauge Rosettes

Any point in a body may be subjected to applied stress along three mutually perpendicular axes but at a surface the possible stress axes are reduced to two, the stress on the third axis perpendicular to the surface being zero. Strain gauges are thus used to measure strain resulting from either uniaxial or biaxial stress.

Uniaxial Stress

Hooke's Law is valid only along the line of action of a uniaxial applied stress and within the elastic limit. In all directions other than the line of action the law does not apply because of the Poisson effect. Induced stress is zero along planes perpendicular to the applied stress, although Poisson strain is present.

Biaxial Stress

Hooke's Law is not valid in any direction on a surface subjected to biaxial stress because Poisson strain occurs in all directions. In such a system the maximum and minimum direct stress values are always mutually perpendicular to each other and are called the maximum and minimum *principal stresses*, the planes on which they act are called the principal planes, and the strains on the principal planes are called the *principal strains*. Shear stresses and strains on principal planes are zero. When biaxial stresses are present the direction of the principal stresses may either be known or unknown. To take the simplest case, where the principal stress and thus the principal strain directions are known the equations for stress along the mutually perpendicular axes x and y are

$$\sigma_x = \frac{E}{1 - \nu^2}(e_x + \nu e_y)$$

$$\sigma_y = \frac{E}{1 - \nu^2}(e_y + \nu e_x)$$

where σ_x and σ_y are the principal stresses, ν is Poisson's ratio, E is Young's modulus, and e_x and e_y are the principal strains. Clearly both principal strains must be known to determine either one or both of the principal stresses. The simplest rosette must thus consist of two mutually perpendicular gauges which must be accurately positioned in the principal stress directions. The direction of the principal stresses may be determined by the brittle lacquer or "stress-coat" technique in which the surface under test is coated with a special lacquer which exhibits a large number of fine cracks when the body to which it is applied is subjected to a given level of stress. The crack pattern which emerges is analogous to the strain field at the surface and from examination the principal strain directions may be determined. A further refinement of the technique involving the use of a separate calibrating strip provides a rough approximation to within $\pm 20\%$ of the maximum principal stress value.

When the directions of the principal stresses are unknown or not known with sufficient certainty it is necessary to use three strain gauges which will provide sufficient strain information for the principal strains and hence the principal stresses to be determined or again to utilize the brittle lacquer technique. The gauges are arranged to form a convenient pattern and to confine the strain measurements to the smallest possible area as ideally measurement at a point on the surface is required at which strain values are constant. Point measurement is impossible but a closer approximation is achieved by using short rather than long gauges where strain gradients in the plane of the gauges are likely to exist. Although any angles between the gauges may be used, in practice the simplest solutions and smallest errors are obtained by using the largest possible angles, and commercial gauges are manufactured with these factors in mind. The basic arrangements of the main types of commercial gauges are shown at Fig. 5.18.

An overlapped two gauge rosette suitable for applications where the principal strain directions are accurately known is shown at 5.18*a*. Overlapping reduces the area covered by the gauges to a minimum. Gauges constructed in this manner are not suitable for use on members subject to severe bending or torsion and having a neutral axis close to the surface, e.g. thin sheet material, as the stress gradient from the neutral axis of the material to the surface is very high and thus the strain on the upper gauge is significantly greater than the strain on the lower gauge. Foil gauges, which are very much thinner than wire gauges, suffer less from this error source but single plane configurations offer the easiest solution.

Two rectangular rosettes are shown at (5.18*b*) and a delta rosette is shown at (5.18*c*). These are particularly useful when the directions of the principal stresses are in doubt. The T-delta rosette shown at (5.18*d*) incorporates one over-lapping gauge which is used as a check on the results obtained from the other three gauges.

5.10 Lead Connections

The lead wires connecting the gauge to the bridge should have a low resistance compared with the gauge so that temperature variations cause only small resistance changes; they should be flexible and insulated to suit operating

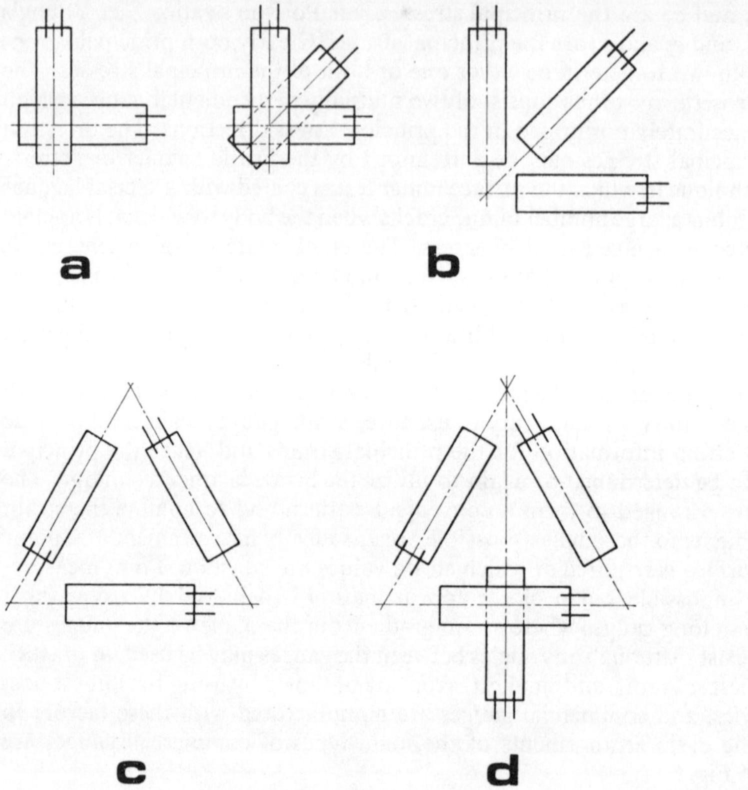

Fig. 5.18 BASIC ARRANGEMENTS OF THE MAIN TYPES OF GAUGE
a two gauge rosette (90°) *b* rectangular rosettes (45°)
c delta rosette *d* T-delta rosette

conditions. Plastic-coated wire is suitable for normal temperature operation but other insulators such as fibreglass, ceramics, etc., are used for higher temperatures. Leads from conventional gauges offer no temperature compensation problems provided that they are at the same temperature as each

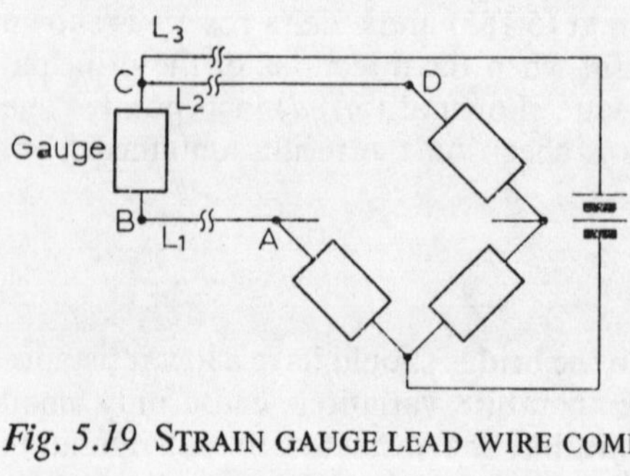

Fig. 5.19 STRAIN GAUGE LEAD WIRE COMPENSATION

other. However, lead compensation is required when using a single self-compensated gauge and this may be achieved by the three lead system shown in Fig. 5.19. Temperature change in the battery lead L_3 does not affect bridge balance.

Gauge leads L_1 (AB) and L_2 (CD) are in adjacent bridge arms by this arrangement and similar resistance changes in both leads will be cancelled. Normally, commercial instruments have at least two external bridge arms and to complete the bridge the second arm should incorporate a dummy gauge maintained at constant temperature so that only changes at the active gauge affect bridge output.

5.11 Sources of Error in Strain Gauge Installations

The following factors will affect the accuracy of strain observations.

Temperature Change

Perfect temperature compensation is unlikely to be achieved for the following reasons. No two gauges have identical electrical resistances, coefficients of expansion or gauge factors, although selection from the same batch reduces the effect of these differences to very small proportions. Apart from these considerations the gauges should be protected from draughts and incidental temperature changes which might affect one gauge and not another. Quite small temperature gradients across the test body can cause significant false strain readings. The effect of a small temperature change is illustrated by the following example.

A 120 Ω gauge, gauge factor 2 and temperature coefficient 12 $\mu\Omega°C/\Omega$ is subjected to a temperature rise of 5°C.

$$\delta R = (120 \times 12 \times 10^{-6} \times 5)\Omega$$

$$\text{Apparent strain } e_a = \delta R/R\lambda$$

$$= \frac{120 \times 12 \times 5}{120 \times 2 \times 10^6} = 30\ \mu\text{m/m or 30 microstrain}$$

A stress of approximately 6 MPa would produce this strain on steel.

The above remarks apply equally to the internal arms of the bridge, and also to the leads.

Hysteresis

The strain gauge should sustain exactly the same amount of strain as the surface to which it is bonded under temperature conditions suited to the bonding cement, and the maintenance of that strain should be independent of the length of time over which the strain is applied. However, if $\delta R/R$ is plotted against $\delta L/L$ to a high strain level a small measure of non-linearity may develop and on relieving strain to zero a hysteresis loop is formed. Normally, repeated cycles will form further, though rapidly narrowing loops (Fig. 5.20). It is usual to cycle the gauges a number of times before use to a higher strain level than actually required to eliminate this effect. Severe hysteresis arises from faulty bonding and will not be eliminated by cycling.

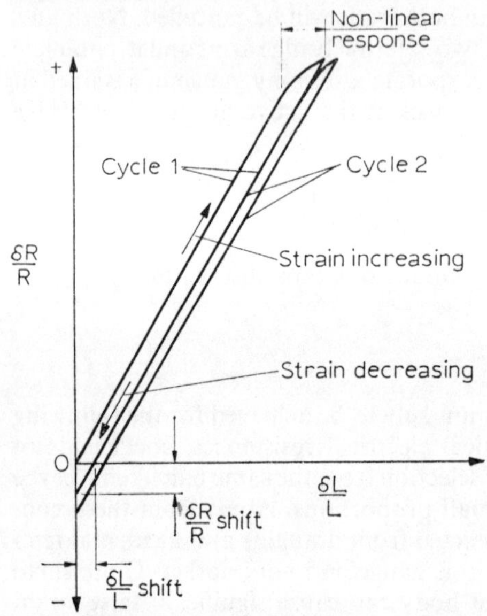

Fig. 5.20 HYSTERESIS LOOPS

Creep

Both the backing cement in which the gauge filament is embedded and the bonding cement holding the gauge in position must be much stronger than the filament itself if strain is exactly to be transmitted to the gauge. If this is not the case then the gauge filament relaxes under strain and strain indication is less than true strain becoming increasingly non-linear as strain increases. Creep arises from a number of sources:

(1) Faulty bonding either at the gauge-surface interface or within the gauge itself, i.e. the backing into which the filament is embedded.
(2) Operating temperature increase which softens the plastic bonds thus allowing the filament to relax. This latter effect largely provides the temperature limitation at which the gauge can satisfactorily operate. Paper-base cold setting cement gauges operate satisfactorily up to 50°C and creep is of the order of 0·5% of the total strain in 24 hours. Thermo-setting base (Bakelite) gauges and cements exhibit little creep (0·05%) up to 50°C and about 0·5% up to 100°C. Operations above these temperatures are usually of short duration and require special precautions.
(3) Unlike hysteresis, creep is time-dependent and sufficient time must be allowed for creep to become evident when the gauges are initially

installed. Static strain tests are more likely to be affected by creep than dynamic tests which are usually of short duration.

(4) Smaller gauges more easily develop significant creep than large gauges because of their smaller bonding area; thus preference should be given to larger gauges wherever possible.

Fatigue

This usually occurs in dynamic strain installations resulting from stress reversals at the point where the connecting leads are attached to the gauge. In all cases leads should be well secured to prevent relative motion between the gauge and the leads. Fatigue can also cause decay in the gauge base or fixing cement leading to filament relaxation.

Humidity and Moisture

Moisture absorption by the gauge backing or fixing cement causes volume changes which may extend or compress the grid. Thus gauges should be applied in dry conditions and any danger of moisture absorption due to changes of humidity should be guarded against by the application of moisture-resistant compounds such as silicone wax, bitumen or other compounds or by means of rubber covers, metal cans, etc. Precise details of waterproofing techniques for both gauges and lead connections are supplied by manufacturers. Another aspect of moisture absorption is that it can lead to a sharp fall in the impedance of the gauge to ground which should normally be at least 100 MΩ and preferably 1 000 MΩ. The reason for this is that the gauge is electrically in parallel with the backing and fixing cement and if these become conductors by developing low resistance then the effective gauge resistance is reduced.

5.12 Semiconductor or Piezoresistive Strain Gauges

Semiconductor gauges are cut from single crystals of silicon or germanium in which are combined exact amounts of special impurities such as boron which impart certain desirable characteristics. Such crystals are known as "doped" crystals.

The essential advantage of semiconductor gauges is their high strain sensitivity which allows very small strains to be measured accurately and,whereas wire and foil gauges have gauge factors of the order of 2, the gauge factors of most semiconductor gauges lie in the range 100 to 200. The change of electrical resistance of semiconductors may be either positive or negative, i.e. a gauge whose electrical resistance increases in response to tensile strain is known as a positive or *p*-type gauge and one whose resistance decreases in response to tensile strain is known as a negative or *n*-type gauge.

The gauge consists of a single rectangular filament about 0·05 mm thick by 0·25 mm wide and gauge lengths range from 1·5 to 12 mm. They are made as thin as possible as the breaking stress of the material rises as the cross-sectional area decreases and also the gauge can be bent to a much smaller radius of curvature without fracture. The element may be bonded to a plastic or stainless steel backing which can be welded in position or it may be unbonded in which case extra care must be taken in handling.

At Fig. 5.21*a* a single filament gauge is shown with leads on either side. A number of lead configurations are available to suit requirements. At 5.21*b* a

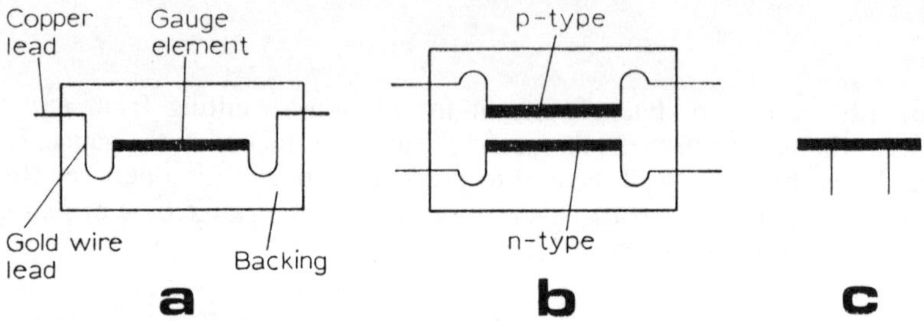

Fig. 5.21 SEMICONDUCTOR STRAIN GAUGES

pair of filaments, one *p* and one *n*, is shown. This type provides temperature compensation because both have positive temperature coefficients of resistance. Backed gauges are bonded to surfaces by the same methods as wire and foil gauges with the precaution that because of the extreme brittleness of the

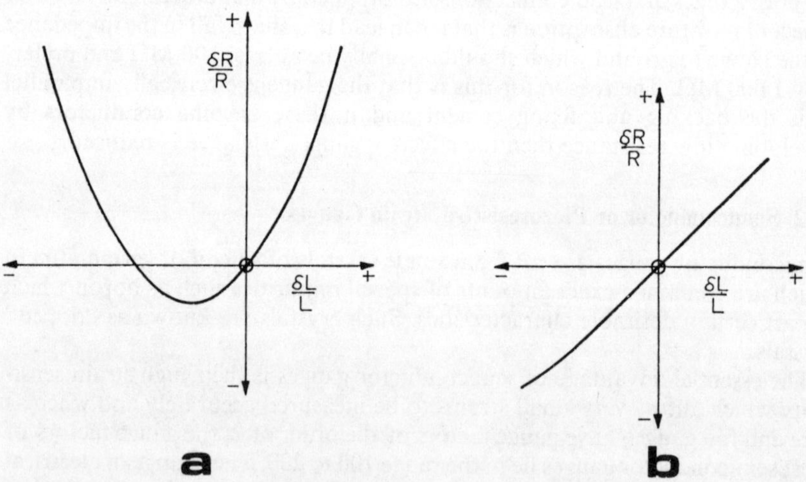

Fig. 5.22 OUTPUT CHARACTERISTICS OF A *p*-TYPE SEMICONDUCTOR STRAIN GAUGE

filaments they should not be squeezed in any way. Unbacked gauges as shown at 5.21*c* are available but are not in wide use.

Lightly doped gauges have very high sensitivity, i.e. high gauge factors and also high electrical resistance of the order of 200 Ω although gauges having resistances of 1000 Ω or even 10 000 Ω are available. Heavily doped gauges having lower electrical resistance also have lower gauge factors of the order of 50. At constant room temperature the change of resistance as a ratio of the unstrained resistance is very closely parabolic, i.e. the output is distinctly non-linear and a lightly doped *p*-type silicon gauge might have a relative resistance change/strain curve as shown in Fig. 5.22*a*.

The slope of the curve at any point is the gauge factor and it is evident that quite small strains either tensile or compressive bring about a change in gauge factor. It will be noticed that the two legs of the curve become more nearly linear as they increase in length and in fact heavily doped materials provide curves in which the gauge factor is virtually constant at constant temperature about the zero point (5.22*b*).

Temperature changes cause considerable resistance and sensitivity changes in semiconductor gauges; again heavily doped gauges are much less affected in this respect, e.g. a normal commercial *p*-type silicon gauge having a gauge factor of 120 could suffer a gauge factor variation of about −20 for a temperature rise of 100°C whereas a heavily doped *p*-type silicon gauge of 50 gauge factor would show virtually no change of factor for a similar rise in temperature.

Temperature Compensation

If semiconductor gauges are to provide accurate results temperature compensation is essential. This can be achieved by the use of self-compensating gauges or by circuit arrangements which provide compensation in a similar manner to that used for wire and foil gauges and in addition *p* and *n* combinations, thermistors, etc., may also be used.

The temperature coefficient of resistance of both *n* and *p*-type gauges is positive but the gauge factor of *n*-type gauges is negative. Thus the positive temperature coefficient and the negative gauge factor can be balanced against each other, and these in turn against the non-linear characteristics of the Wheatstone bridge at high resistance variations to provide temperature compensation within a limited temperature range. There are numerous circuit arrangements which can be used to provide near-linear outputs from semiconductor strain gauges. Where circumstances permit, similar gauges, one in compression and one in tension, can form two adjacent arms of the bridge; this is called "push-pull" operation. In such circumstances the gauge non-linearities are in opposition and largely tend to cancel each other. Similarly, *p* and *n* type pairs such as those shown at Fig. 5.21*b* may be used, and in this case both gauges experience the same stress condition and are connected into adjacent bridge arms. A similar result can be obtained if separate *p* and *n* gauges are used providing their non-linearities are carefully matched and they both experience the same temperature changes. The loss of sensitivity in gauges having negative coefficients of gauge factor can be compensated by providing a constant current bridge supply. One method is to connect a thermistor in series with the bridge supply so that as the temperature of the gauges and the thermistor increases the fall in resistance of the thermistor allows an increase in voltage across the bridge, thus maintaining a constant current supply to the bridge.

Field of Application

Because of their high sensitivity compared with wire and foil gauges, semi-conductor gauges can be used to measure low strains making unnecessary the use of signal amplification. They are particularly useful when measuring strains on rotating bodies as variations of electrical resistance occur at the interface of the slip rings and brushes used to convey the signal to the stationary strain measuring instruments. Such variations impose unwanted signals on the output which may be a considerable proportion of the total output when wire and foil gauges are used, thus making it difficult to identify the signal arising from strain.

5.13 Slip (Collector) Rings and Conducting Brushes

When strain gauges are attached to a moving body such as a rotating shaft or a turbine blade the gauge resistance changes must be conveyed to the stationary

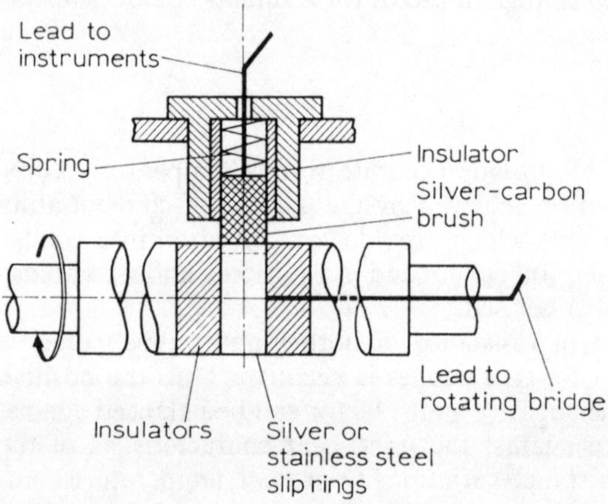

Fig. 5.23 SLIP RING AND CONDUCTING BRUSH

measuring system by an intermediate mechanism. This mechanism consists of rotating slip rings and stationary brushes operating in the same manner as the armature and brushes of a car dynamo. The leads from the gauges which must be very securely held in position are led to the centre of rotation and then soldered to rings incorporated in the rotating mechanism or to a separate shaft attached to the mechanism which carries the rings. A spring loaded stationary conducting brush or brushes in permanent contact with the rotating ring carries the current to the measuring system (Fig. 5.23).

The main problem to be overcome with this method is variation of electrical resistance at the interfaces of the rings and brushes. The change of resistance of an average strain gauge subjected to a strain equivalent of 7 MPa stress on steel is about 0·01 Ω but the contact resistance variation or "noise" may easily be equal to this amount on balance of all the rings and brushes in use and if they were connected in series to the gauges the strain signal might be completely submerged in contact resistance variations. A combination of the following three methods is used largely to reduce if not completely to eliminate this effect.

(*a*) The circuit is arranged so that the rings and brushes are connected in parallel with the gauge (Fig. 5.24).
(*b*) High resistance gauges are used to obtain a proportionately higher resistance change for a given strain (the qualities of semiconductor gauges are of value in this respect).
(*c*) Brush pressures and ring and brush materials are chosen to afford the best operating conditions. For general use up to 5000 r/min, silver, silver plated or stainless steel rings and silver–graphite brushes at pressures between 140 and 210 kPa are widely used.

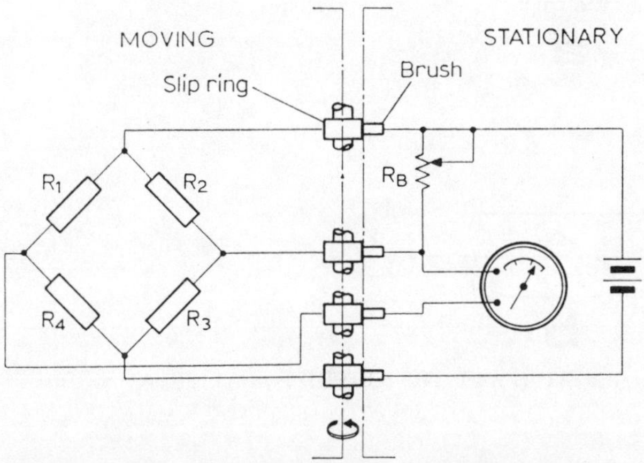

Fig. 5.24 STRAIN MEASUREMENT ON ROTATING ELEMENTS

The system shown in Fig. 5.24 can be used for null balance operation as a large balancing resistor R_B is incorporated in the circuit. Performance at high speed can be improved by incorporating two or more brushes in parallel on each ring, each having springs of a different spring rate so that the effect of resonances or other vibration sources will not cause all the brushes to lift from the ring at the same time. When not in use the brushes may be lifted from the rings.

Slip rings and strain gauge leads may also be brush painted on to surfaces using special silver paint. This is rather similar to the printed circuit technique and careful insulation with insulating paints is essential. This method may be useful when normal rings and brushes are difficult to apply. In all cases slip rings must have a very high quality surface finish and concentricity of rotation is essential, the latter being extremely important for high speed operation (10 000 to 20 000 r/min).

Circuits other than that illustrated at Fig. 5.24 are used when one or two strain gauges only can be applied to the moving part but the method of signal transfer to the stationary measuring instrument remains the same.

An essential requirement when measuring strain on moving parts is to avoid fatigue conditions at the leads. Normally this presents little difficulty on rotating parts but when linear or oscillating motion is involved as is often the case in machine tools, e.g. strain gauges attached to the tool on a shaping machine, then a pivoted arm oscillating with the tool can be used to which the leads are attached. The leads are fixed along the arm to a point near the centre of rotation and then generously coiled and supported before entering the stationary measuring instrument. In other circumstances linear sliding contacts can be used in a similar manner to that used for rotational motion.

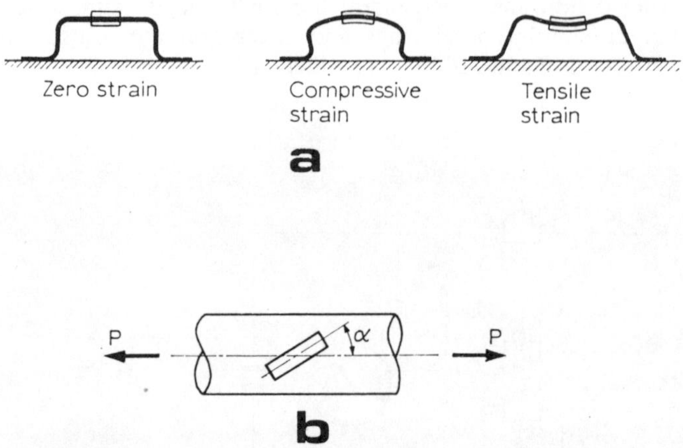

Fig. 5.25 CLIP GAUGE, SHOWING AT *b* THE METHOD FOR ATTENUATING GAUGE OUTPUT

An entirely different approach is to install suitable radio or television equipment on the moving part and broadcast the strain signals to a remote receiver where they can be analysed. This technique is called telemetering and of course involves considerable expense.

5.14 Plastic Strain

Most strain measurements are made within the elastic limit of metals but there are occasions when measurement well into the plastic range of a material is required and even to rupture. Resistance strain gauges if they are of reasonable length measure strain well beyond the elastic limit for all metals recording strains up to 40 000 μm/m without serious error but strains for very ductile metals can be as high as 25%, i.e. 250 000 μm/m and for such purposes special gauges may be used. Such post-yield gauges have the following characteristics and operating requirements: (i) highly ductile wire or foil, (ii) a plastic gauge factor equal to the elastic gauge factor, (iii) a fixing cement which prevents slip and is of sufficient ductility to resist cracking and rupture, and (iv) a backing material, usually 0·025 mm thick nylon, which resists cracking or rupture at high strains. Gauges fulfilling these requirements are commercially available to measure strains up to 150 000 μm/m.

The two following methods avoid the use of special gauges for measuring plastic strain by attenuating the strain on the gauge so that its output is less but proportional to the strain on the member.

THE CLIP GAUGE This consists of a spring steel bridge which is fixed to the body under test (Fig. 5.25*a*).

Gauges are cemented to the upper and lower sides of the bridge and under strain the legs deflect much more than the bridge. Output from the gauges can be varied by varying the dimensions of the clip. Initial calibration is required.

The second method is most easily appreciated by reference to a uniaxial stress field. In this case the maximum strain occurs along the axis of maximum stress; consequently, a gauge attached to the member at some convenient angle to the maximum strain axis will be subject to output attenuation proportional to the true strain (Fig. 5.25*b*). A similar technique can be used on a biaxial stress field when the positions of the principal axes are known.

5.15 Miscellaneous Applications

Stress Measurement

The attenuation technique can be used to modify the output of a gauge so that it is proportional to stress rather than strain for the measurement of principal stresses.

Because both principal strains must be known to determine either one or both principal stresses the angle to the chosen principal plane to which the gauge is set must be such that the gauge is affected in correct proportion by both principal strains (Fig. 5.26). It can be shown that the angle required is dependent only on Poisson's ratio for the material, as in the case of principal stress shear stress is zero, thus

$$\tan\theta = \sqrt{\nu} \qquad \text{or} \qquad \theta = \tan^{-1}\sqrt{\nu}$$

where ν is Poisson's ratio.

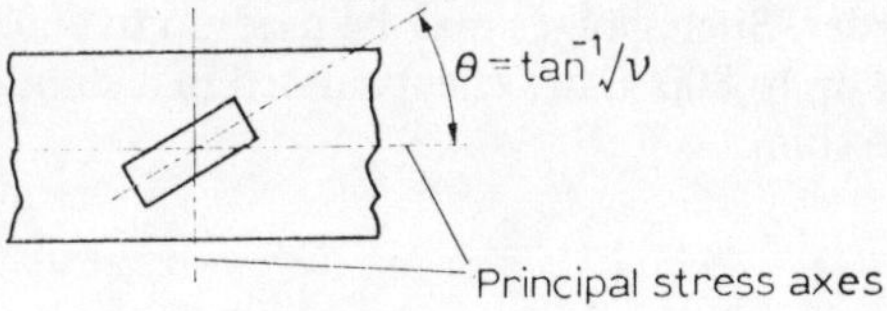

Fig. 5.26 THE STRAIN GAUGE AS STRESS GAUGE

Stress Gauges

These gauges also provide an output proportional to stress. Their grids are arranged so that they respond to the two orthogonal strains associated with the stress, i.e. the strain along the gauge axis and the strain perpendicular to that axis. Two types are available, one having two mutually perpendicular grids whose lengths are proportional to Poisson's ratio (Fig. 5.27*a*), and are suited only to materials having the same ratio. These gauges can be used to measure stress along any axis including the principal stress axis by using leads (2) and (3) but in fact are unnecessary for that application as the attenuation technique applied to a normal gauge is completely satisfactory. The gauge becomes a normal strain gauge when leads (1) and (3) are used. The second type has a vee-shaped grid. Each arm of the vee is affected by shear strain, one arm being in compression and the other in tension due to shear, thus the effect of shear strain is eliminated. The angle θ of each arm of the gauge to the stress axis is equal to $\tan^{-1}\sqrt{\nu}$ as before (Fig. 5.27*b*).

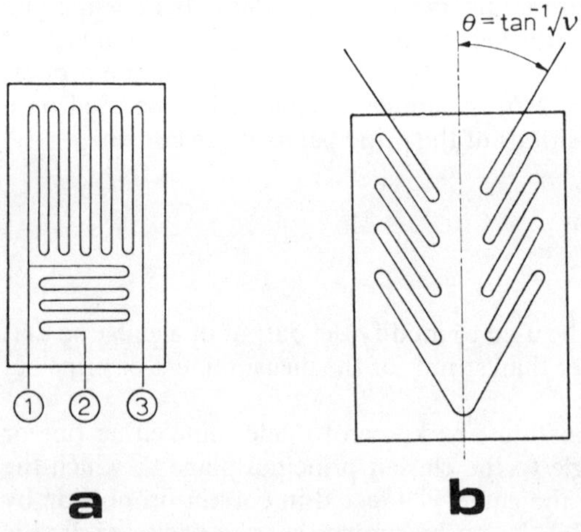

Fig. 5.27 COMMERCIAL STRESS GAUGES

Weldable Gauges

These gauges are bonded on to a thin (0·125 mm) stainless steel shim which can then be welded in position. They are used for high-temperature strain measurement where it is essential that the quality of the bond to the tested surface should be of the highest order. Such gauges may be used up to 400°C. Gauges for dynamic measurement up to 800°C are encapsulated in a stainless steel tube which is mounted on the shim.

"*Non-linear*" *Materials*

Many alloys, e.g. cast iron, do not have linear stress–strain characteristics and it is necessary to investigate their properties in this regard by mechanical testing so as to determine the probable response of a gauge bonded to the surface. The necessary information must include a close approximation to the direct stress–strain relationship and also Poisson's ratio.

Gauges can be applied to non-metallic materials such as glass, plastics, ceramics, wood, concrete, etc., and in each case the behaviour of the material under stress must be investigated. Such considerations as surface finish and characteristics, e.g. porosity, must be assessed and environmental effects must also be investigated, e.g. moisture absorption by wood. A knowledge of the effect of adhesive cements must be obtained if gauges are to be applied to plastics as some of these may dissolve the surface.

5.16 Zero Drift in Strain Instruments

Wheatstone Bridge

Under d.c. operation bridge balance is virtually independent of the supply voltage but this condition is true only if the internal arms of the bridge are subjected to the same temperature conditions and have the same thermal coefficients as the gauges themselves.

Instrument Instability

Some battery-operated instruments incorporate an audio frequency oscillator, a rectifier and an amplifier, and these provide a source for zero drift if the battery is allowed to decay as their performance depends on a stable battery voltage.

Zero drift increases in importance with time, thus short time readings will be little affected. Where the time between readings is long (weeks, months), then it is essential either to eliminate the effect of drift at each reading by returning the tested structure to zero strain and rebalancing the bridge if necessary or, if this is not possible, to detect it and thence correct the apparent strain readings to obtain true strain values. A simple solution to the problem when a mains supply is available is to interpose a power regulator between the mains and the instrument, thus ensuring a constant voltage and current. Such regulators replace the battery and can provide alternating or direct current.

A separate dummy bridge may be used as a reference. Here gauges are attached to a piece of metal which remains unstrained. Suitably temperature compensated and balanced, such a bridge should remain at zero strain; if it does not then the strain readings are taken to be zero drift and the active gauge readings adjusted accordingly.

The bridge galvanometer senses both the direction and magnitude of current flow when the bridge is unbalanced. If all the resistances in the bridge and the leads were initially equal, then reversing the position of the active and dummy gauges in the bridge should provide an equal and opposite reading on the galvanometer, e.g. if R_1 and R_4 were originally connected at A then the reverse would be shown when they were connected at B (Fig. 5.28*a*).

If zero drift does not occur, then a real strain–time and a "reversed" strain–time curve plotted on the same axes should be mirror images of each other and the mean of both readings should be zero, the zero line produced representing the initial real strain reading (5.28*b*). If zero drift occurs, then the mean of the two readings will depart from the zero line and these departures when plotted provide the datum from which zero drift at any time may be derived (5.28*c*).

It is unlikely that the resistances and bridge leads will be in perfect balance and in this case the two initial readings will be displaced from each other on the strain–time graph. The mean of the two readings then fixes the position of the zero line. The mean values of the readings are then plotted as before but because the real and "reversed" strain curves are displaced the true zero strain line will be displaced by an amount equal to half the distance between the two initial readings (5.28*d*).

It must be clearly understood that the methods outlined will detect zero drift only in the instrumentation and not in the gauges themselves. In practice, drift in the instrumentation is usually very small in comparison with gauge drift which is by far the most likely source of error.

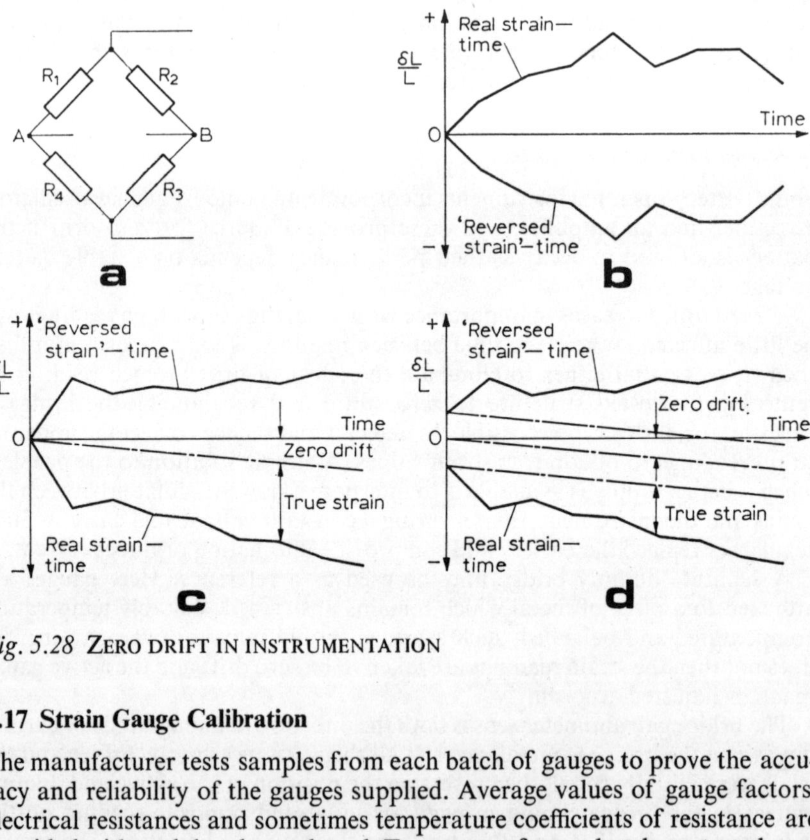

Fig. 5.28 ZERO DRIFT IN INSTRUMENTATION

5.17 Strain Gauge Calibration

The manufacturer tests samples from each batch of gauges to prove the accuracy and reliability of the gauges supplied. Average values of gauge factors, electrical resistances and sometimes temperature coefficients of resistance are provided with each batch purchased. Two gauges from a batch are tested on a symmetrically supported and loaded rectangular steel beam (Fig. 5.29*a*).

The exact position of the gauges is unimportant provided that they lie within the vee or ball supports A and B with their axes parallel to the beam axis as the bending moment to which the beam is subjected under load is constant within this length L and the tensile and compressive strains developed at the beam surfaces are equal but opposite in sign. A calibrated dial test indicator positioned at a point half way between A and B records the beam deflection at the centre of the span. Before loads are applied the dial indicator is set at zero and the bridge balanced. Equal loads are applied at W and values of resistance change $\delta R/R$, allowing for augmented output factors, and deflections y are recorded. Having rebalanced the bridge, further loads are applied until a suitable number of readings are obtained. It can be shown that

Surface strain $e = 4ty/L^2$

where t = beam thickness
y = beam deflection
L = length of beam subjected to constant bending moment.

It can also be shown that the gauge factor $\lambda = \delta R/Re$.

From the values obtained a graph of $\delta R/R$ and e can be plotted, the tangent of the resultant linear curve providing the gauge factor (Fig. 5.29*b*). This may

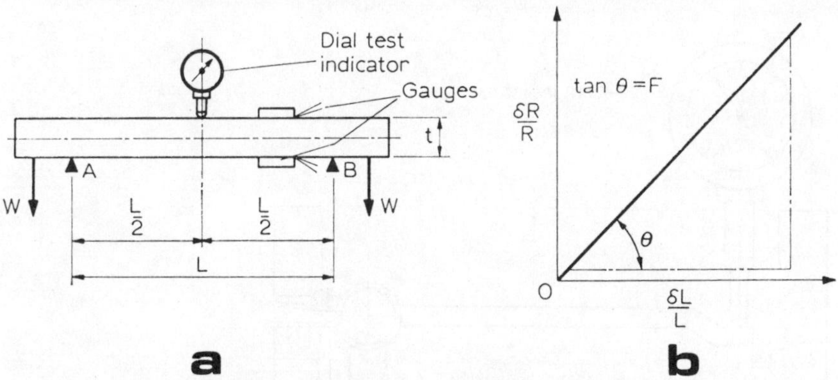

Fig. 5.29 STRAIN GAUGE CALIBRATION BEAM ARRANGEMENT

be compared with the manufacturer's value. An examination of the apparatus suggests that the value *t* should be taken to the centre lines of the gauges' grids but as this extra distance is negligible compared with the beam thickness it may be ignored in this case without detriment to the results. It should be noted that constant temperature conditions are essential during the test.

5.18 Extensometers

Prior to the invention and development of resistance strain gauges the main method used for determining strain was the extensometer. Extensometers are still used for certain purposes, e.g. on tensile test pieces and for structural applications in civil engineering where long gauge length and robust instruments are required. The main limitation of extensometers is their size which makes application difficult except in easily accessible positions; further, the elements to which they are attached must be geometrically simple such as rods of sufficient length and uniformity of cross-section to accommodate the instrument. Their use is limited largely to static strain measurement although dynamic measurement is possible with some types. Extensometers are sometimes referred to as strain gauges but in fact, as their name implies, they are extension-measuring devices because the gauge length of the instrument must be known before strain can be computed.

Two similar methods, the knife edge and the conical penetrator, are used to ensure that the instrument responds in unison with the increase or decrease in length of the member to which it is attached. These are embedded into the surface of the member which consequently suffers minor damage introducing a stress concentration and hence an error source. Three types are illustrated in Fig. 5.30, mechanical, optical and pneumatic.

Mechanical Extensometers

The Berry extensometer shown at Fig. 5.30*a* is used for structural applications in civil engineering under approximately linear strain conditions over long

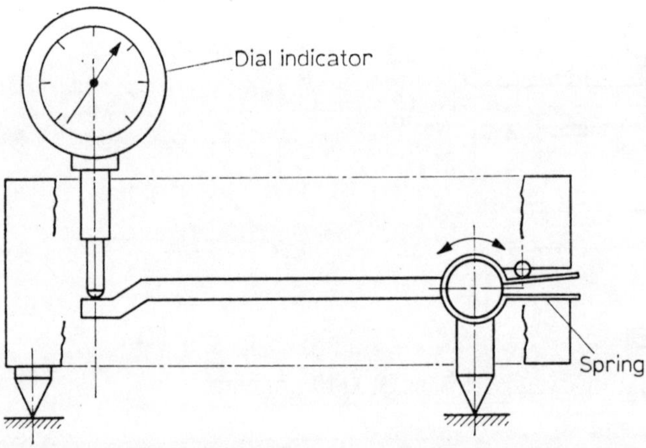

Fig. 5.30a BERRY EXTENSOMETER

gauge lengths up to 200 mm. An interesting feature of this gauge is the kinematically located spring loaded pivot spindle which eliminates bearing clearance.

The Huggenberger (5.30*b*) is another well known mechanical extensometer of which a number of models are obtainable, the most widely used being capable of gauge lengths of approximately 12 mm or 25 mm. This instrument depends entirely on very low friction knife edged pivots and may be reset to zero after each extension reading if desired. Magnifications up to 2000 can be obtained depending on the model.

The principle of the Johansson extensometer is shown at (5.30*c*); here a twisted metal torsion tape is used to which a light pointer is attached at its centre. One end of the tape is attached to the fixed knife edge and the opposite end to the moving knife edge which is flexibly attached to the instrument body by a leaf or flexure spring. Extension causes the tape to unwind thus rotating the pointer against the scale. Instrument friction is reduced virtually to zero, and magnifications of up to 5000 with gauge lengths of 1 to 50 mm are obtainable depending on the model. The torsion tape device is often referred to as the Johansson Microkator principle. An advantage of this gauge is its low profile.

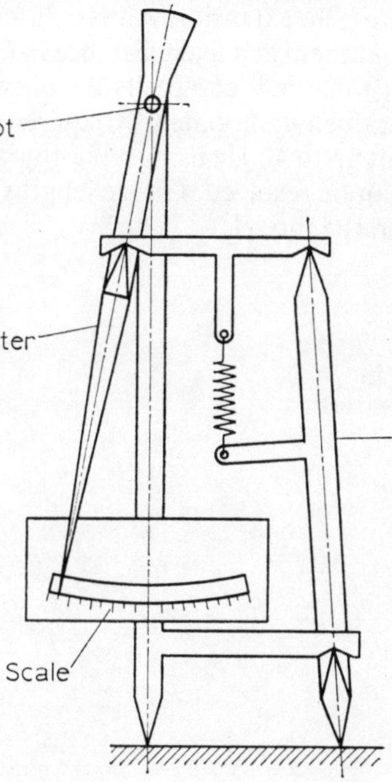

Fig. 5.30b HUGGENBERGER EXTENSOMETER

Optical Extensometer

Another low profile gauge is the Tuckerman optical extensometer which requires an auto-collimator to take the extension readings (Fig. 5.30*d*). The gauge consists of a fixed and a moving knife edge to which a mirror is attached.

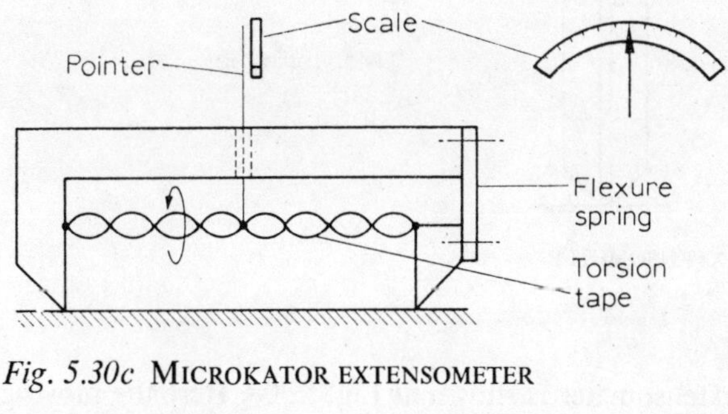

Fig. 5.30c MICROKATOR EXTENSOMETER

A graticule image is transmitted by the auto-collimator to the mirror which in turn reflects the image to the roof prism and thus back to the auto-collimator. Having initially established the position of the reflected image at zero extension any strain will rotate the knife edge and thus the mirror, displacement of the reflected image in the auto-collimator indicating the extension. Auto-collimators have the advantage that there is no need to set them at a fixed distance from the reflecting mirror and roof prism; thus gauge readings can easily be taken at a considerable distance (metres) from the strained component if required. This gauge can be used for dynamic measurement up to 40 Hz using a photographic recorder, and strains as small as 2 μm/m can be resolved. Gauge lengths vary from about 6 mm to 250 mm depending on the model.

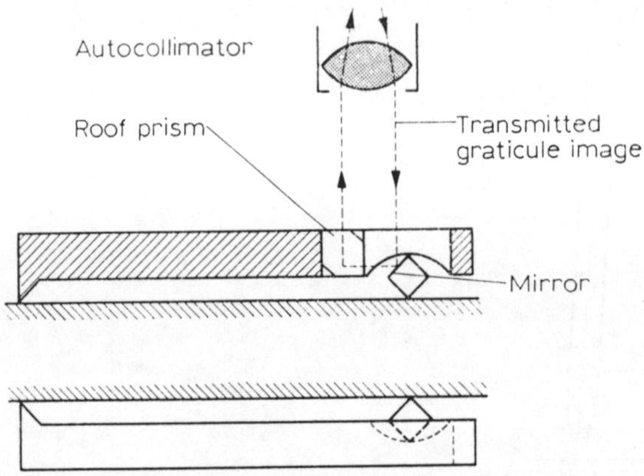

Fig. 5.30d TUCKERMAN OPTICAL EXTENSOMETER

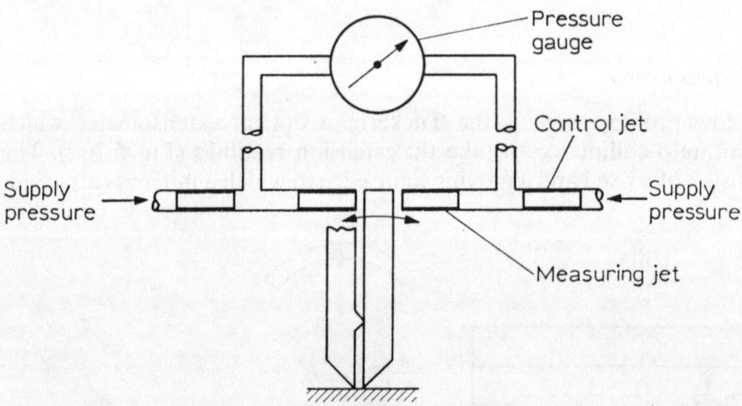

Fig. 5.30e PNEUMATIC EXTENSOMETER

Pneumatic Extensometer

A differential pressure extensometer is shown at Fig. 5.30*e*. Here the moving knife edge operates as a flapper between two outlet jets, the effective area of each jet being equal to the area of the cylinder between the jet and its corresponding flapper face. Both jets are supplied with a nonfluctuating and equal-air flow at constant pressure. The effect of strain is to rotate the flapper and thus reduce the area of one jet while increasing the area of the other. Pressure on the constricted side rises towards the supply pressure and on the other towards atmosphere, the differential pressure being recorded on the pressure gauge. Provided that the gaps between the jets and the flapper are small, close linearity is obtained. Magnifications up to 100 000 and gauge lengths as small as 1 mm are obtainable. The pressure gauge can be sited at a distance from the sensing element.

The Acoustic Extensometer

A wire under tension vibrates at its natural frequency. The natural frequency changes if for instance the temperature or tension in the wire increases or

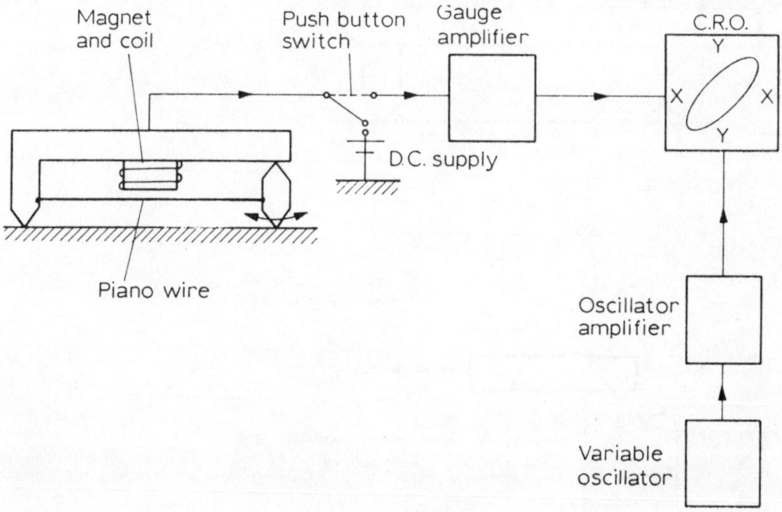

Fig. 5.31 ACOUSTIC STRAIN GAUGE SYSTEM

decreases. When all conditions that might affect the strained wire remain steady an increase or decrease in tension is the only factor which will change the vibrational frequency; such changes of frequency are proportional to changes in tension and thus to extension or strain in the wire. In order to use the vibrational frequency of the gauge wire as a source of strain information it is necessary to provide a previously calibrated vibrational system such as a variable oscillator against which it may be compared. An acoustic gauge system is shown at Fig. 5.31.

Steel piano wire about 0·3 mm diameter is mounted between the fixed and moving knife edges of the gauge and mounted above the wire is a permanent magnet round which is wound a coil. The push button arrangement allows a pulse of direct current to flow to the coil which sets the wire vibrating at its natural frequency for a short period transmitting a small alternating current to the gauge signal amplifier and thence to the X-plates of an oscilloscope. The Y-plates of the oscilloscope are supplied with a reference signal from the variable oscillator (range 200 to 2000 Hz) which is then adjusted until a circular or elliptical Lissajous figure appears on the screen when both the gauge and oscillator frequencies are matched and the vibrational frequency of the gauge wire is then known. The variable oscillator can be checked by replacing the gauge signal by the signal from a standard tuning fork.

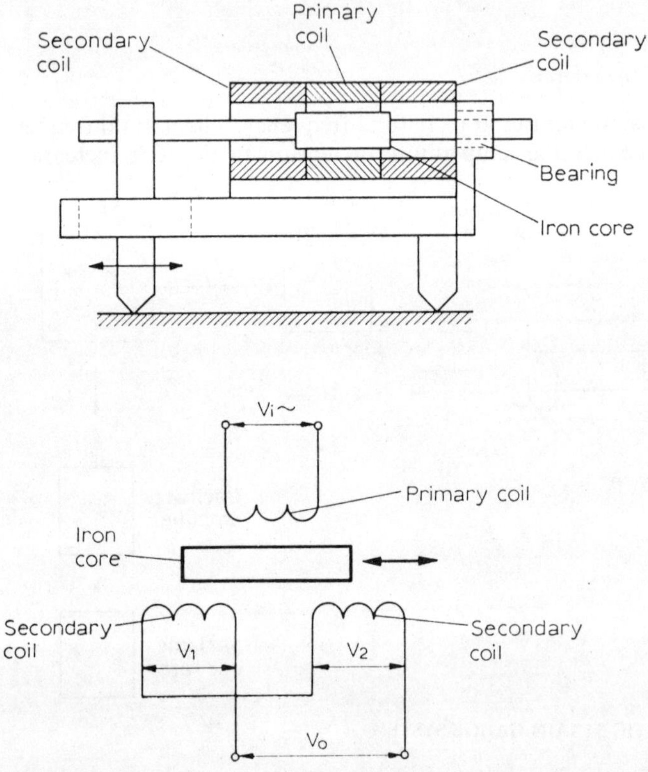

Fig. 5.32a INDUCTANCE STRAIN GAUGE

The strain in a wire varies as the square of the frequency, thus

$$\frac{\delta L}{L} = \frac{4L^2\rho}{AE} \times f^2$$

where L = wire length (m)
A = wire cross-sectional area (m^2)
E = Young's modulus of the wire (Pa)
f = vibrational frequency (Hz)
ρ = density of the material (kg/m^2)

It is thus an easy matter to compute the strain caused by a change of frequency from the gauge but digital readout versions are now available.

Acoustic gauges are used mainly in civil engineering and have the advantages of robustness and long-term reliability. High sensitivity of the order of

0·5 μm/m is readily obtainable and gauge lengths from 20 to 340 mm have been used. Because the system measures frequency, gauge lead and switch resistances do not affect the output. The gauges are easily covered for use externally on all types of structures and instead of knife edges studs may be fixed into the structure to support the wire, e.g. on steel girders. These gauges have operated successfully when mounted in a protective tube and immersed in concrete and, providing the covers are carefully sealed, operation under water offers few problems. A disadvantage is that quite considerable tension is required to provide a sufficiently taut wire, thus these gauges are not suitable for light structures.

The Inductance Extensometer

There are various ways in which linear movements can be transformed into inductance variations. A well tried method is the linear variable differential

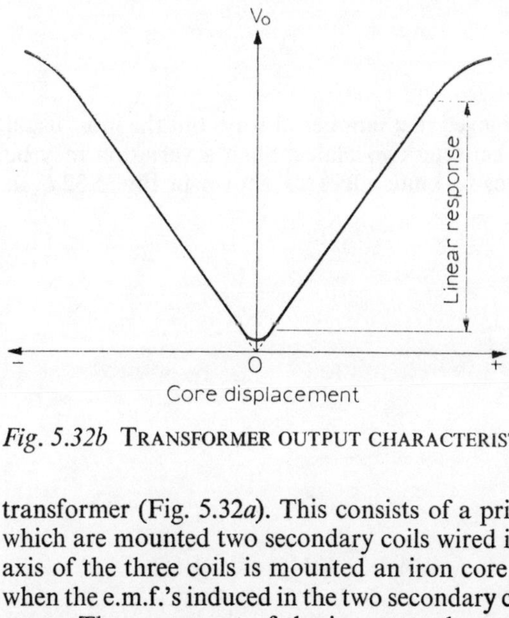

Fig. 5.32b TRANSFORMER OUTPUT CHARACTERISTICS

transformer (Fig. 5.32*a*). This consists of a primary a.c. coil on each side of which are mounted two secondary coils wired in series opposition. Along the axis of the three coils is mounted an iron core on a non-magnetic rod. Only when the e.m.f.'s induced in the two secondary coils are unequal will an output occur. The movement of the iron core due to extension of the tested body causes the induced e.m.f.'s in the secondary coils to vary and because of their series opposition connection their combined output will be the difference, if any, of the e.m.f.'s induced.

In the practical differential transformer there is always a capacitive effect between the primary and secondary coils which results in a small output voltage even when the induced e.m.f.'s in the secondary coils are in equal opposition, but this is normally less than 1% of the maximum voltage. The output characteristics of the transformer are shown at 5.32*b*.

Commercial inductance transformers used for strain measurement have base lengths between 10 and 50 mm although base lengths as small as 0·5 mm are available for special purposes. The driving force required is less than 0·01 N and thus dynamic strain can be measured provided that the frequency of the primary coil voltage is at least ten times the measured frequency.

An inductance bridge arrangement employing a single coil and moving iron can be used (5.32*c*) and is similar in operation to a resistance bridge. Each arm of the bridge can be active and temperature compensation can be p.·ovided.

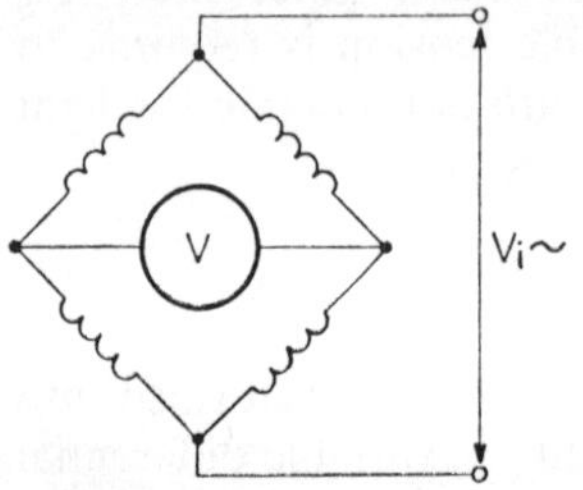

Fig. 5.32c INDUCTANCE BRIDGE

The Capacitance Extensometer

Variable capacitance can be achieved in a number of ways but the most usual method is to vary the air gap between two plates. Such a variation may be achieved by attaching the plates to knife edges as shown in Fig. 5.32*d*. In

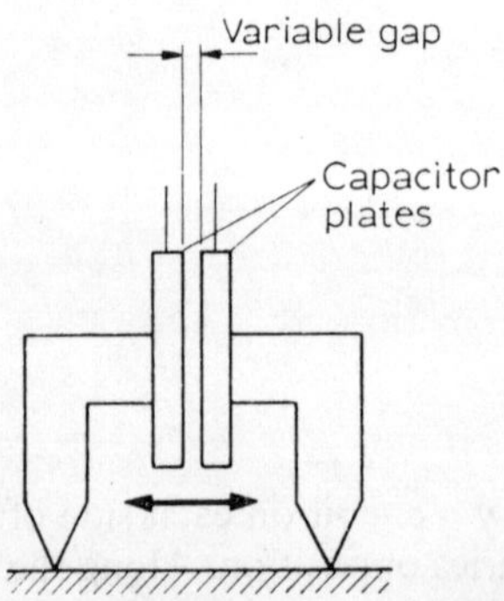

Fig. 5.32d CAPACITANCE BRIDGE

practice this type of gauge is little used for strain measurement as circuit and temperature problems generally outweigh the advantages of the method.

5.19 Photoelasticity

Photoelasticity is an indirect method of stress analysis which is extremely useful at the design stage. A study is made of the stress conditions in a scale model of the component made from a suitable transparent material such as Araldyte CT.200 and from this the stresses in the original metal component can be deduced, both qualitative and quantitative results being obtained.

When the model is stressed, placed in the path of a beam of polarized light and the light is arranged to fall on a screen, an interference pattern may be obtained. Inspection of the interference pattern indicates areas of high stress concentration and, equally important, areas of very low stress where the material is not being used to its best advantage. Full quantitative results of peak stresses at both the boundaries and interior of the model can easily be obtained and a simple process of scaling provides the actual stress likely to be present in the metal component.

Using this technique many design faults can quickly and easily be located and suitable modifications made to the original design.

5.20 Polarized Light

A beam of light is assumed to travel in wave formation in any number of directions transverse to the axis of propagation. If the waveform can be confined to one plane, then the light is said to be plane polarized. Certain crystals

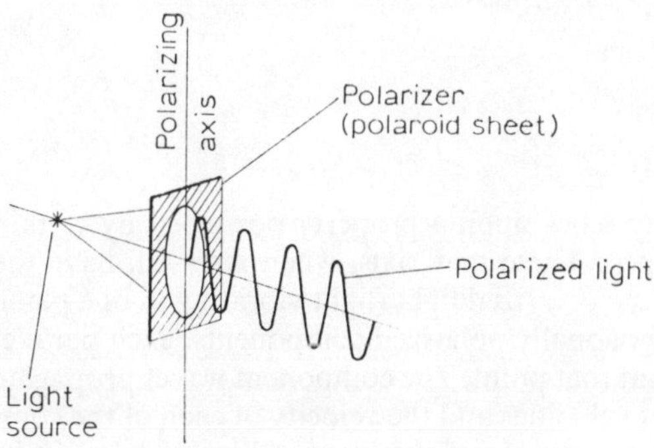

Fig. 5.33 PLANE POLARIZED LIGHT

such as tourmaline and herapathite are said to have only one optic axis, that is light can only pass through the crystal in one direction, and in the plane perpendicular to the optic axis light is absorbed. The polarizing medium used in photoelasticity is polaroid which consists of crystals of herapathite embedded in celluloid, and rays of light transmitted by the Polaroid sheet (the polarizer) vibrate in a plane parallel to the polarizing axis as shown in Fig. 5.33.

Elliptically and Circularly Polarized Light

If two plane-polarized waves vibrating in planes at right angles have a phase difference of α wavelengths it can be shown that the resulting motion represents the equation of an ellipse and would represent elliptically polarized light. If the amplitude of the two waves are equal and the phase difference $\alpha = \pm\frac{1}{2}\pi$ radians then the resulting motion represents circularly polarized light.

A plane polariscope is shown in Fig. 5.34*a* where the axes of the polarizer and a second polaroid sheet, known as the analyser, are crossed (perpendicular to each other). No light will be transmitted in this case as the emergent rays from the polarizer vibrate in a plane at 90° to the axis of the analyser.

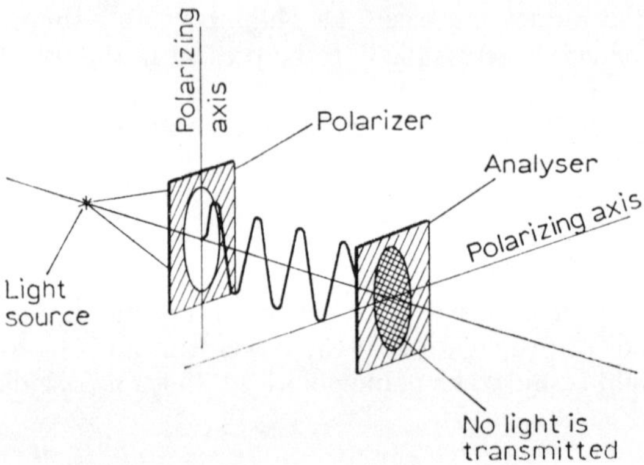

Fig. 5.34a THE PLANE POLARISCOPE

5.21 Principles of Photoelasticity

The technique of photoelasticity relies upon a property possessed by certain materials known as birefringence. These materials, when stressed, have the property of splitting a ray of plane polarized light (light vibrating in one plane only) at any point into two orthogonally polarized components, each parallel to a direction of principal stress at that point. The component waves propagate through the material at different velocities and the velocity of each of the components is proportional to the magnitude of the principal stress lying in its plane of vibration:

$$V_1 \propto \sigma_1 \qquad \text{and} \qquad V_2 \propto \sigma_2$$

where V_1 and V_2 are the velocities of the respective components and σ_1 and σ_2 are the principal stresses. The principal stresses are defined as the maximum σ_1 and minimum σ_2 values of direct stress that can be obtained at a point in a material. The principal stresses act along planes known as principal planes which are 90° apart. It may also be shown that the difference in the principal stresses at a point is equal to twice the maximum shear stress at that point:

$$\sigma_1 - \sigma_2 = 2\tau_m$$

where τ_m = the maximum shear stress at the point.

The emergent rays from the photoelastic model will have suffered a relative phase change or "relative retardation" proportional to $(V_1 - V_2)$. However, as $(V_1 - V_2)$ is in turn proportional to $(\sigma_1 - \sigma_2)$ the relative retardation of the emergent rays will be proportional to the difference in the principal stresses $(\sigma_1 - \sigma_2)$ at that point. The emergent rays can be recombined into a single plane by means of an analyser, crossed with the polarizer (the planes of the respective polarizing axes are at 90°), such that only the components of the two rays parallel with the analyser axis will pass through. This set-up is known as a plane polariscope and is shown in Fig. 5.34*b*.

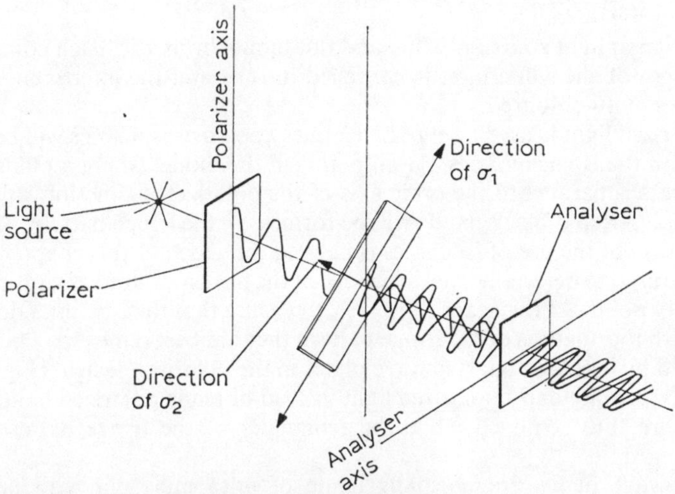

Fig. 5.34b PHOTOELASTIC SPECIMEN IN THE PLANE POLARISCOPE

Fig. 5.35 MONOCHROMATIC FRINGE PATTERN OF A DISC SUBJECTED TO DIAMETRAL LOAD (DARK FIELD)

The rays emerging from the analyser will interfere optically, i.e. cancel each other in some places and reinforce in others. If the light is allowed to fall on a screen an interference pattern or "fringe pattern" is obtained. The amount of interference at any point is proportional to the difference in the principal stress values at that point ($\sigma_1 - \sigma_2$) and hence to the maximum shear stress [$\tau_m = \frac{1}{2}(\sigma_1 - \sigma_2)$].

If the original light source is monochromatic (e.g. mercury green) the fringes will appear as black lines on a green background, the black fringes corresponding to regions where the rays are 180° out of phase. These fringes are known as isochromatics and a typical fringe pattern for a disc subjected to a diametral load is shown in Fig. 5.35.

When the original light source is white and not monochromatic, each composite wavelength of the white light is cancelled in turn and the interference pattern becomes multi-coloured.

If plane polarized light is used a set of black lines known as isoclinics will be superimposed on the isochromatics. At all points in the model where a plane of principal stress is parallel to the optic axis of the polarizer, extinction will occur at the analyser, and black bands will be formed on the fringe pattern. A detailed discussion of the use of isoclinics is beyond the scope of this chapter, but they may be used to determine the directions of the principal stresses within the model at any point. At this stage, however, to ensure that the isoclinics do not interfere with the analysis of the fringe pattern they are best removed. This may be achieved by inserting quarter-wave plates in the polariscope field (Fig. 5.36) which produce circularly polarized light instead of plane polarized light. The isoclinics are thus removed while the remainder of the fringe pattern remains unchanged.

The quarter-wave plates are normally made of mica and split a plane polarized ray of light into two orthogonally polarized components. The thickness of the plate is arranged so as to produce a phase difference, or relative retardation, of $\frac{1}{2}\pi$ radians, or one quarter of a wavelength, between the

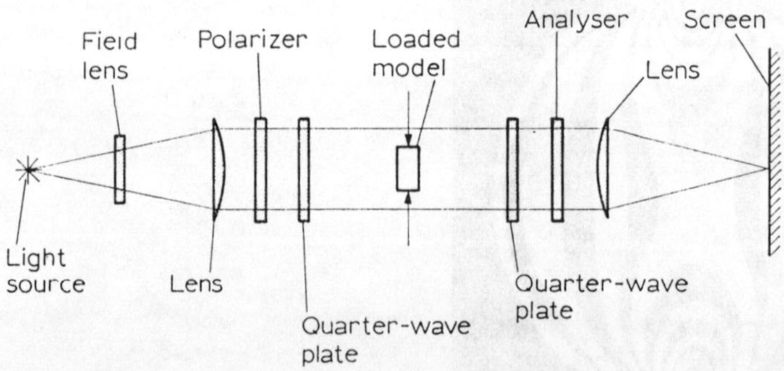

Fig. 5.36 POLARISCOPE ARRANGEMENT FOR CIRCULARLY POLARIZED LIGHT

component rays. If the axes of the plate are arranged at 45° to an incoming plane polarized ray, then the amplitudes of the two component rays will be equal and circularly polarized light will result.

The quarter-wave plates may be used in two ways, e.g. the plates can be adjusted to two relative positions in the instrument:

(*a*) "Crossed", when the "fast" plane (the component ray with the greater velocity is in this plane) is positioned opposite the "slow" plane (Fig. 5.37*a*).

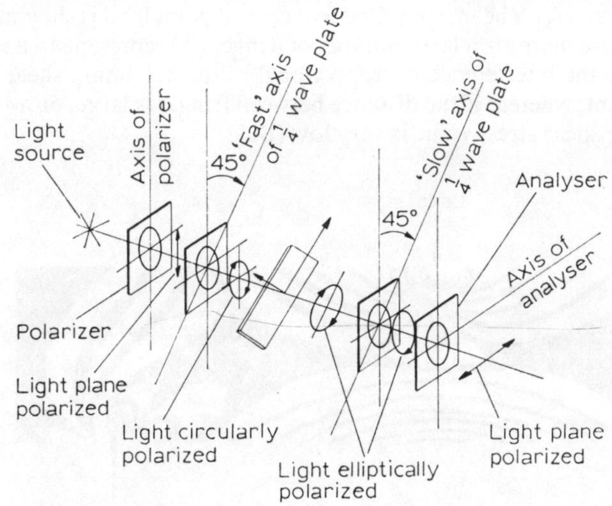

Fig. 5.37a CROSSED CIRCULAR POLARISCOPE

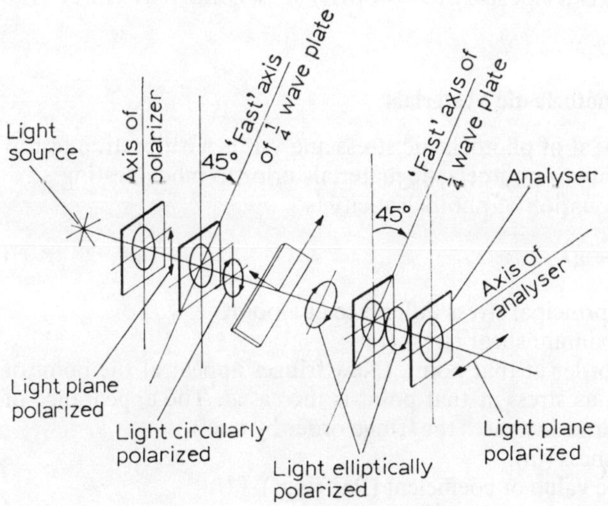

Fig. 5.37b PARALLEL CIRCULAR POLARISCOPE

(*b*) "Parallel", when the two "fast" planes are arranged to coincide and the two "slow" planes likewise (Fig. 5.37*b*).

The crossed-circular polariscope gives extinction of the light at the viewing screen when no photoelastic model is present (dark field) whereas the parallel condition produces an overall addition of half a wavelength and rotates the resultant plane of vibrations through 90° giving maximum brightness (light field).

A typical fringe pattern for a beam subjected to a central point load is shown in Fig. 5.38. Areas where there are a large number of fringes (A) represent areas of high shear stress as the interference is proportional to the maximum shear stress value at any point, whereas if the distance between fringes is large, or no fringes exist at all, the shear stress value is very low (B).

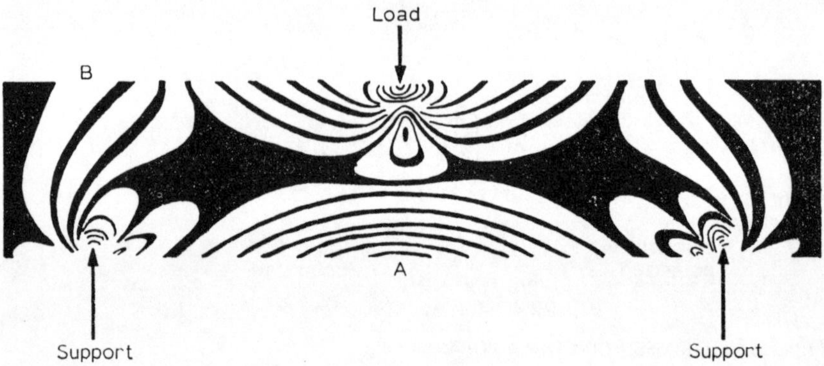

Fig. 5.38 FRINGE PATTERN FOR SIMPLY SUPPORTED BEAM LOADED AT THE CENTRE

5.22 Calibration of Photoelastic Materials

An essential requirement of photoelastic stress analysis is a calibration test to determine the sensitivity of photoelastic materials prior to model testing.

The fundamental equation of photoelasticity is

$$\sigma_1 - \sigma_2 = nf/t = nF \qquad \text{(i)}$$

where $\sigma_1 - \sigma_2$ is the principal stress difference at a point
$= 2 \times$ the maximum shear stress (Pa)
$n =$ The fringe order at that point. (New fringes appear at the point of observation as stress at that point is increased. The appearance of each new fringe increases the fringe order by one.)
$t =$ model thickness (m)
$F =$ model fringe value or coefficient (Pa/fringe)
$f =$ material fringe value or coefficient (Pa/fringe/m or N/m fringe).

The material or model fringe coefficients are defined as the stress value required to produce one fringe order at any point in the material or model respectively.

Thus, if the material or model fringe value is known the principal stress difference, and hence the maximum shear stress at any point, can easily be calculated from the fringe pattern. At a boundary the principal stress normal to the boundary, σ_2, is equal to zero and thus fringe values at the boundary can be used to calculate the boundary stress σ_1.

The following calibration techniques are in general use:

(*a*) Diametral loading of a disc.
(*b*) Simple tension.
(*c*) Four-point bending.

The three systems are used to determine the material or model fringe coefficients. Under a particular loading condition the value of fringe order is obtained at a point in the model where the stress difference ($\sigma_1 - \sigma_2$) is known. This procedure is repeated for a series of different loads and a graph of load versus fringe order plotted. As an example the diametral loading of a disc method is now discussed in more detail.

Consider a thin disc of photoelastic material subjected to two concentrated loads P on a diameter as shown in Fig. 5.39.

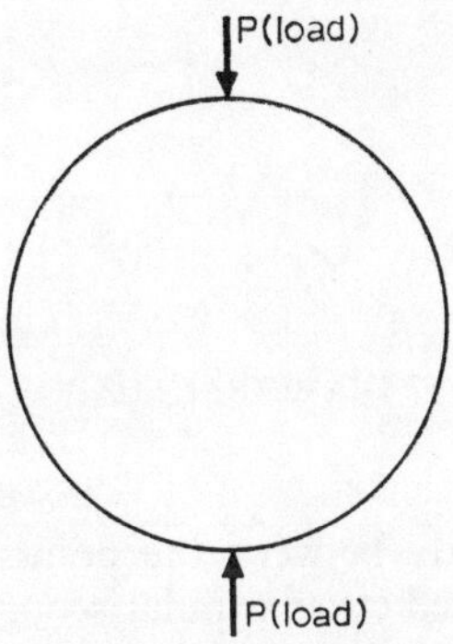

Fig. 5.39 DISC SUBJECTED TO DIAMETRAL LOAD (COMPRESSIVE)

It can be shown that the principal stresses set up at the centre of the disc are given by

$$\sigma_1 = 2P/\pi t D \qquad \text{and} \qquad \sigma_2 = -6P/\pi t D$$

where t is the thickness of the disc and D is the diameter.

$$\sigma_1 - \sigma_2 = 8P/\pi t D \tag{ii}$$

but from equation (i), $\sigma_1 - \sigma_2 = nf/t = nF$
and using equations (i) and (ii),

$$f\,(\text{the material fringe value}) = \frac{8}{\pi D} \cdot \frac{P}{n} \tag{iii}$$

or

$$F\,(\text{the model fringe value}) = \frac{8}{\pi D t} \cdot \frac{P}{n} \tag{iv}$$

In practice the load on the disc is increased so as to produce successive fringe orders at the centre using the crossed circular polariscope set-up thus producing a dark field or background. The ratio P/n in equations (iii) and (iv) is given by the slope of the graph of load P against fringe order n which is a straight line through the origin (Fig. 5.40).

Further points may be determined for the graph without increasing the load range by using a light field background (polarizer and analyser axes parallel) when the half-order fringes are obtained. The points on the graph

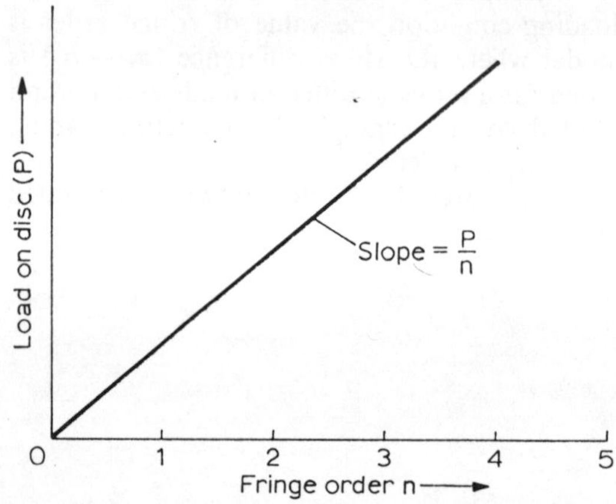

Fig. 5.40 GRAPH OF LOAD AGAINST FRINGE ORDER FOR THE DIAMETRAL LOADING OF A DISC IN COMPRESSION

corresponding to the half-order fringes will lie exactly between the points corresponding to the full-order fringes (using the crossed circular polariscope set-up).

5.23 Three-dimensional Photoelasticity

The above discussion refers mainly to two-dimensional models, i.e. thin models where the stresses do not vary considerably throughout the thickness. The method can be adapted, however, to deal with three-dimensional models using the "frozen stress" technique (Fig. 5.41).

The figure shows a spring placed in a beaker of water and loaded with a mass m. The water may then be frozen such that the stress remains in the

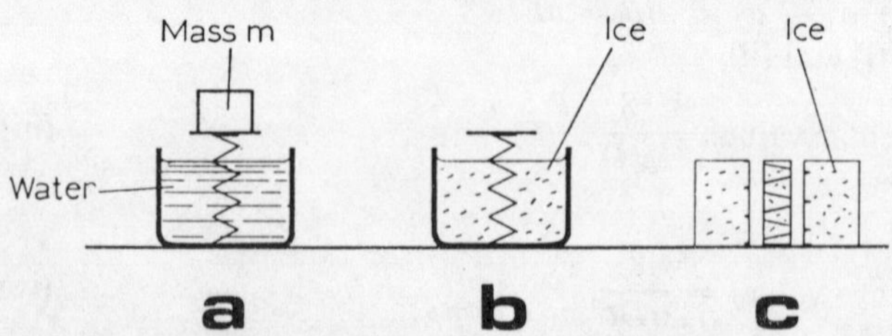

Fig. 5.41 THE "FROZEN STRESS" PRINCIPLE
a Loaded spring *b* Load removed *c* Ice and spring sliced for examination

spring when the mass is removed. If it were possible to cut the ice without melting it the spring and ice could now be sliced into thin strips for examination by the method discussed for two-dimensional models.

A similar effect is obtained when loaded photoelastic models are heated to a temperature above the "softening" temperature of the material and then slowly cooled. The stress pattern is found to be "frozen" in the model which may be sliced into thin strips for investigation as described previously.

In order to eliminate thermal stresses strict control of the cooling rate is essential and a one degree temperature gradient across the model is desirable. With models of 250—1250 mm in thickness this can mean a cooling rate of between 0·5 and 2·5°C/hour.

5.24 Photoelastic Strain Gauges

Transparent materials such as glass and plastic when strained are doubly refracting or birefringent. This effect when observed under polarized light

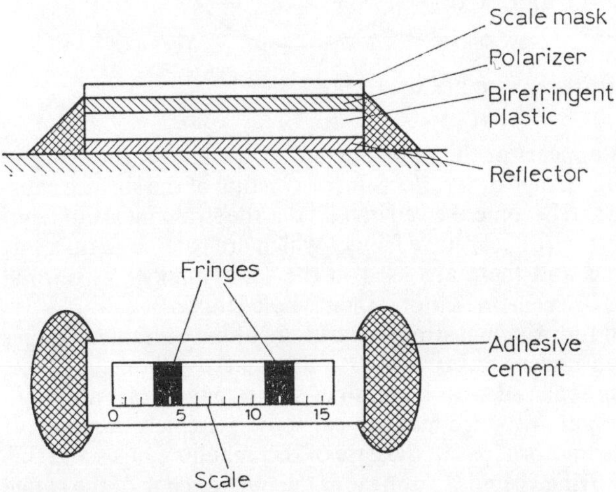

Fig. 5.42 RECTANGULAR PHOTOELASTIC STRAIN GAUGE

reveals interference fringes in the material which are analogous to the strain induced. Relief of strain eliminates the fringes. Permanent fringes can be induced by heating the material while under stress and maintaining the stress until it has cooled. The "frozen" fringe pattern so formed is permanent and change of stress simply causes fringe displacement. Rectangular and circular gauges based on this principle are commercially available. The gauges are of sandwich construction consisting of a piece of birefringent plastic approximately 2·5 mm thick on one side of which is a reflective coating and on the other a material which has the characteristics of a polariscope. Mounted above this is a calibrated and numbered strain scale (Fig. 5.42).

Gauges are mounted to surfaces with cements similar to those used for wire gauges although, because they are much less flexible, the surfaces to which they can be attached must be near to flatness. To eliminate the Poisson effect and thus ensure that strain along the gauge axis only is recorded, the gauges are cemented at their ends and not along their whole length. Two gauges can be used to determine principal strains in a biaxial stress field. Low tensile strain transmitted to the gauge displaces both fringes to the right and compressive strain causes displacement to the left. By noting the displacement of a particular fringe on the scale the strain magnitude can be determined. High strains cause fringes to pass off the scale but each time a fringe disappears it is replaced

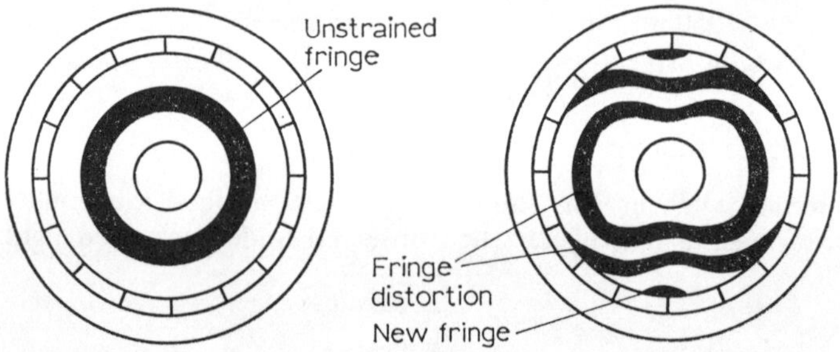

Fig. 5.43 CIRCULAR PHOTOELASTIC STRAIN GAUGE

by a new fringe which appears at the opposite end of the gauge, thus careful note must be made of the fringe order. Bending or twisting of the surface causing the fringes to rotate and become skewed invalidates the strain measurement.

Gauges for measuring pure shear have fringes which are rotated to an angle of 45° to the gauge axis and these are used in the same manner as normal gauges except that they are cemented along their whole length.

Circular gauges used for principal strain measurement have circular fringes which become distorted under strain; they also move across the gauge face, disappearing and being replaced with increase in strain magnitude, although the visible fringe pattern at any stage may appear to be extremely complex. It is essential to note the fringe order as in the case of rectangular gauges and this is done each time a new fringe begins to appear at the outside edge of the gauge (Fig. 5.43). This procedure is followed on both strain axes and, with the aid of a chart supplied by the manufacturer, the principal strains and thence the principal stresses may be determined.

Photoelastic strain gauges are direct reading devices and require no instrumentation. They are large, the minimum gauge length obtainable being about 20 mm, and the scale requirements limit their narrowness. They are inflexible and are thus suited only to flat or nearly flat surfaces. Their sensitivity is of the order of 40 to 75 μm/m and they have a maximum operating temperature of 50°C. It would appear that their main applications would lie in the field of sensitive transducer elements.

Photoelastic stress gauges are also manufactured.

6 Measurement of Time, Speed, Acceleration, and Frequency

6.1 Mechanical and Electrical Timers

Usually a mechanical or electrical hand-started timer is used in engineering work. The most common mechanical timers are the clock and watch based on the oscillations of a flywheel making one oscillation every fifth of a second. Stop watches and clocks do not measure absolute time but measure the time for various events to occur. Since mechanical timers are normally started by hand the accuracy of the timing operation involves not only the accuracy of the timer itself but also the speed of reaction of the operator. If, however, the same operator starts and stops the watch and his time lag is approximately constant the timing error will be reduced. The timing error due to the human operator may well be insignificant when times greater than one minute are measured. If, however, the time measured is of the order of a few seconds the error will be a factor of major importance.

If time intervals of less than twenty seconds are to be measured accurately it may be necessary to provide some means of starting and stopping the timer which does not depend on human muscular response. Mechanical trips can be used to start and stop the watch and highly sensitive timers are available in which the timing mechanism is running steadily, with a quick-acting clutch mechanism to start and stop the hand. If errors less than 0·05 s are required then electrical timers are usually employed.

Electrical timers use as their standard the frequency of oscillation of an alternating current which is fed to them. The frequency may be fixed, as is the case with electric clocks which base their time indication on the 50 Hz mains frequency, or varied as the operator requires to suit the accuracy of the work. The higher the frequency fed to the timer and the more regular the oscillation the more accurate is the timing. Starting and stopping these devices is normally achieved by electric pulses. The number of cycles occurring between starting and stopping is recorded and displayed usually on a digital readout.

6.2 The Measurement of Speed

Speed measurement may be divided into angular velocity measurement and linear velocity measurement.

The angular velocity of a rotating machine is usually expressed in radians per second (rad/s) or revolutions per minute (r/min).

Linear velocity is the distance moved per unit of time and is expressed as metres per second (m/s).

6.3 Tachometers

A tachometer is an instrument which measures directly the angular velocity of a rotating shaft. Tachometers may be broadly divided into mechanical and electrical instruments and are considered below.

6.4 Mechanical Tachometers

Revolution Counter and Timer

The simplest method available in this category is to use a revolution counter and a stop watch. This method is quite satisfactory for measuring low speeds and some instruments are commercially available which consist of a revolution counter with a built-in timing device (Fig. 6.1).

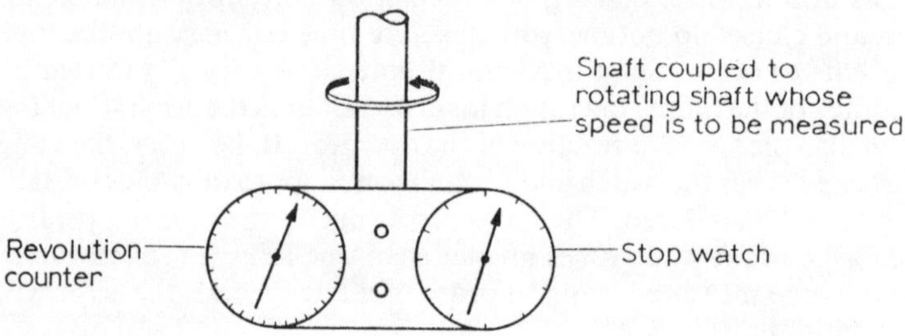

Fig. 6.1 REVOLUTION COUNTER WITH BUILT-IN TIMING DEVICE

The Slipping Clutch Type

This tachometer shown in Fig. 6.2 consists of an indicator shaft driven by an input shaft through a slipping clutch. A pointer is attached to the indicator shaft which is rotated against the torque of a spiral spring. As the speed of the input shaft is increased, the greater the torque applied to the indicator shaft which moves the pointer over a calibrated scale.

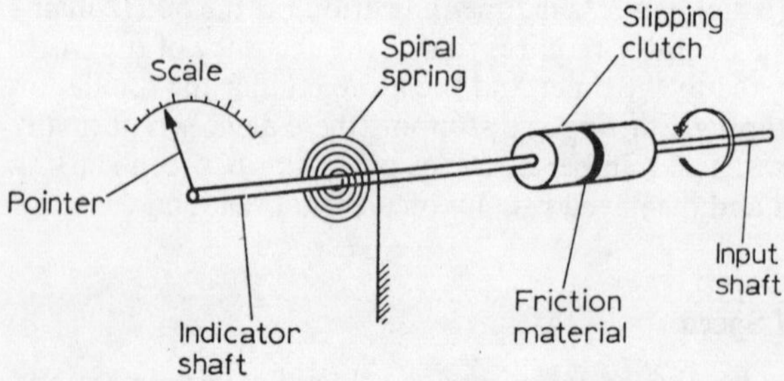

Fig. 6.2 SLIPPING CLUTCH TYPE TACHOMETER

The Centrifugal Force Type

The centrifugal force type (Fig. 6.3) operates on the Watt governor principle. Two masses are mounted on leaf springs which are attached at one end to a driven shaft. The other ends of the springs are attached to a grooved collar which can slide on the shaft and move a pointer.

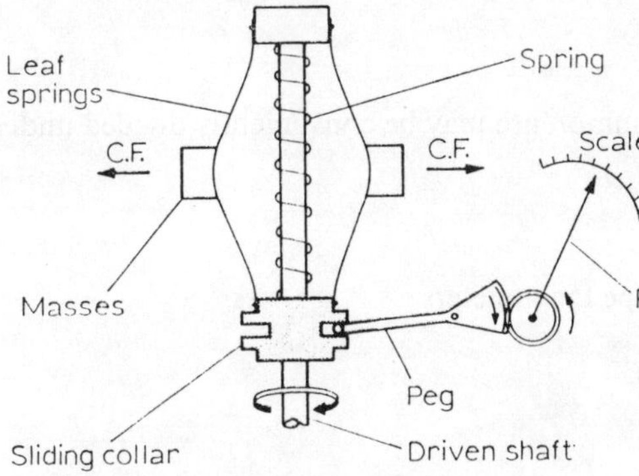

Fig. 6.3 CENTRIFUGAL FORCE TACHOMETER

As the speed increases the centrifugal force on the masses increases (i.e. centrifugal force $= mr\omega^2$ where m = mass, r = radius from centre of motion, ω = angular velocity) thus causing the masses to move outwards and the sliding collar to move up the shaft. The movement of the collar is transmitted through a quadrant and pinion to a pointer. This instrument may be used to measure rotational speeds up to 40 000 r/min with a maximum accuracy of ±1%.

The Resonance or Vibrating-reed Tachometer

The vibrating-reed tachometer consists of a set of thin cantilever members each member having a different natural frequency of vibration (Fig. 6.4*a*). The reeds are lined up in order of their natural frequency of vibration and when the frame of the tachometer is placed in mechanical contact with the frame

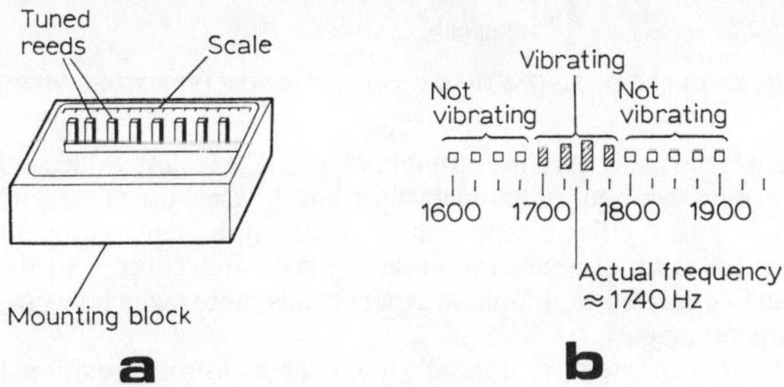

Fig. 6.4 VIBRATING-REED TACHOMETER

of a rotating machine the reed with natural frequency closest to the frequency of the machine vibration will respond the most readily. The reed vibration is indicated against a calibrated scale (Fig. 6.4*b*).

An advantage of this instrument is that the meter need only be in contact with a non-moving part of the machine. They may be used over the range 600–10 000 r/min and the accuracy may be better than $\pm\frac{1}{2}\%$.

6.5 Electrical Tachometers

The electrical tachometers in common use may be conveniently divided under the following headings:

(1) Inductive tachometers
(2) Tachogenerators
(3) Commutated capacitor-type tachometers
(4) Digital counters
(5) Stroboscopic tachometers.

(1) *Inductive Tachometers*

THE EDDY CURRENT OR MAGNETIC DRAG TYPE This instrument, shown in Fig. 6.5, has a rotating permanent magnet which is driven by the machine shaft. Eddy currents are induced in the aluminium drag cup and produce a torque which rotates the cup against the torque produced by a spiral spring. The pointer attached to the cup indicates rotational speed on a calibrated scale.

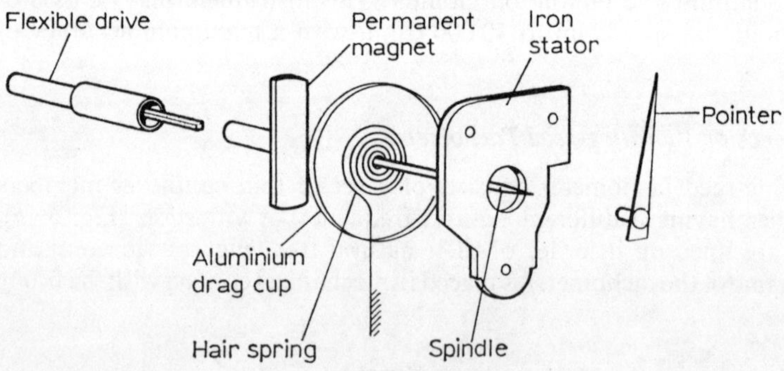

Fig. 6.5 THE EDDY CURRENT OR MAGNETIC DRAG TACHOMETER (EXPLODED VIEW)

This instrument is normally used in automobiles when the scale is calibrated in linear velocity (miles per hour or kilometres per hour). When used for speed measurement in aircraft the flexible mechanical drive can be replaced by an electrical drive in the form of a small 3-phase a.c. generator coupled to the machine shaft and connected to a 3-phase synchronous motor which in turn drives the tachometer magnet.

The eddy current tachometer is normally used for measuring rotational speeds up to 12 000 r/min at an accuracy rather less than $\pm 3\%$.

THE A.C. DRAG-CUP GENERATOR TYPE This instrument has an aluminium cup which is rotated in a laminated iron magnet system. The stator core carries two windings which have their axes at right angles as shown in Fig. 6.6.

When an alternating current is applied to one of the windings, eddy currents are induced in the aluminium cup. When the cup is rotated a cross-magnetizing flux is set up which induces an e.m.f. proportional to the speed of rotation at the output coils. This output voltage may be measured using a voltmeter calibrated in r/min. This type of tachometer is widely used in control systems where a

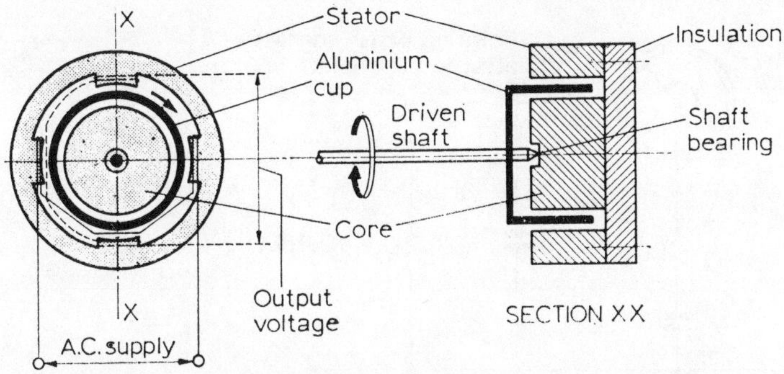

Fig. 6.6 THE A.C. DRAG-CUP GENERATOR TYPE TACHOMETER

linear output voltage at a constant frequency is necessary. The instrument may be used for measuring rotational speeds up to 5000 r/min at an accuracy within $\pm\frac{1}{2}\%$.

(2) *Tachogenerators*

Tachogenerators are also used extensively in control systems. Many control systems require a voltage signal proportional to the rotational speed of a shaft to be compared with the system input voltage. This can easily be achieved by means of a tachogenerator fixed to the output shaft of the system.

D.C. TACHOGENERATORS These instruments are small d.c. generators with a permanent field. The instrument may be fixed to the rotating shaft, and the output voltage from the tachogenerator is proportional to the rotational speed. The output voltage may be measured using a moving coil instrument calibrated in r/min.

The d.c. tachogenerator may be used for measurements up to 5000 r/min at an accuracy within $\pm 2\%$ or in the case of the more expensive models $\pm 0{\cdot}1\%$.

A.C. TACHOGENERATORS A.C. tachogenerators are small brushless alternators having a rotating multi-polar permanent magnet. The alternator is driven by the rotating shaft and the output voltage is measured by a voltmeter calibrated in r/min.

As in the d.c. tachometer the output voltage is proportional to speed, but the frequency also varies with speed and affects the accuracy of the instrument. The instrument may be used up to 5 000 r/min at an accuracy within $\pm 2\%$.

(3) *Commutated Capacitor-type Tachometers*

The tachometer consists of a rotating switch which alternately charges and discharges a capacitor (Fig. 6.7). The capacitor C_1 is charged from a constant voltage d.c. supply and discharged through the resistor R_2 by rotation of the

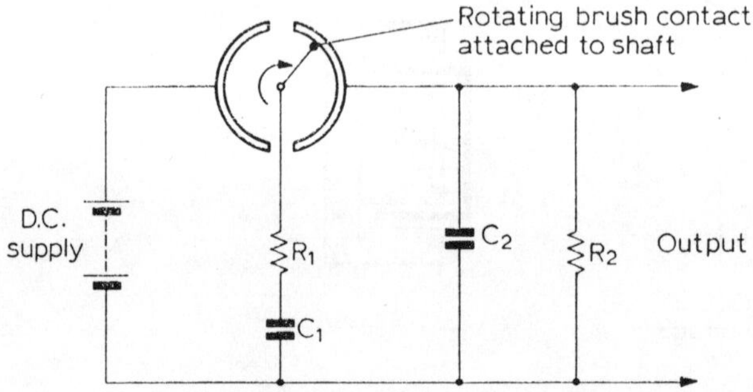

Fig. 6.7 COMMUTATED CAPACITOR-TYPE TACHOMETER

switch. The mean discharge current and hence the potential across R_2 is proportional to the speed of rotation of the shaft. The value of $C_2 = 100C_1$ and the resistance R_1 must be small but large enough to limit the charging current and hence reduce burn at the switch contacts. This instrument is used within the range 200–10 000 r/min.

(4) *Digital Signal Pick-up Tachometers*

The electrical tachometers discussed so far give an analogue indication of speed whereas these instruments give a pulsed or digital output signal. The digital output can be used for direct indication of speed, to transmit measurements for computation, or for applications in a control system. As there is no physical connection between the transducer and the shaft little or no load is

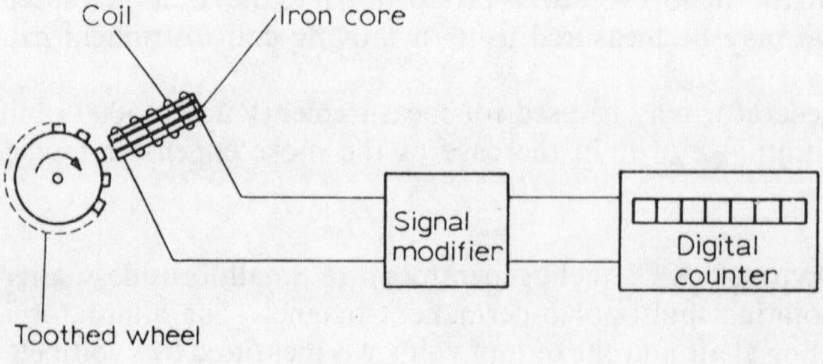

Fig. 6.8 INDUCTIVE PICK-UP TACHOMETER

applied to the machine shaft which is a distinct advantage of this type of instrument. Various forms of digital pick-up may be used and three of the most popular are discussed below.

INDUCTIVE PICK-UPS A typical system using an inductive pick-up is shown in Fig. 6.8. Each tooth on the wheel changes the reluctance of the magnetic circuit as it passes thus producing an e.m.f. change in the form of a pulse. The pulses are fed into a digital counter which counts the number of pulses for some pre-set time interval. The function of the signal modifier is to shape the digital signal to give a distinct series of pulses of usable amplitude.

CAPACITIVE-TYPE PICK-UPS Capacitive pick-ups have a vane attached to the machine shaft. As the vane rotates between the fixed capacitor plates a change in the capacitance to earth ratio is produced. The output is fed into a pulse shaper and a digital counter as shown in Fig. 6.9.

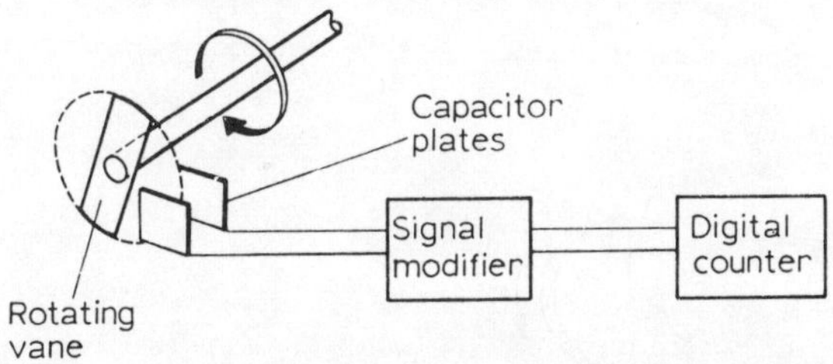

Fig. 6.9 CAPACITIVE PICK-UP TACHOMETER

PHOTOELECTRIC PICK-UPS This instrument consists of a photoelectric cell which is activated by light pulses deflected from a rotating machine shaft. The shaft has an intermittent reflecting surface and when a beam of light is focused on to the shaft a number of pulses of light are obtained. The frequency of the light pulses is proportional to the speed of the shaft and hence the frequency of the electrical output pulses from the photoelectric cell will also be proportional to the shaft speed. Again the pulses are shaped to provide a suitable input for a digital counter (Fig. 6.10)

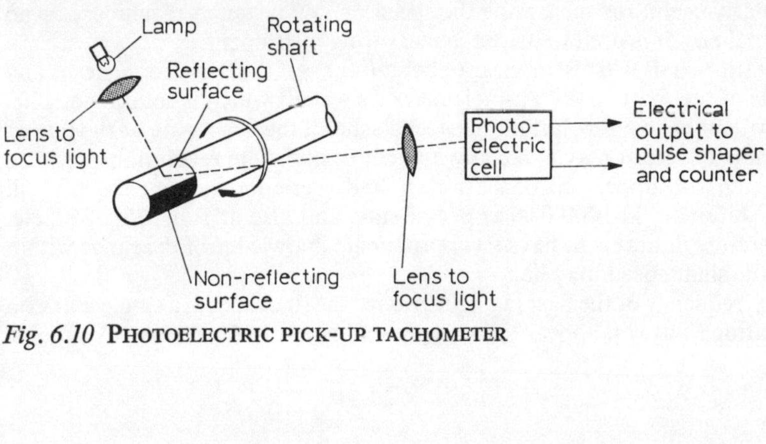

Fig. 6.10 PHOTOELECTRIC PICK-UP TACHOMETER

Digital measuring systems may be used for speeds up to 3 000 000 r/min and accuracy is only limited by the accuracy of the counting device.

(5) *Stroboscopic Tachometers*

The instrument operates on the principle that if a repeating event is only viewed when at one particular point in its cycle it appears to be stationary.

If a mark is made on a shaft and the shaft is slowly rotated under a steady light, the angular velocity may be found by watching the mark and counting the number of revolutions during a definite time interval. As the rotational speed increases, however, it becomes difficult to see the revolutions of the mark.

If a flashing light is substituted for the steady light and the frequency and duration of flashing is adjusted so that there is one very short flash per revolution then the observer will see the mark on the shaft apparently standing still. This is shown in Fig. 6.11, where the rotating disc is illuminated at one point during each revolution.

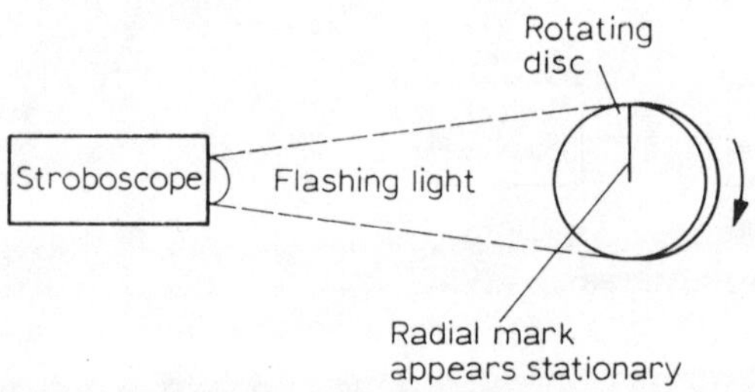

Fig. 6.11 STROBOSCOPIC TACHOMETER

Commercially available stroboscopes usually incorporate a neon-tube which flashes at a rate set on a calibrated control switch, to read angular velocities between 600 and 20 000 r/min. It is possible to obtain the speed of a rotating shaft without making contact with it and the stroboscope is therefore particularly useful for measuring the speed of low-powered machinery as no additional power is used to drive the measuring instrument.

A rotating shaft will also appear to be stationary if the flash frequency is any multiple of the shaft speed. A single mark on a shaft which is rotating at 1000 r/min will make two revolutions between flashes if the flashes are at the rate of 500 r/min. The shaft may make any number of complete revolutions between flashes and still appear to be stationary and therefore the above shaft will appear stationary at 1000 flashes per minute, and also at 500, $333\frac{1}{3}$, 250, etc. It is therefore desirable to have an approximate knowledge of the range within which the shaft speed may lie.

If the frequency of the flashing light is twice the shaft speed, a single mark on the rotating shaft will appear to be two standing marks 180° apart. A suitable technique for determining the shaft speed when the general speed range is unknown is as follows. Gradually increase the frequency of flashing from a low value until the rotating member appears to be stationary and note the frequency. The flash frequency is then doubled and if there is still only one apparent stationary image the frequency is again doubled. This procedure is continued until two images appear 180° apart. When these two images first appear the flash frequency is twice the speed of rotation.

6.6 Measurement of Acceleration and Vibration Amplitude

Principles of the Seismic Instrument

The basic functional form of the seismic instrument is as shown in Fig. 6.12. The base is connected to the vibration source whose characteristics are to be

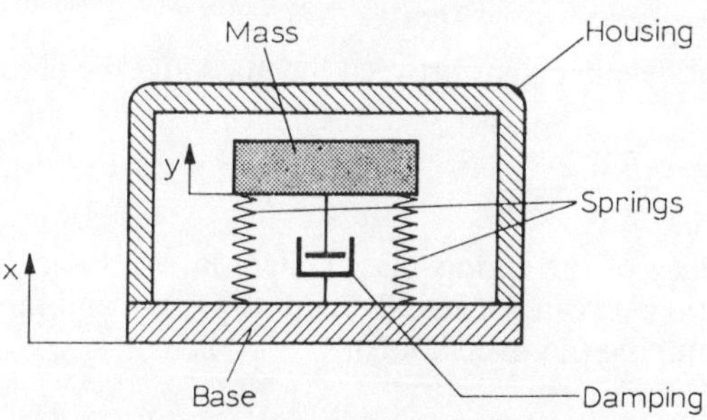

Fig. 6.12 BASIC SEISMIC MASS INSTRUMENT

Relative movement between the mass and the base $z = y - x$

measured. The mass tends to remain fixed in its equilibrium position so that the vibration motion is registered as a relative displacement between the mass and the base. This displacement is then sensed and measured by an appropriate transducer.

The seismic instrument may be used for either displacement or acceleration measurements by proper selection of mass, spring, and damper combinations. In general, a large mass and a soft spring are desirable for vibration displacement measurements, while a relatively small mass and a stiff spring are used for the measurement of acceleration.

If a harmonic input is applied to the base the movement is given by the expression

$$x = X \sin \omega t$$

where x is the displacement at time t from the centre of the movement
X is the maximum displacement or amplitude
ω is the frequency of the harmonic input.

A typical harmonic input is shown in Fig. 6.13.

This is simple harmonic motion of amplitude X and frequency $\omega/2\pi$ Hz. It should also be noted that the acceleration of the base is given by $\omega^2 x$ and thus the maximúm acceleration is $\omega^2 X$.

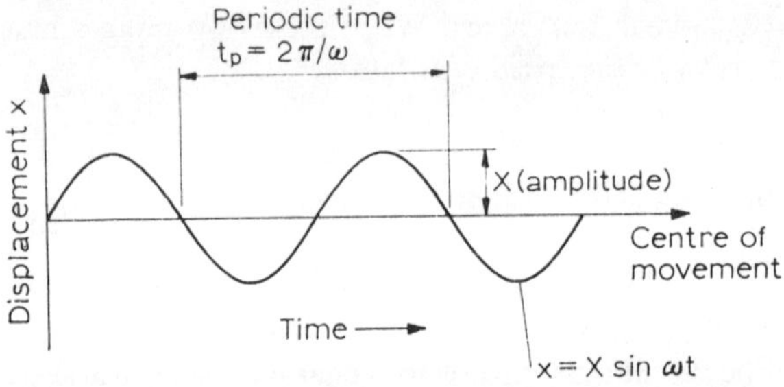

Fig. 6.13 HARMONIC INPUT APPLIED TO BASE OF SEISMIC INSTRUMENT

It can be shown that the relative movement between the mass and the base is given by

$$Z = \frac{\omega^2 X}{\omega_n{}^2\{[1 - (\omega/\omega_n)^2]^2 + [2\zeta\omega/\omega_n]^2\}^{1/2}} \qquad \text{(i)}$$

where ω_n is the natural frequency of the spring–mass system in rad/s and ζ is the damping ratio, i.e. the ratio between the actual damping present and the amount required for critical damping (no overshoots).

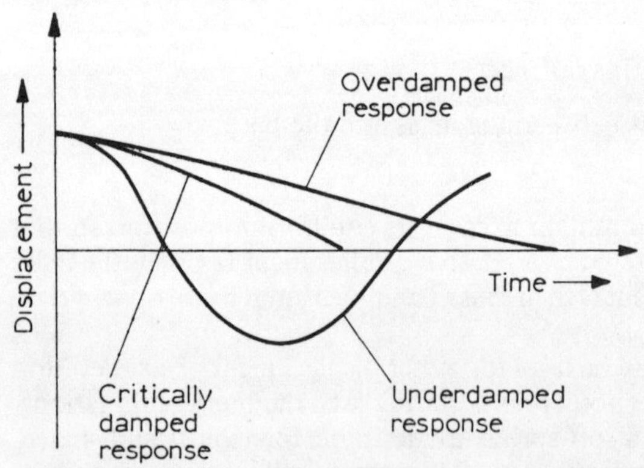

Fig. 6.14 TYPICAL DAMPED HARMONIC RESPONSE CURVES

The Measurement of Displacement

If the frequency applied to the base of the seismic instrument, ω, is much higher than ω_n then the term $[2\zeta\omega/\omega_n]^2$ in equation (i) may be ignored compared with $[(\omega/\omega_n)^2]^2$ and an approximate expression for Z becomes

$$Z = \frac{\omega^2 X}{\omega_n{}^2\{[(\omega/\omega_n)^2]^2\}^{1/2}} = \frac{\omega^2 X}{\omega_n{}^2(\omega/\omega_n)^2} = X$$

Thus the relative movement between the mass and the base is approximately equal to the movement of the base. At frequencies higher than $2\omega_n$ this system can be used to measure the amplitude of vibration by means of a small displacement transducer between the mass and the base. A typical instrument is shown in Fig. 6.15.

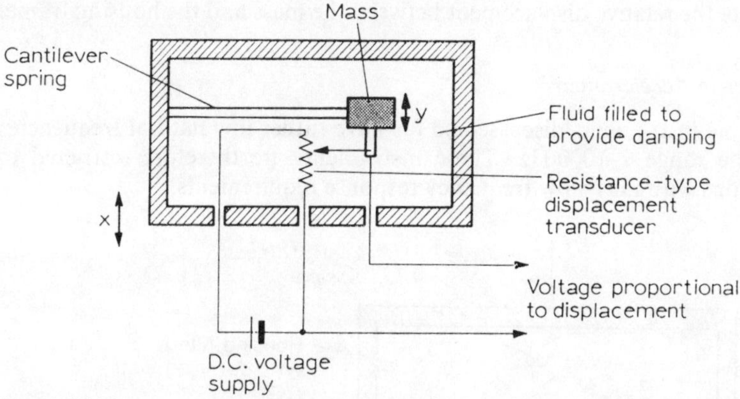

Fig. 6.15 TYPICAL SEISMIC MASS INSTRUMENT WITH CANTILEVER SPRING

The Measurement of Acceleration

If the input frequency ω is much smaller than the natural frequency ω_n then the expression for Z becomes

$$Z = \frac{\omega^2 X}{\omega_n^{\ 2}\sqrt{1}}$$

and this assumption remains valid for $\omega \leqslant 0{\cdot}4\omega_n$.

However $\omega^2 X$ is the maximum acceleration of the base and therefore

$$Z = \frac{1}{\omega_n^{\ 2}} \times \text{maximum acceleration}$$

Thus the instrument shown in Fig. 6.15 could be used to measure the maximum acceleration of a vibrating part for values of ω up to about $0{\cdot}4\omega_n$.

Strain Gauge Accelerometer

The electrical resistance strain gauge may also be used for a displacement-sensing device in a seismic instrument as shown in Fig. 6.16.

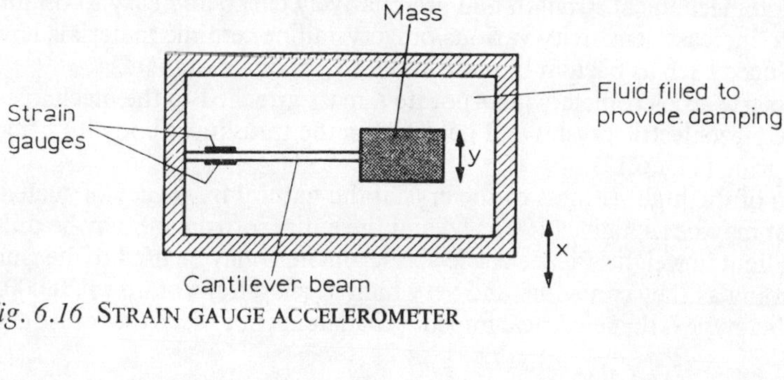

Fig. 6.16 STRAIN GAUGE ACCELEROMETER

The seismic mass is mounted on a cantilever beam and a resistance strain gauge is mounted on each side of the beam to sense the strain in the beam resulting from the vibrational displacement of the mass. Damping for the system is provided by the viscous liquid which fills the housing. The outputs from the strain gauges are connected to an appropriate bridge circuit which is used to indicate the relative displacement between the mass and the housing frame.

Piezoelectric Accelerometers

The seismic instruments discussed so far have rather low natural frequencies within the range 2–1000 Hz. These instruments are therefore restricted to applications with fairly low frequency response requirements.

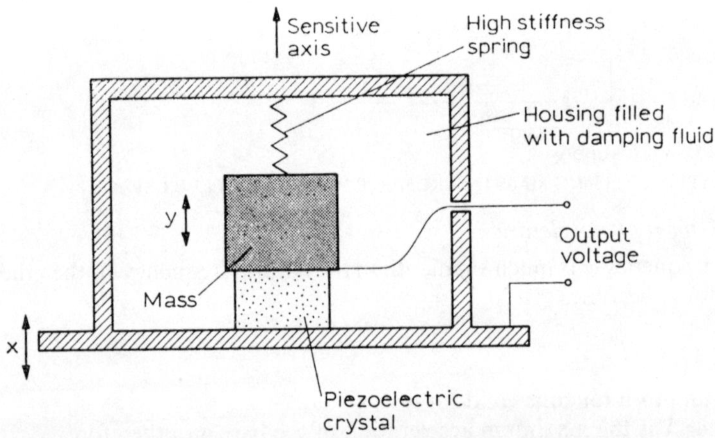

Fig. 6.17 PIEZOELECTRIC CRYSTAL ACCELEROMETER

The seismic instrument frequently employs a piezoelectric transducer for high frequency measurements. The basic element in a piezoelectric transducer is a block of crystalline material capable of generating an electrical potential when subjected to mechanical strain along a preferred axis. Examples of common materials possessing this effect are quartz and Rochelle salt (potassium sodium tartarate) and ammonium dihydrogen phosphate. These materials have a fairly high mechanical strength and are relatively cheap and easy to obtain. In order to increase sensitivity various polycrystalline ceramic materials have been produced such as barium titanate and lead zirconate titanate.

Piezoelectric accelerometers incorporate a mass attached to the mechanical axis of the piezoelectric crystal and held within the transducer body by a high stiffness spring (Fig. 6.17).

Because of the high stiffness of the crystal the natural frequency of such an instrument may be as high as 100 kHz, and the entire instrument may be quite small and light in weight. Piezoelectric accelerometers may be used to measure shock loading as they can withstand very high accelerations of up to 20 000*g*. One disadvantage is that electrical impedance matching between the transducer and readout circuitry is usually a critical matter requiring careful design considerations (see Chapter 1).

Inductive Accelerometers

Inductive accelerometers employ a pair of coils and a magnetically-coupled mass (Fig. 6.18). When an acceleration is applied the mass is displaced and alters the magnetic coupling path which changes the inductive ratio of the coils. Inductive accelerometers offer a high electrical output for essentially continuous resolution and are used to measure static or dynamic accelerations.

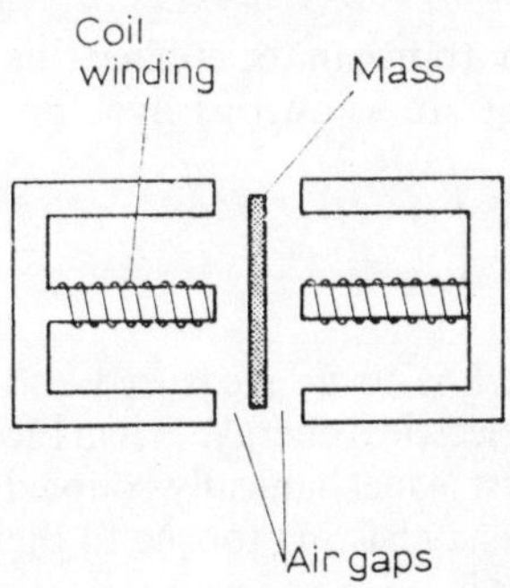

Fig. 6.18 INDUCTIVE ACCELEROMETER

These instruments must be excited with alternating current and be reactively and resistively balanced. Inductive accelerometers have a low frequency response and suffer from the disadvantage that they are susceptible to magnetic fields.

6.7 The Effect of Temperature on Seismic Instruments

Seismic vibration instruments may be influenced by temperature effects. Devices which employ a variable-resistance transducer to sense the relative displacement between the mass and the instrument housing will require a correction to account for resistance changes due to temperature.

Also most seismic instruments are filled with a viscous fluid, e.g. silicone oil, to provide damping. Obviously the viscosity of the damping fluid will vary with temperature and a fluid with a small change in viscosity with temperature would be desirable. However most fluids which fall into this latter category, e.g. gases, have low viscosities as well and are therefore unsuitable for use as damping fluids. One way of eliminating this problem is to introduce an electrical resistance heater into the fluid to maintain the temperature at a constant value regardless of the surrounding temperature.

7 Measurement of Force, Torque and Power

The accurate measurement of power is of major importance in the engineering industry. Instruments used for power measurement are known as dynamometers and may be divided into two main groups:

(*a*) Absorption dynamometers
(*b*) Transmission dynamometers.

Absorption dynamometers absorb the power which is to be measured and dissipate the energy extracted in the form of heat or electrical energy. When the absorption dynamometer is used the engine under test is mechanically coupled directly to the dynamometer or brake, which applies a resisting torque to the common shaft. The magnitude of the resisting torque is measured using instruments connected to the dynamometer and hence the power output of the engine may be calculated.

A *transmission dynamometer* enables the torque on the engine shaft to be determined when it is driving a normal load, no power being absorbed by the dynamometer. Transmission dynamometers work on the principle that when a torque is applied to a shaft, and is resisted by an equal and opposite torque, the shaft twists through a small angle which is proportional to the applied torque (provided that the elastic limit of the shaft material is not exceeded). The applied torque may be calculated from a knowledge of the angle of twist for a given length of shaft and if the rotational speed of the shaft is known the power transmitted may be calculated.

Force measurement is closely associated with torque and power measurement since power is defined as the rate of doing work and work is defined as the product of force and distance. The accurate measurement of power therefore requires that force, distance and time may be similarly measured.

7.1 Force Measurement

Force is defined as the product of mass and acceleration. The unit of force is the newton where 1 newton is the force required to give a mass of 1 kilogramme an acceleration of 1 metre per second per second. That is

$$1\ \text{N} = 1\ \text{kg} \times 1\ \text{m/s}^2$$

Instruments for force measurement are available over a very wide range and fall into three main groups:

(*a*) Mechanical methods.
(*b*) Hydraulic and Pneumatic methods.
(*c*) Electrical methods.

In general, electrical methods are most useful when remote measurement of force is required. Electrical methods are highly sensitive and can detect very small forces but may be unreliable as they are influenced by stray magnetic fields and temperature changes. Hydraulic methods are used mainly to generate and measure large forces and apply the hydraulic ram principle. Referring to

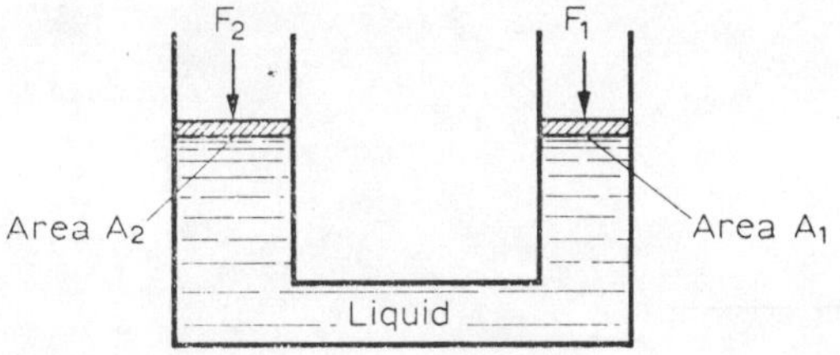

Fig. 7.1 PRINCIPLE OF THE HYDRAULIC RAM

$$F_1 = F_2A_1/A_2$$

Fig. 7.1, a large force acting on the piston of area A_2 may be balanced by a much smaller force acting on the piston of area A_1, e.g.

$$F_1/A_1 = F_2/A_2 \qquad \text{therefore } F_1 = F_2\frac{A_1}{A_2}$$

where $A_2 \gg A_1$.

7.2 Mechanical Methods of Force Measurement

The Simple Lever

The simple lever provides a force-measuring device by equating moments. An unknown force at one end of a pivoted lever may be balanced by applying a known force at the other end to provide equilibrium. When the lever is in equilibrium the moments of the two forces about the pivot point will be equal; thus the unknown force may easily be calculated from the lever ratio and the balancing force.

An example of a simple lever system where the lever arms are equal is the chemical balance. In order to reduce the number of masses required for balance

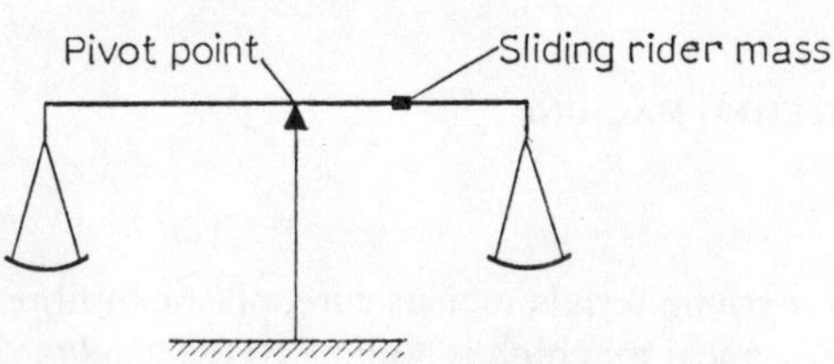

Fig. 7.2 CHEMICAL BALANCE

an adjustable lever arm may be used with a rider mass which can be moved along one lever arm until balance is obtained (Fig. 7.2).

The simple lever principle may be applied to a tensile/compressive testing machine as shown in Fig. 7.3. The lower end of the vertically mounted test

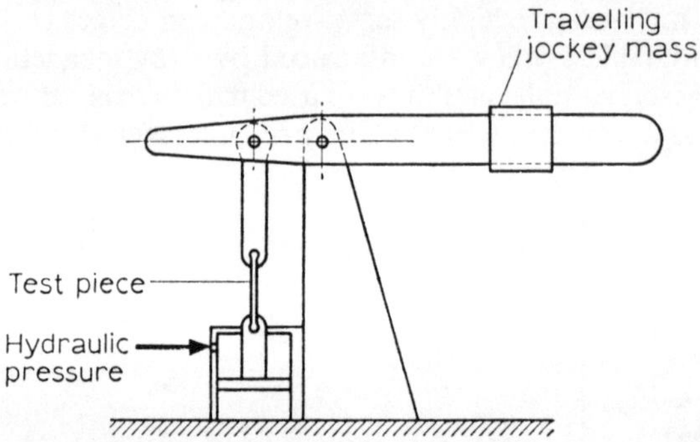

Fig. 7.3 LEVER-TYPE TENSILE TESTING MACHINE

specimen is attached to some form of loading, either a screw jack or a hydraulic ram, while its upper end is gripped by a swivel and attached to a steelyard to measure load.

This type of machine would normally be used for testing materials such as low strength wire, plastic, fibre, etc.

The Compound Lever

The simple lever may be compounded by introducing a series of pivot points and additional levers. A more compact machine is therefore capable of exerting greater loads and this principle is applied to a number of testing machines (Fig. 7.4).

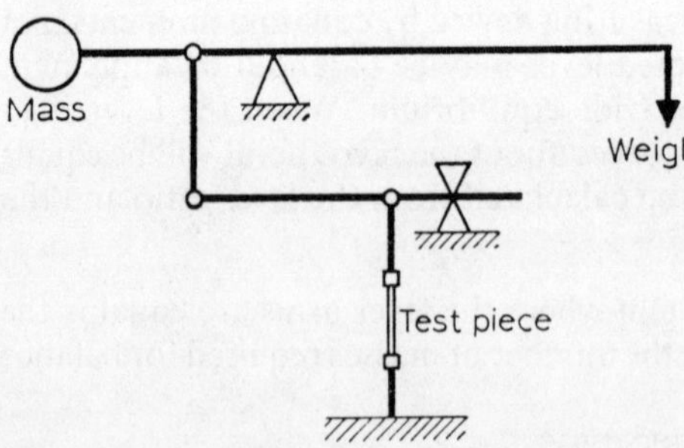

Fig. 7.4 COMPOUND LEVER-TYPE TESTING MACHINE

The Pendulum

Many machines for testing low-strength materials such as wire, plastic or fibre, incorporate a simple pendulum. A typical machine is similar to the steelyard testing machine discussed previously but with the steelyard replaced by a pendulum. The load may be applied with a screw or motor thus causing the pendulum to rise along a quadrant scale. Fig. 7.5 shows a direct-reading balance using the simple pendulum principle.

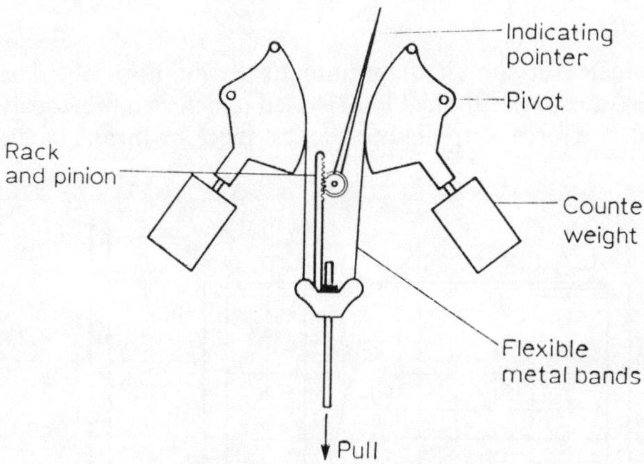

Fig. 7.5 DIRECT-READING BALANCE USING THE SIMPLE PENDULUM PRINCIPLE

The Spring and Proving Ring

If a force is applied to an elastic member such as a spring or a steel bar, then the member will deform. Provided that the force is not excessive then the elastic deformation of the bar or spring may be measured, and this deformation is proportional to the applied force. For example, for a simple spring

$$F = kx$$

where F = the applied force, k = a constant, x = the spring displacement.

The proving ring (Fig. 7.6) is a ring of known physical dimensions and

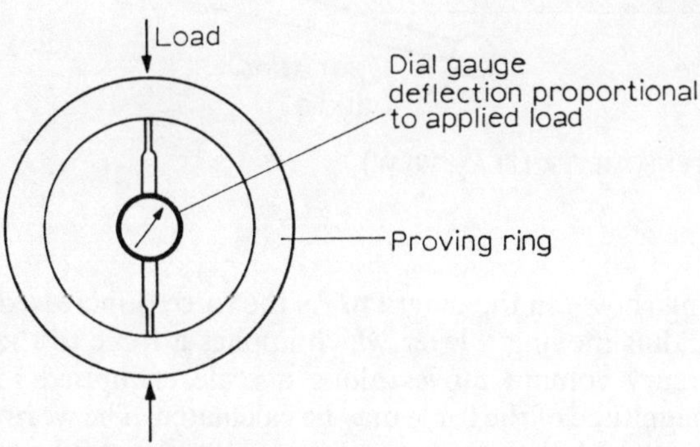

Fig. 7.6 THE PROVING RING

mechanical properties. An external tensile or compressive force applied across the ring diameter causes distortion which is proportional to that force. The amount of distortion can be measured using a dial gauge, micrometer, or strain gauges.

The Hounsfield Tensometer

The Hounsfield Tensometer is shown diagrammatically in Fig. 7.7. The extension of the test piece due to the applied load is used to deflect a previously calibrated spring beam. The force is applied to the test piece by means of the

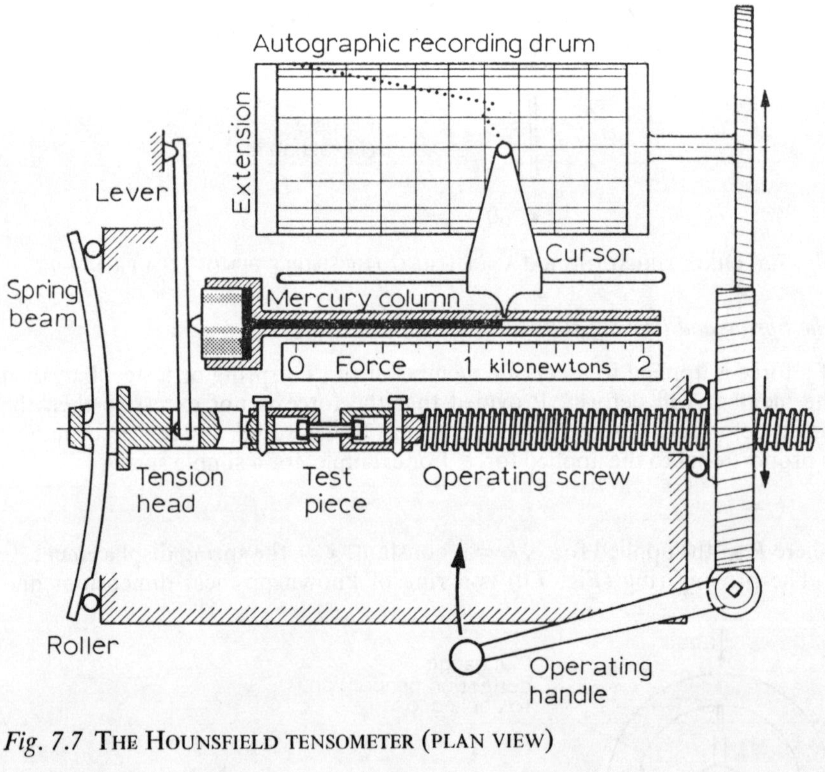

Fig. 7.7 THE HOUNSFIELD TENSOMETER (PLAN VIEW)

operating handle and gearing shown in the diagram. As the force is increased the spring beam will bend, thus moving a lever which applies a force to the mercury column. The mercury column moves along a scale calibrated in kilonewtons and thus the magnitude of the force may be calculated. The worm gear rotates the recording drum such that the distance through which the drum rotates is proportional to the extension of the test piece. The cursor is used to follow the mercury column as the force on the test piece is increased and in this way a record of force against extension is obtained. A set of spring beams provide a range up to 20 kN which is the maximum load of the apparatus.

7.3 Hydraulic and Pneumatic Methods of Force Measurement

Hydraulic pressure may be used to measure force as shown in Fig. 7.8. A chamber or capsule with a diaphragm and containing oil is connected to a Bourdon tube or some other suitable pressure-measuring device. When a force is applied to the diaphragm a pressure is developed in the chamber equal to the

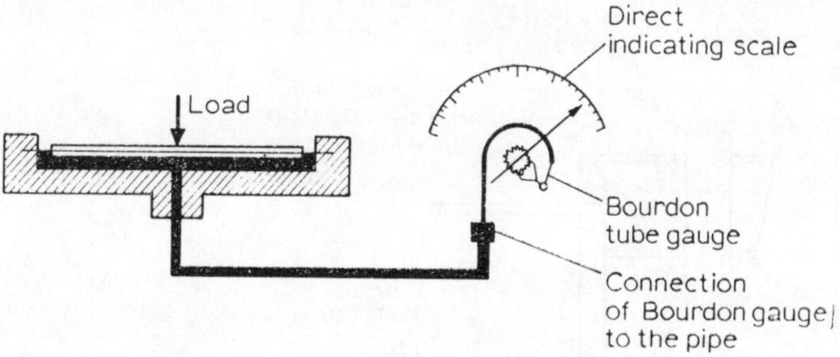

Fig. 7.8 HYDRAULIC LOAD CELL

applied force divided by the effective diaphragm area. This pressure is indicated by the Bourdon gauge which may be calibrated directly in force units. The apparatus is similar to the dead-weight pressure tester discussed in Chapter 8.

A pneumatic load cell in which air pressure is used to measure force is shown in Fig. 7.9. This device is similar to the conventional pneumatic relay

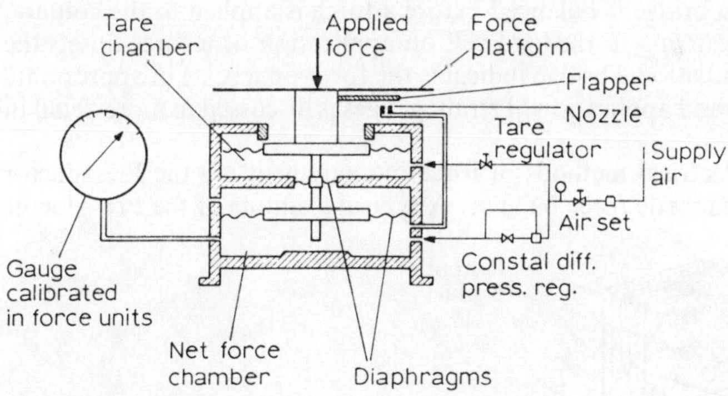

Fig. 7.9 PNEUMATIC LOAD CELL

operating on the force-balance principle and employing a nozzle and flapper unit. The weight of the force platform and fittings when no external force is applied is balanced by adjusting the pressure in the tare chamber. When an external force is applied the flapper will move nearer to the nozzle thus restricting the passage of air from the nozzle to the atmosphere. The pressure in the net force chamber will thus increase as the external force is increased and this pressure is directly proportional to the applied force. The force may conveniently be measured using a Bourdon tube gauge calibrated in force units as shown in the diagram.

7.4 Electrical Methods of Force Measurement

A typical strain gauge force transducer is shown in Fig. 7.10. In this system a column of steel suitably proportioned to provide the desired strain from a known load is either stretched or compressed by the load. Strain gauges are attached to the column, their electrical resistance being altered according to

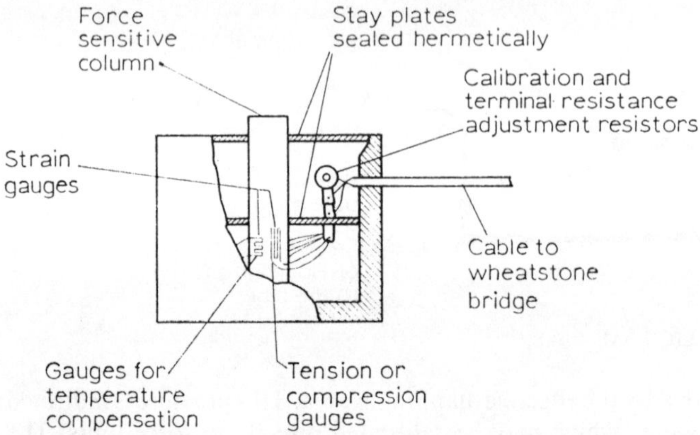

Fig. 7.10 STRAIN GAUGE FORCE TRANSDUCER

the load. The strain gauges are connected to the appropriate arms of a Wheatstone Bridge. The bridge is balanced before a force is applied to the column, and the stress occurring in the load cell on application of a force upsets the balance of the bridge. The bridge indicates the force on a scale in appropriate units. The theory and application of strain gauges is discussed in more detail in Chapter 5.

Two further electrical methods of force measurement are the Pressductor and the Electromagnetic force balance. A typical example of the Pressductor

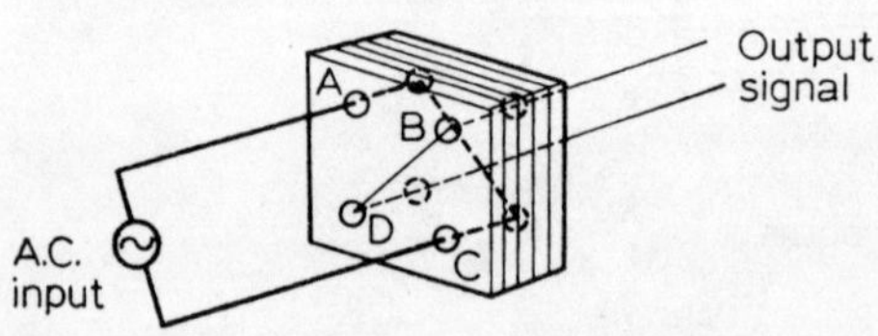

Fig. 7.11 PRESSDUCTOR
Input is fed through A and back through C.
Output lead is fed through B and D.

is shown in Fig. 7.11. The Pressductor uses the principle that when a force is applied to a material, e.g. a steel supporting member, an internal stress pattern is set up within the member. This stress will disturb the magnetic flux distribution within the member if a magnetic field is established. The Pressductor consists of a block of bonded laminations with four holes drilled through its thickness. A wire is fed through one hole diagonally across the back of the laminations and returns through another hole. This wire is fed with alternating current. A second wire is then fed through the other diagonal holes and connected to a suitably instrumented circuit.

Provided that no stress is applied to the block, a symmetrical pattern of magnetic flux is established within the block and no voltage is induced into the output wire. On application of stress the flux pattern is disturbed and a voltage is induced in the output wire, this voltage being proportional to the applied

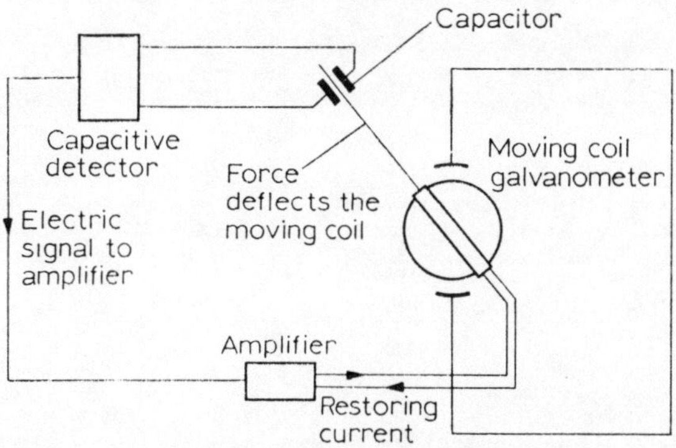

Fig. 7.12 PRINCIPLE OF ELECTROMAGNETIC FORCE BALANCE

force on the block. This type of instrument may normally be used for measuring forces up to 220 N and above.

When forces less than 10 N are to be measured the electromagnetic force balance principle can be used (Fig. 7.12). If a force is applied which deflects the moving coil from its zero position a signal is transmitted from the capacitive detector to the amplifier. The amplifier transmits a current proportional to this signal sufficient to produce a restoring force on the moving coil which is placed in a steady magnetic field. This restoring current is proportional to the deflecting force.

7.5 Power Measurement

As stated in the introduction to this chapter instruments used for power measurements are known as dynamometers and may be divided into absorption dynamometers and transmission dynamometers. Absorption dynamometers absorb the mechanical energy after measuring it whereas transmission dynamometers provide an indication of the force or torque which is passing through the dynamometer.

Absorption dynamometers may be divided into three major groups as follows:

(*a*) Mechanical methods
(*b*) Hydraulic methods
(*c*) Electrical methods.

7.6 Absorption Dynamometers: Mechanical Methods

One of the most simple forms of measuring power is the rope brake shown in Fig. 7.13. A loop of rope is wrapped round the flywheel or brake drum, one end of the rope being connected to a spring balance while the other end is connected to a mass. When the flywheel is rotated a frictional force is developed and the rope tightens, thus inducing a force on the spring balance.

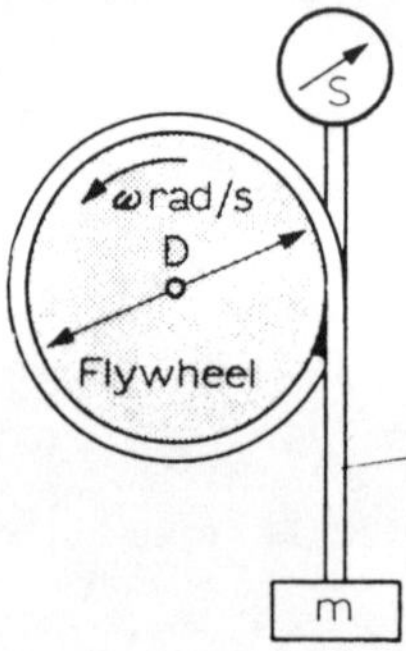

Fig. 7.13 ROPE BRAKE

The torque on the flywheel is equal to the resultant force acting on the flywheel multiplied by the moment arm:

$$T = \tfrac{1}{2}D(mg - S)$$

where T is the torque (N m)
D is the flywheel diameter (m)
mg is the applied weight (N)
S is the force on the spring balance (N).

The power output from the driving engine can now be determined by multiplying the flywheel torque by its angular velocity in rad/s, i.e.

$$\text{Brake power} = T\omega \text{ watts}$$

where ω is the angular velocity of the flywheel in rad/s.

The band brake shown in Fig. 7.14 operates on the same principle as the rope brake. A band of brake lining material (A) passes over the upper part of a flywheel F as shown. Attached to the ends of this band are two equaiiy loaded mass carriers. At B a metal strap connects the band to a piece of belt which passes over and is attached at C to a small wheel D. The pointer E is fixed to the wheel D, and passes over the scale S. The power can be calculated from the following relationship:

Brake power or the power output at engine shaft
$= \omega r \times$ (scale reading in newtons)

where ω = angular velocity in rad/s and r = flywheel radius.

Two further mechanical brakes commonly used are the Prony brake and the Appold brake. In the former the conversion of energy is by friction of a brake band or brake shoe on a brake drum. The braking action is due to friction between individual brake shoes or between a flexible brake band and the drum. Heat is dissipated by cooling the brake with water when in continuous use. The

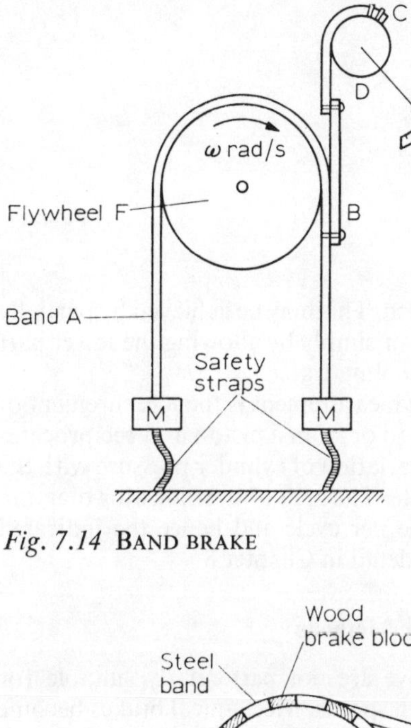

Fig. 7.14 BAND BRAKE

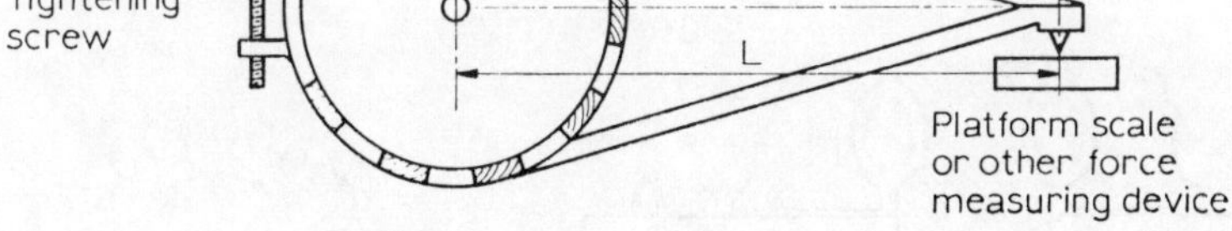

Fig. 7.15 PRONY BRAKE

power output from the driving engine can be measured by determining the speed of rotation and the torque required to hold the brake arm still (Fig. 7.15).

A limitation of the Prony brake is the variation of the coefficient of friction between the brake blocks and the wheel. The Appold brake compensates for this by incorporating a lever which applies a known tension to the band of brake blocks (Fig. 7.16).

Any increase in friction drag causes the lever to move down and hit the lower stop. The stop causes the lever to pivot about C thus reducing the applied tension in the band. A reduction in friction drag will cause the lever to hit the upper stop and hence the applied tension in the band is increased.

In all the dynamometers so far discussed the work done against friction is dissipated as heat and the brake drums or flywheels become very hot. Efficient

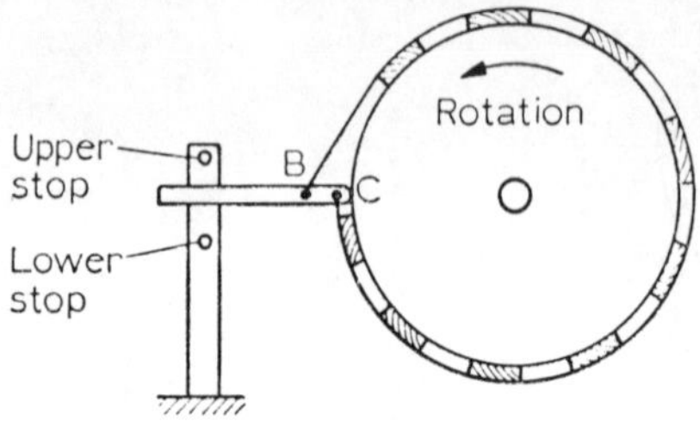

Fig. 7.16 APPOLD BRAKE

means of cooling must, therefore, be applied. This may be achieved by partially filling the brake drum with cooling water or simply by allowing the lower part of the flywheel or brake drum to run in cooling water.

A further mechanical method of power measurement is the measurement of indicated power or the power transferred to or from a piston as it reciprocates within a cylinder. A trace is taken of the variation of cylinder pressure with the cylinder volume during the operating cycle. The area of the indicator diagram produced is a measure of the work done per cycle and hence the indicated power. Engine indication is discussed in detail in Chapter 8.

7.7 Absorption Dynamometers: Hydraulic Methods

The mechanical methods described above are not particularly suitable for testing high-power engines running at high speeds. Mechanical brakes become cumbersome in larger sizes and flexible control is not obtained.

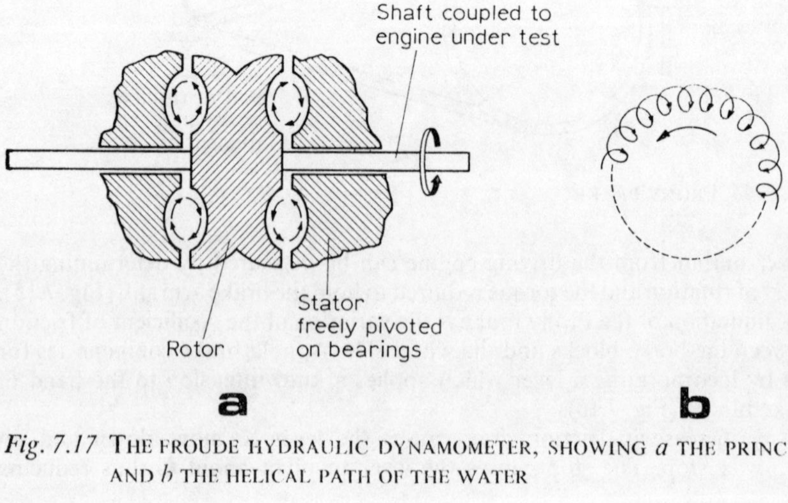

Fig. 7.17 THE FROUDE HYDRAULIC DYNAMOMETER, SHOWING *a* THE PRINCIPLE AND *b* THE HELICAL PATH OF THE WATER

The hydraulic dynamometer consists essentially of a rotor mounted on the dynamometer shaft which is coupled to the shaft of the engine under test. The rotor is enclosed by a watertight stator casing mounted on trunnions and thus freely pivoted (Fig. 7.17*a*). The stator casing is connected to the dynamometer arm and balance system and is in two halves, one of which is situated on either side of the rotor. The rotor and stator are fitted with a system of cup-shaped pockets, those on each half of the stator being of opposite pitch to those which they face on the rotor and thus forming spherical chambers when they match.

The stator and rotor pockets are so shaped that when the dynamometer is in action the path of the water is a helix as shown in Fig. 7.16*b*. The vortex produced gives rise to a braking effect between rotor and stator, the stator tending to turn in the direction of rotation of the rotor.

The tendency of the stator to rotate is resisted by the dynamometer balancing system which is connected to the stator.

Control of Load

The load applied to the engine is controlled by interposing sluice gates in the clearance space between the stator and rotor pockets. The gates may be moved using threaded rods which are operated by a hand wheel. The effect of blanking off part of the pocket area is to reduce the braking effect between rotor and stator and hence to reduce the load applied to the engine shaft. Smooth control of load over a wide range can be obtained using this method. Typical size range from a few kW up to 35 MW.

Advantages of the Hydraulic Dynamometer over Mechanical Brakes

(1) The hydraulic dynamometer has a constant supply of water running through it so that the braking medium is also the coolant.
(2) The brake power of much larger, higher speed engines can be measured using a hydraulic dynamometer.
(3) Small braking loads such as friction forces acting at the shaft glands will cause an extra small torque to be applied to the stator. This torque, however, will be included in that measured on the balancing system.
(4) The hydraulic dynamometer may be protected from hunting effects by means of a dashpot damper.

Two further hydraulic dynamometers are the fan brake and the hydrostatic dynamometer. The former consists of a shaft having a series of radial arms fixed to it. The arms carry plates which are fixed normal to their direction of motion. On rotation of the brake a resisting torque is produced which can be shown to be proportional to the area of the blades, the speed squared, the density of the ambient air and the cube of the blade radius arm (Fig. 7.18).

In the hydrostatic dynamometer (Fig. 7.19) the engine input is connected to a hydraulic pump unit which is mounted on trunnions. The reaction to the applied torque is sensed by means of a transducer fitted on the pump unit base plate. The pump unit is mounted on an oil reservoir tank and a gear-type make-up pump is also fitted in the circuit. This device has the advantage that it may be used as a motor for engine starting and running in. In this case the engine on test may be rotated by supplying power from the make-up pump but if the dynamometer is designed purely for absorption then the make-up pump

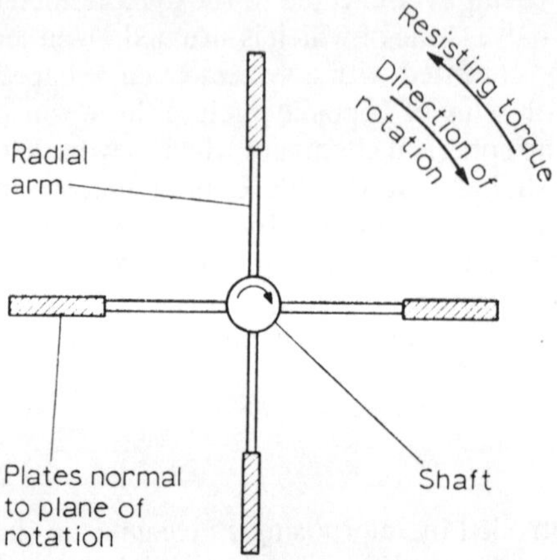

Fig. 7.18 FAN BRAKE

supplies only the system losses. Pressure and flow control valves may be incorporated in the hydraulic circuit so that the load characteristics of the dynamometer can be changed with suitable adjustment.

Apart from the start-up and motoring facility the hydrostatic dynamometer has the following additional desirable features:

(1) High torque at low engine speeds.
(2) Accurate torque readings over the range of the machine.
(3) Two directional operation.
(4) The ability to be programmed on a computer.

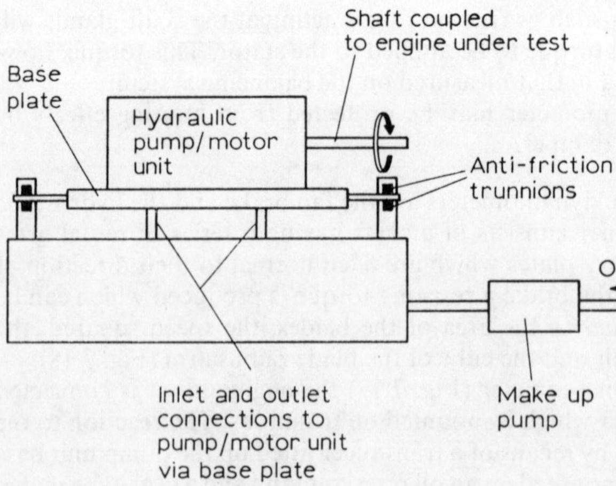

Fig. 7.19 THE HYDROSTATIC DYNAMOMETER

7.8 Absorption Dynamometers: Electrical Methods

If an engine is coupled to a dynamo and an electrical load is applied to the dynamo then the engine must produce power in order to drive the dynamo. The power output from the engine may be estimated if the load on the generator is known. Account must be taken of frictional effects, windage, and electrical losses in the generator, and measurement accuracy may not be very good.

Two electrical dynamometers which operate on a similar principle to the Froude hydraulic dynamometer are the eddy current dynamometer and the all-electric dynamometer. The *eddy current dynamometer* consists of a toothed rotor coupled to the shaft of the engine under test, and running inside the

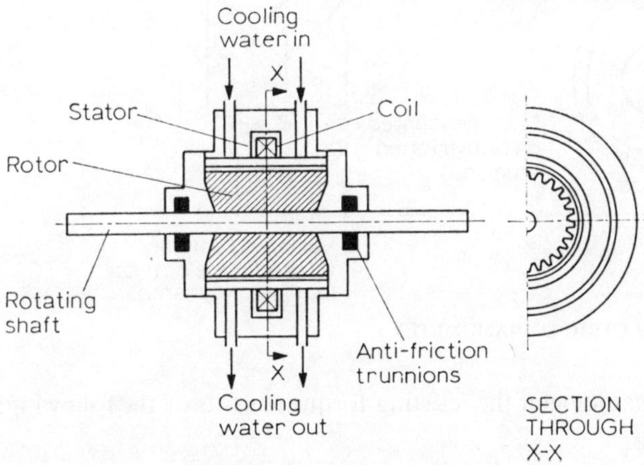

Fig. 7.20 THE EDDY CURRENT DYNAMOMETER

smooth bore of the stator, the clearance between stator bore and rotor teeth being very small. The stator carries circumferentially wound field windings and these, when excited by a small d.c. current, set up a magnetic flux which links the rotor and stator. The magnetic flux concentrates near the rotor teeth and, when the rotor rotates, eddy currents are induced in the stator shell near the surface. The eddy current fields react with the main field and set up a force on the rotor which tends to resist the torque applied by the engine. The stator is mounted in anti-friction trunnions and the torque measuring arm is fitted to the stator. A diagram of the eddy current dynamometer is shown in Fig. 7.20.

The power produced by the engine under test is dissipated as heat which is removed from the dynamometer by a flow of cooling water through the space between rotor and stator, the water being carried round and through the machine by the motion of the rotor teeth.

The following torques act on the stator casing and are all measured by the torque measuring device:

(1) The main torque due to the generation of eddy currents.
(2) The torque due to slight hydraulic effects (cooling water).
(3) Friction torque of shaft bearings.

The *all-electric dynamometer* consists of an electric dynamo, the shaft of which is coupled to that of the engine under test and the stator of which is connected to a balancing system. The stator itself is mounted on anti-friction trunnions. The electromagnetic reaction between the rotating member (armature) and the stator causes a torque to be transmitted between the armature and stator. Control of the engine load is effected by loading the dynamo electrically and thus increasing the electromagnetic reaction between the armature and the stator (see Fig. 7.21).

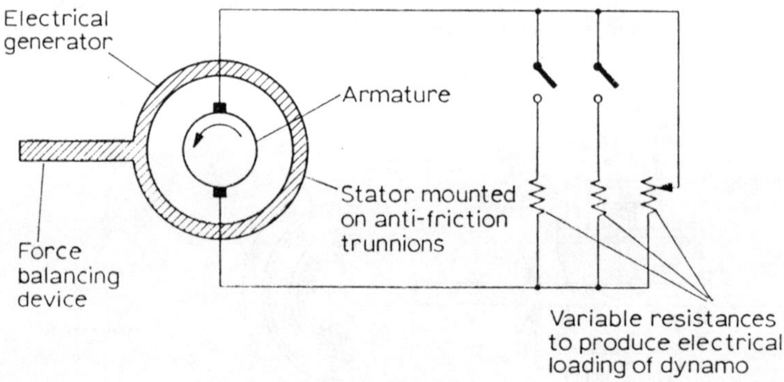

Fig. 7.21 THE ALL-ELECTRIC DYNAMOMETER

On this type of dynamometer the resisting torque consists of the following:

(1) Friction torque.
(2) Resistance torque due to windage.
(3) Torque due to the electromagnetic reaction.

The sum of all these torques is measured by the balancing device which therefore gives an accurate determination of the torque applied to the engine.

An absorption dynamometer which is used to measure electrical power in an alternating current circuit is the Wattmeter. The current is passed between two fixed coils in series while a moving coil acts as the Wattmeter voltage coil and carries a current proportional to the voltage of the circuit across which it is connected. A high non-inductive resistance is connected in series with the voltage coil. The deflection of the moving coil depends on the currents flowing in the coils and the power factor.

In an alternating current circuit the instantaneous power is given by

$$p = ei$$

where p = instantaneous power
e = instantaneous voltage
i = instantaneous current.

Therefore if both the current and voltage waves are sinusoidal, and the current is lagging in phase by an angle ϕ, then

$$e = E_{max} \sin \omega t$$

and

$$i = I_{max} \sin (\omega t - \phi)$$

Therefore the instantaneous power is given by $p = ei$ and substituting θ for ωt

$$p = E_{max} I_{max} \sin \theta \sin (\theta - \phi)$$

The mean power is

$$p = \frac{1}{2\pi} \int_0^{2\pi} E_{max} I_{max} \sin \theta \sin (\theta - \phi)\, d\theta$$

$$= \frac{E_{max} I_{max}}{2\pi} \int_0^{2\pi} \tfrac{1}{2}\{\cos \phi - \cos (2\theta - \phi)\}\, d\theta$$

$$= \frac{E_{max} I_{max}}{4\pi} \left[\theta \cos \phi - \tfrac{1}{2} \sin (2\theta - \phi)\right]_0^{2\pi}$$

$$= \frac{E_{max} I_{max}}{2} \cos \phi$$

$$= EI \cos \phi$$

where E and I are the r.m.s. values of current and voltage, and $\cos \phi$ is known as the power factor.

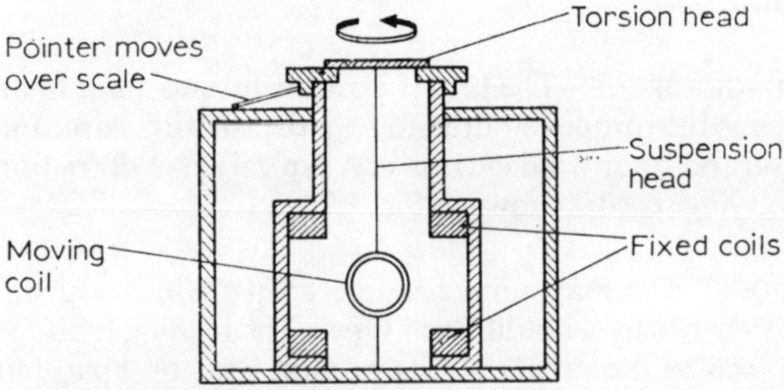

Fig. 7.22a TORSION-HEAD WATTMETER

Wattmeters may be divided into two classes:

(1) Suspended coil, torsion instruments.
(2) Pivoted coil, direct indicating instruments.

In the former class the moving, or voltage, coil is suspended from a torsion head by a metal suspension which serves as a lead to the coil. The moving coil is situated entirely inside the fixed current carrying coils. The torsion head carries a scale, and when in use the moving coil is brought back to the zero position by turning this head, the power being calculated by multiplying the number of divisions turned through by a constant for the instrument. A typical suspended coil Wattmeter is shown in Fig. 7.22*a*.

In the direct indicating instrument the moving coil is carried on a pivoted spindle and the movement is spring-controlled. The moving system carries a pointer and a damping vane which moves in a sector-shaped box; Fig. 7.22*b* shows a typical instrument of this type.

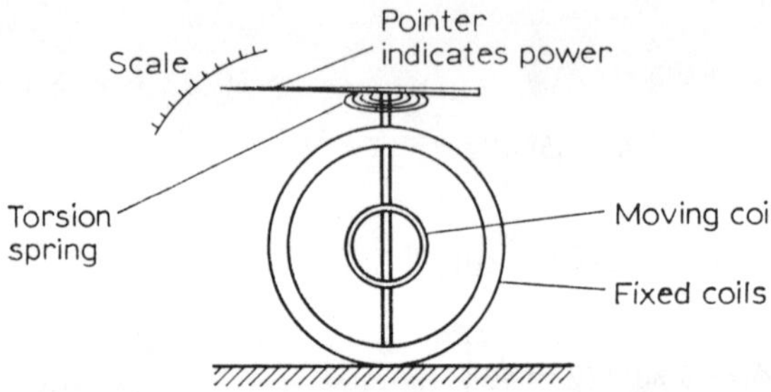

Fig. 7.22b PIVOTED-COIL DIRECT-INDICATING WATTMETER

7.9 Comparison of Absorption Dynamometers

Different dynamometers may easily be compared if the useful range of each is first determined. The useful range is limited by a number of factors which can be combined to form a dynamometer characteristics diagram. These factors are now considered in turn:

(*a*) FRICTION AND WINDAGE TORQUE The lowest possible torque input to an absorption dynamometer is the torque required to turn the rotating parts and overcome bearing friction and rotor windage losses. A typical curve of friction and windage torque at varying speed is shown in Fig. 7.23*a*.

(*b*) LOADING LIMIT TORQUE The maximum possible torque which a dynamometer can exert will vary widely with different types. The loading limit for a Prony brake is determined by the strength of its weakest structural part but electric and hydraulic dynamometers can exert zero torque at zero speed, and the maximum possible torque will increase with an increase in speed. This is shown in Fig. 7.23*b*.

(*c*) MECHANICAL LIMIT TORQUE AND CENTRIFUGAL LIMIT The maximum torque which a dynamometer may exert is limited by the mechanical strength of the dynamometer frame or by the range of the force-measuring device, whichever is the smaller. Therefore a top limiting mechanical torque independent of speed is introduced. Also a further strength limitation is introduced due to the maximum allowable safe operating speed which will limit the usable speed regardless of torque. These are indicated in Fig. 7.23*c*.

(*d*) COOLING LIMIT If a dynamometer is to absorb energy for an appreciable length of time there must be adequate cooling capacity to absorb this energy

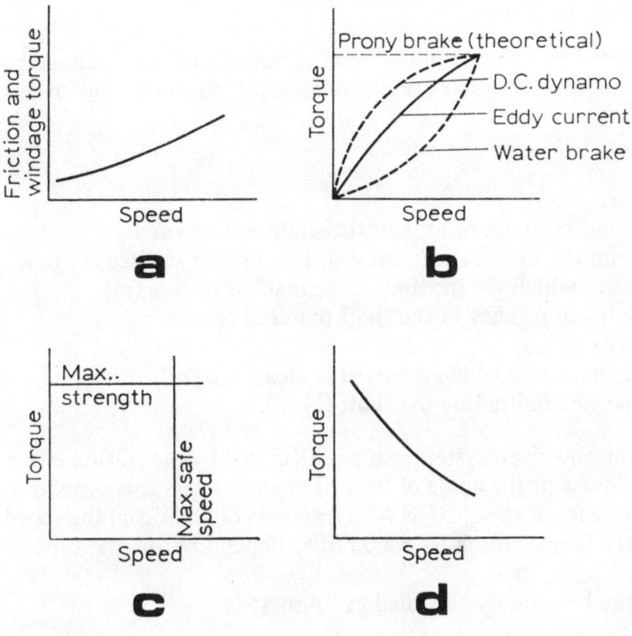

Fig. 7.23

a Friction torque *b* Maximum torque
c Mechanical limits *d* Limits of cooling

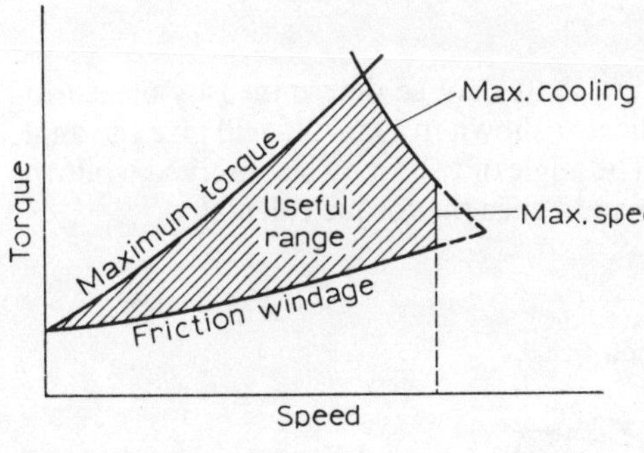

Fig. 7.24 THE USEFUL RANGE OF AN ABSORPTION DYNAMOMETER

in the form of heat. The cooling limit is shown in Fig. 7.23*d* and will result in lower maximum allowable torque at higher speeds since power is torque times angular velocity.

All the limiting conditions discussed above may be combined on one dynamometer characteristics diagram from which the useful range of the dynamometer may be determined (see Fig. 7.24).

7.10 Transmission Dynamometers

If the torque transmitted by a shaft is measured the power can be calculated by multiplying the torque by the angular velocity of the shaft. It can be shown that

$$\frac{T}{J} = \frac{\tau}{r} = \frac{G\theta}{L}$$

where J = polar second moment of area of the shaft section (m^4)
τ = the maximum induced shear stress at the outside surface (Pa)
r = the radius at which the maximum shear stress occurs (m)
G = the modulus of rigidity of the shaft material (Pa)
θ = angle of twist (rad)
L = length of shaft over which the twist is measured (m)
T = the torque transmitted by the shaft (N m).

Thus the torque for any given system can be calculated by measuring either the maximum shear stress or the angle of twist. Torsion meters can be used to measure power ranging from about 20 MW to fractions of 1 kW and the speed of operation can vary from 2 rad/s to 5000 rad/s, depending on the type of instrument used.

Torsion meters may be broadly classified as follows:

(*a*) Mechanical torsion meters
(*b*) Electrical torsion meters
(*c*) Optical torsion meters.

7.11 Mechanical Torsion Meters

The torque transmitted by a rotating shaft may be determined by measuring the angle of twist. The rotating indicator shown in Fig. 7.25 will give the angle of twist over a given length of bar. The angle of twist may be observed while the bar is rotating using a stroboscope and hence the torque calculated.

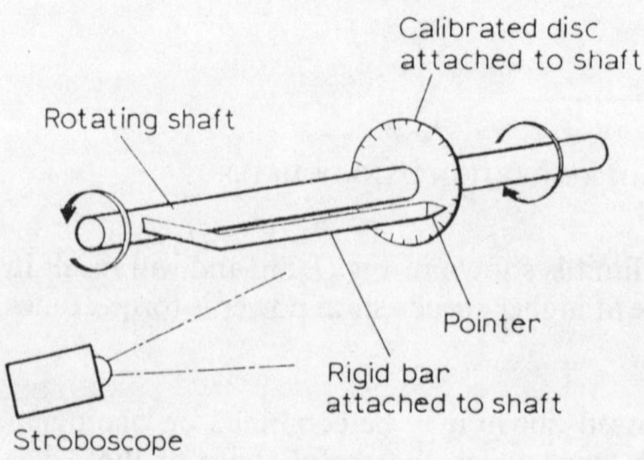

Fig. 7.25 STROBOSCOPIC TORQUE METER

The power transmitted may now be calculated by multiplying the torque by the rotational speed of the bar, i.e.

$P = T\omega$ watts

where T = torqe (N m) and ω = angular velocity (rad/s).

A simple torque meter suitable for applications where the driven shaft is near to and parallel with the drive shaft is shown in Fig. 7.26. The applied

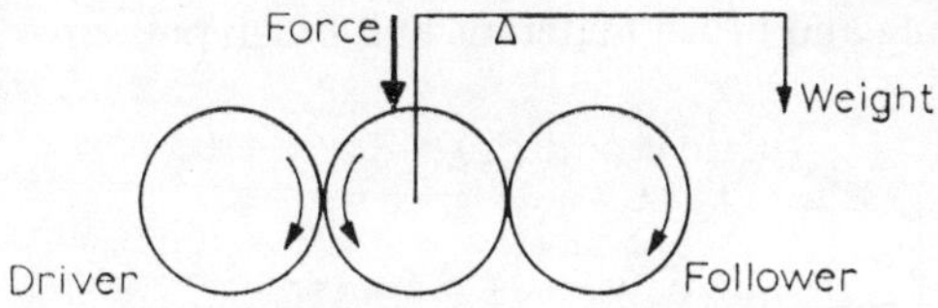

Fig. 7.26 SIMPLE TORQUE METER

torque can be resolved as a force at the pitch circle diameter of the drive gear and for equilibrium this force is resisted by a force equal in magnitude but opposite in direction. The resisting force is applied through a lever system using weights.

7.12 Electrical Torsion Meters

Perhaps the most widely used electrical torsion meter is the strain gauge torque meter, the principle of which is shown in Fig. 7.27. The strain gauges

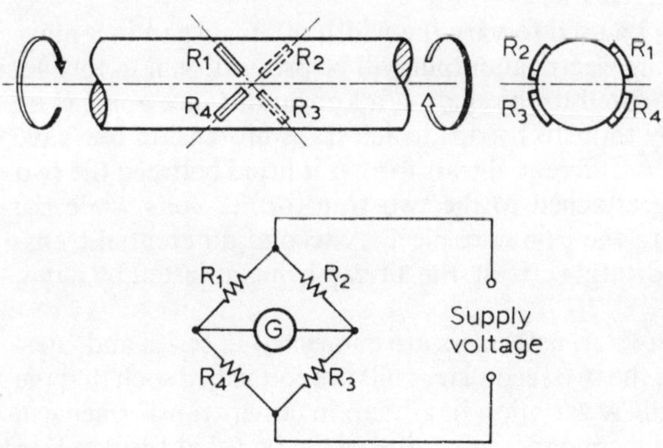

Fig. 7.27 STRAIN GAUGE TORQUE METER

are mounted mutually at 45° to the axis of the shaft. Under an applied torque the tensile and compressive principal strains can be measured and hence the applied torque may be calculated. The strain gauge output is brought out through slip rings and brushes to a strain bridge indicator. A major difficulty has been that the magnitude of the signal leaving the gauge is low and often the "noise" level in variation of contact resistance is of the same order of magnitude as the signal itself.

Circuits and techniques have now been developed so that slip-rings need not be a source of significant inaccuracy and three methods have been employed to overcome slip-ring contact resistance variations. One method involves construction of an electrical circuit so that the slip-ring contact resistance is not in series with the gauge itself but is placed at some point in the circuit where its effect is minimized. A second, less effective method is to use high-resistance strain gauges so that the resistance change of the gauges will be large compared with those occurring between slip-rings and brushes. The third technique requires careful selection of slip-ring and brush materials and brush pressures

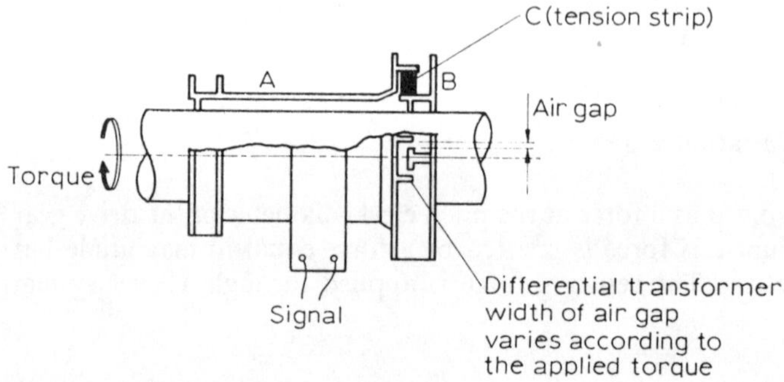

Fig. 7.28 ELECTRICAL TORSION METER

for optimum performance. These techniques are covered in more detail in Section 5.13.

If the applied torque can be used to vary the width of an air gap in a magnetic circuit then the resulting electrical output will be proportional to torque. Two castings, A and B, are fitted to the shaft at a known distance apart (Fig. 7.28). A is attached to B by thin steel strips which transmit tension but have little resistance to bending. A differential transformer is fitted between the two castings, one casting being attached to the two transformer coils while the second casting is attached to the iron core piece. A second differential transformer is included in the indicating circuit, the air gap being adjusted by hand, using a micrometer screw.

The primary coils of the two transformers are connected in series and energized with a.c. current while the two secondary coils are connected such that the induced e.m.f.s are opposed. A variation in air gap in either transformer will cause current to flow in the secondary circuit due to the fact that the e.m.f. in one secondary coil is increased while that in the other secondary coil is decreased by an equal amount. The flow of current in the secondary circuit may be measured using a suitable instrument.

The air gap of the indicator unit may be adjusted using the micrometer screw until no current flows in the secondary circuit. The air gaps in the two transformers must now be equal. Hence the applied torque is directly proportional to the width of the air gap or the micrometer reading.

In practice the power is calculated by multiplying the micrometer reading by the shaft speed and a constant for the meter.

A similar meter to the above uses the principle of varying capacitance as a measure of shaft twist, and hence applied torque is available. The arrangement is similar to the above but in this instrument the plates of a condenser are connected to the castings clamped to the shaft. Any twist in the shaft will change the relative positions of the condenser plates and hence the output signal will vary. This output signal can then be used as a measure of applied torque. Hence

Power transmitted = output signal × constant for meter

7.13 Optical Torsion Meters

Using a similar apparatus to the electrical torsion meters described above, optical methods may be used to measure the angle of twist of a shaft. The

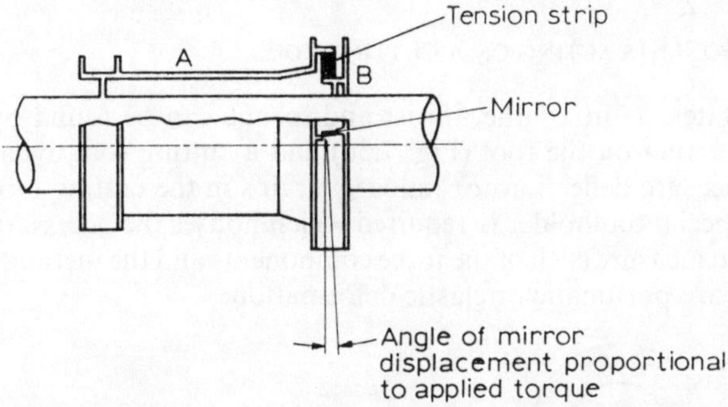

Fig. 7.29 OPTICAL TORSION METER

apparatus shown in Fig. 7.29 is basically the same as that in Fig. 7.28 except that in addition to the tension strips there are two links which have torque mirrors incorporated. Hence any relative movement of the two castings due to applied torque results in partial inclination of the two mirrors. If a beam of light from a source is reflected by the mirrors on to a graduated scale, a movement on the scale will represent a measure of applied torque.

If the mirrors are arranged back to back there will be a reflection from each mirror every half revolution which is quite sufficient if the torque remains constant during each revolution, e.g. a steam turbine. In applications where the torque varies during a revolution, e.g. a reciprocating engine, a second system of mirrors giving four readings per revolution would be desirable.

The deflection of the mirrors is directly proportional to the applied torque. Therefore

$$\text{Power} = \frac{\text{scale reading (m)}}{\text{a constant}} \times \text{angular velocity (rad/s)}$$

where the value of the constant will be specific to the meter.

7.14 The Power Absorbed in Machining Operations

In most machining operations the motion of the cutting tool will be both angular and linear with respect to the work. The power absorbed in the cutting process can be divided into two sections:

(*a*) the power absorbed by the torque exerted on the rotating member
(*b*) the power absorbed by the thrust needed to produce linear motion.

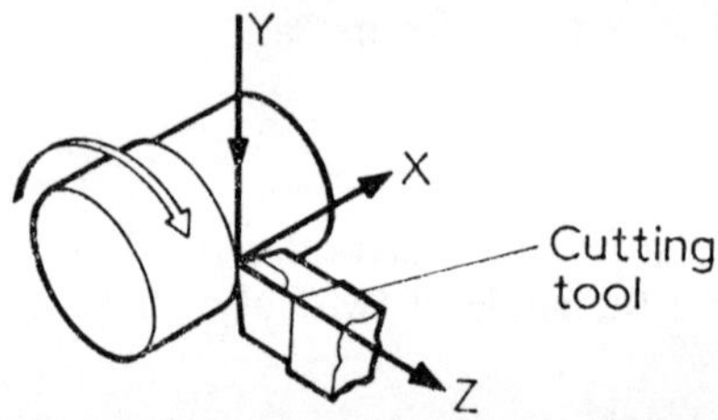

Fig. 7.30 FORCE COMPONENTS ACTING ON A CUTTING TOOL

When the work rotates, as in a lathe, thrust and torque can be found by measuring the forces acting on the tool (Fig. 7.30) and a cutting tool dynamometer is used to measure deflections or induced strains in the cutting tool support member. A special toolholder is required which houses the necessary transducers required to measure each of the force components and the methods most commonly used are pneumatic or elastic deformation.

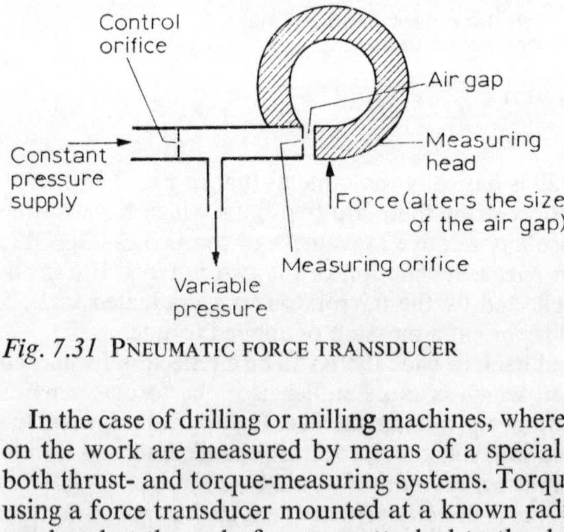

Fig. 7.31 PNEUMATIC FORCE TRANSDUCER

In the case of drilling or milling machines, where the tool rotates, the forces on the work are measured by means of a special workholder which houses both thrust- and torque-measuring systems. Torque measurement is achieved using a force transducer mounted at a known radius which detects the force produced at the end of an arm attached to the dynamometer body which is mounted in bearings. A typical pneumatic force transducer which produces a pressure signal proportional to applied force is shown in Fig. 7.31.

A force applied to the measuring head will deflect the measuring head thus closing the air gap and increasing the variable pressure signal which may be measured using a suitable pressure-measuring device.

8 Measurement of Pressure

Pressure measurement can be broadly divided into two groups as follows:

(*a*) The measurement of static pressure.
(*b*) The measurement of fluctuating pressure.

The term "static pressure" is widely used in the measurement of pressure, liquid level, and rate of flow. When a fluid in a process is still or "static", the static pressure at a point in the fluid is the pressure exerted by the height of fluid above that point. If a pipe full of a fluid in motion is considered, the static pressure at a tapping in the pipe wall may be found by connecting a suitable pressure-measuring device to the tapping.

When it is required to measure a rapidly fluctuating pressure (e.g. the cylinder pressure in an internal combustion engine) the methods used for static pressure measurement are not suitable. In this case a pressure transducer, which converts the pressure into a signal which may be recorded, is used.

8.1 Static Pressure Measurement

Static pressure measurement may conveniently be considered under the following headings:

(*a*) Differential pressure measurement.
(*b*) Gauge pressure, which is the pressure measured above the local atmospheric pressure.
(*c*) Absolute pressure, which is the total pressure measured from zero pressure as the datum point. When the absolute pressure exceeds the local atmospheric pressure it may be considered to be the sum of the gauge pressure and local atmospheric pressure.
(*d*) Vacuum measurement. In this case the pressure to be measured is less than the local atmospheric pressure.

8.2 Differential Pressure Measurement

One of the most popular devices for measuring differential pressure is the U-tube manometer (Fig. 8.1). The glass U-shaped tube is filled about half full with liquid, usually either water or mercury. Generally speaking water is most commonly used as the measuring medium but if higher pressures or more compact gauges are required mercury is used. A scale is fixed between the limbs and the difference in level between the limbs A and B due to a pressure differential $(P_1 - P_2)$ may be measured. If the higher pressure P_1 is applied to A and the lower pressure P_2 to B, the liquid in A will be forced down, that in B will rise, and the action will continue until pressure P_1 is

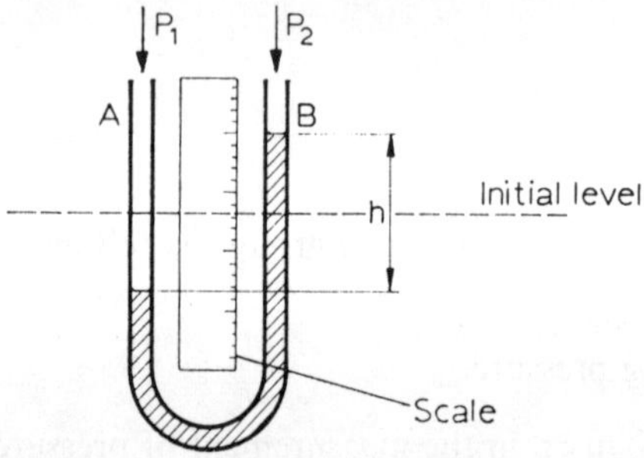

Fig. 8.1 THE U-TUBE MANOMETER

balanced by the sum of P_2 and the pressure due to the height of liquid column h between the two levels. Then

$$P_1 = P_2 + \rho g h$$

where P_1 and P_2 are absolute pressures (N/m² or Pa)
h = difference in levels (m)
ρ = density of manometer fluid (kg/m³)
g = acceleration due to gravity (m/s²)

Or alternatively

$$P_1 - P_2 = \rho g h$$

where $P_1 - P_2$ is the pressure difference applied across the manometer.

If limb B of the manometer is open to the atmosphere then in this case P_2 is the atmospheric pressure. Therefore the above equation becomes

$$P_1 - \text{atmospheric pressure} = \rho g h$$

Hence in this case the manometer is measuring the difference between the absolute pressure and the local atmospheric pressure. This is known as the *gauge pressure*.

Range of the Simple U-tube Manometer

The simple U-tube manometer is normally used between 10×10^{-3} m water and 0–1·25 m mercury (0–162 000 N/m² or 0–162 kPa).

An industrial development of the simple U-tube manometer for measuring high pressures is shown in Fig. 8.2. Here one limb is of much larger diameter than the other. Also the limbs are often constructed of steel.

A differential pressure is applied to the manometer, the lower pressure P_2 to the wide limb and the higher pressure P_1 to the narrow limb. The liquid in the wide limb rises and that in the narrow limb falls until the following pressure balance is obtained:

$$P_1 = P_2 + \rho g H = P_2 + \rho g(h + d)$$

where H = total difference in levels (m)
h = the difference between the level in the small diameter limb and the zero level (m)
d = the difference between the level in the large diameter limb and the zero level (m).

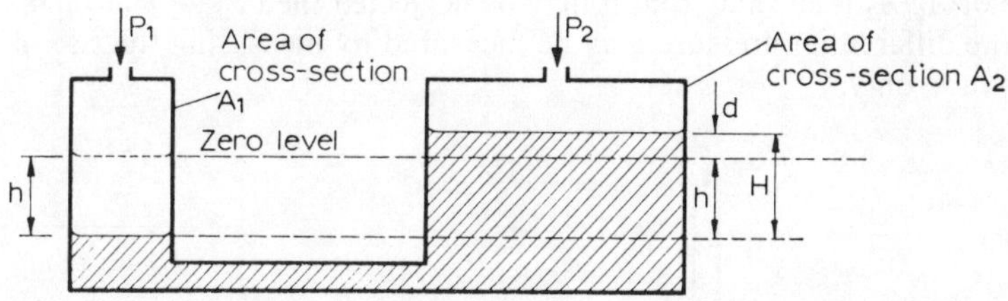

Fig. 8.2 INDUSTRIAL U-TUBE MANOMETER

Also the volume of liquid which has left limb A (the narrow limb) must be equal to that which has passed into limb B (the wide limb), i.e.

$$A_1 h = A_2 d \qquad \text{hence } h = A_2 d / A_1$$

where A_1 = the cross-sectional area of the small diameter limb (m^2)
A_2 = the cross-sectional area of the large diameter limb (m^2).

Substituting

$$P_1 = P_2 + \rho g d\left(1 + \frac{A_2}{A_1}\right)$$

or

$$P_1 - P_2 = \rho g d\left(1 + \frac{A_2}{A_1}\right)$$

or

$$d = \frac{P_1 - P_2}{\rho g(1 + A_2/A_1)}$$

The distance therefore that the liquid rises in the wide tube is proportional to the differential pressure ($P_1 - P_2$). In many industrial manometers of this type a metal float is included in the wide chamber, the float being connected via a linking mechanism to a pointer or pen recorder. Therefore as the level of the liquid varies in response to changing pressure differentials so the float will ride on the surface of the liquid and record the change in level or differential pressure.

The Cistern Manometer

The working of the above instrument would not be affected if the narrow tube were inserted directly into the wide limb as shown in Fig. 8.3.

If a pressure differential $(P_1 - P_2)$ is applied to the manometer then

$$P_1 - P_2 = \rho g(h + d)$$

Also $A_1 d = A_2 h$, and therefore $d = \frac{A_2}{A_1} h$.

$$P_1 - P_2 = \rho g h \left(1 + \frac{A_2}{A_1}\right)$$

If the value of A_2/A_1 is so small that it may be neglected then $P_1 - P_2 = \rho g h$. Therefore the differential pressure may be measured by measuring the rise in the narrow tube only.

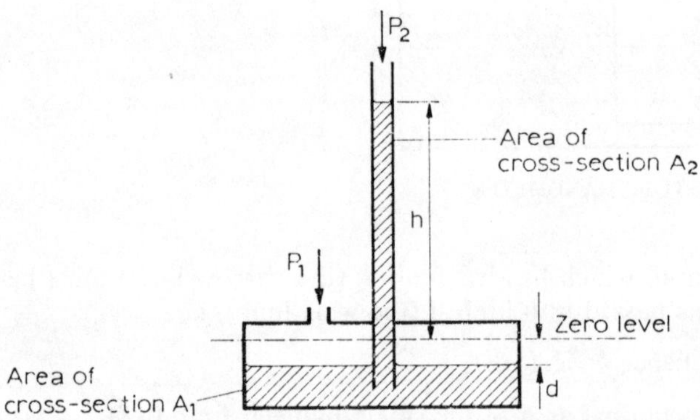

Fig. 8.3 THE CISTERN MANOMETER

In practice the cistern diameter (wide limb) is made so large compared with the narrow tube diameter that the drop in level, *d*, from the zero level in the cistern is negligible.

The Inclined Tube Manometer

A special development of the U-tube manometer is the inclined tube manometer. The advantage of this device is that it gives an increased length of scale compared with the simple U-tube for the same differential pressure. One limb consists of a glass tube inclined at an angle to the horizontal or vertical, and the other of a wide chamber or cistern. The sloping tube carries a scale adjacent to it. A typical inclined tube device is shown in Fig. 8.4.

$$P_1 - P_2 = \rho g(h_1 + d \sin \alpha)$$

$$A_1 h_1 = A_2 d \qquad \text{i.e. } h_1 = \frac{A_2}{A_1} d$$

$$P_1 - P_2 = \rho g d \left(\frac{A_2}{A_1} + \sin \alpha\right)$$

and as in the previous example if A_1 is large compared with A_2 the ratio A_2/A_1 may be neglected. Therefore

$$P_1 - P_2 = \rho g d \sin \alpha$$
$$d \sin \alpha = h$$

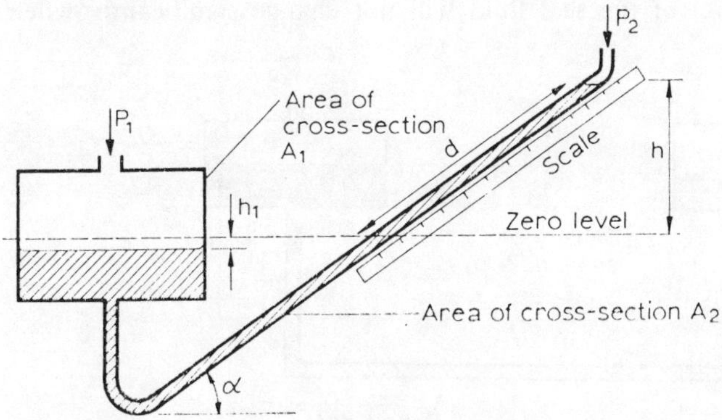

Fig. 8.4 INCLINED U-TUBE MANOMETER

where h is the height which would have been registered if an ordinary U-tube had been used.

If the tube is inclined at an angle of 30° to the horizontal, i.e. $\sin \alpha = \frac{1}{2}$, then

$$d \times \tfrac{1}{2} = h \qquad \text{or} \qquad d = 2h$$

Therefore a scale length of about double the normal value is obtained by inclining the tube at 30° to the horizontal.

The Use of Sealing Fluids with Manometers: the Two Liquid U-tube Manometer

When the pressures of certain liquids are measured using a U-tube manometer it may be necessary to separate the liquid from the manometer liquid by using a sealing fluid. For example if the two liquids would undergo a chemical reaction when mixed, a seal fluid is required. Seal fluids must not mix with the manometer fluid or the fluid being measured and therefore they must be

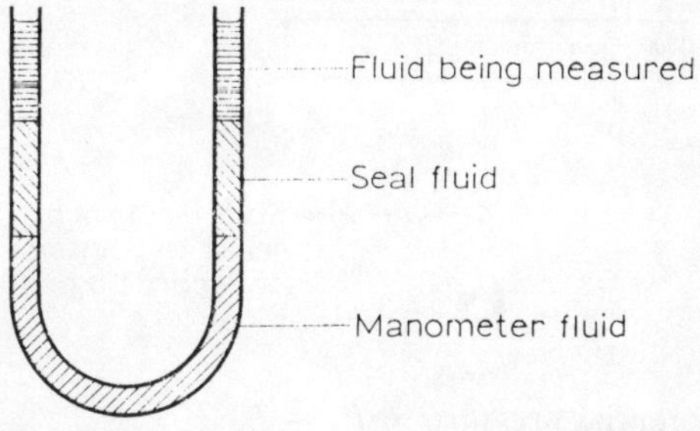

Fig. 8.5 U-TUBE MANOMETER WITH SEAL FLUID

lighter than the manometer fluid but heavier than the measured fluid (see Fig. 8.5). Exactly the same amount of seal fluid is added to both columns of the manometer so that they cancel out.

Seal chambers are normally used with seal fluids. These are simple reservoirs having a diameter much larger than the diameter of the manometer tube so that the level of the seal fluid will not change significantly when

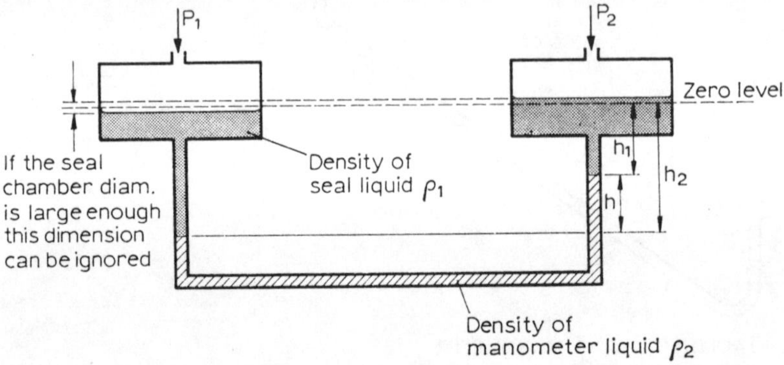

Fig. 8.6 U-TUBE MANOMETER WITH SEALING FLUID AND SEAL CHAMBERS

pressure is applied. Fig. 8.6 shows a typical U-tube manometer together with sealing fluid and seal chambers.

$$P_1 + \rho_1 g h_2 = P_2 + \rho_1 g h_1 + \rho_2 g h$$

Therefore

$$\begin{aligned} P_1 - P_2 &= \rho_1 g(h_1 - h_2) + \rho_2 g h \\ &= -\rho_1 g h + \rho_2 g h = (\rho_2 - \rho_1) g h \end{aligned}$$

Diaphragms, Diaphragm Stacks, and Bellows Gauges

The *single diaphragm* in its most simple form is a thin flat plate of circular shape. The plate is fixed round its edge and when a differential pressure occurs across it the plate will deflect as shown in Fig. 8.7*a*. Only for relatively small

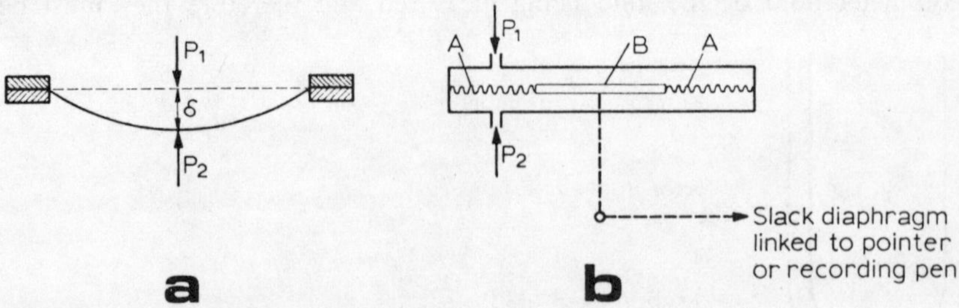

Fig. 8.7

a A simple diaphragm: differential pressure = $P_1 - P_2$
b The slack diaphragm differential pressure gauge

movements is there a linear connection between pressure and deflection and some modification of this simple case is desirable for practical instrumentation.

For the measurement of low pressures an extremely flexible diaphragm is required. The diaphragm assembly is made in the form of a ring of fabric with a disc of metal or other rigid material at the centre. The principle of a fabric or "slack" diaphragm differential pressure gauge is shown in Fig. 8.7*b*.

A is the fabric and B is the solid centre piece which is linked to a pointer or recording pen. With the slack diaphragm gauge it is unnecessary for the

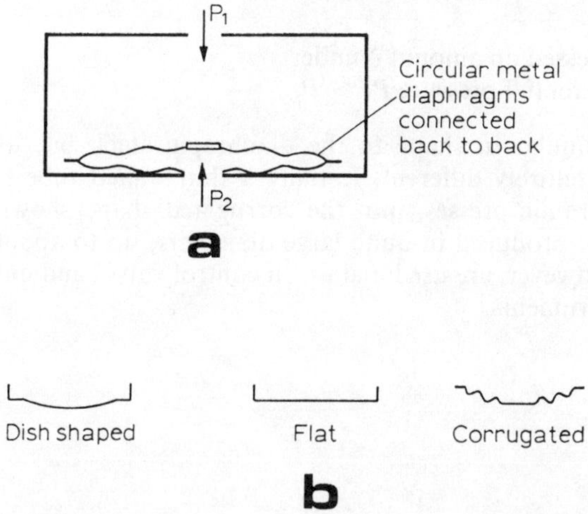

Fig. 8.8

a Metallic capsule pressure gauge
b Typical diaphragm shapes

diaphragm assembly to be circular and quite often it may be elongated. A variety of materials have been used for the slack diaphragm, the most popular being leather, plastic, rubberized fabric, nylon and silk. Gauges working on this principle can be used up to 1 m of water or about 10 kPa.

An alternative form of pressure-measuring instrument is the *metallic capsule*. The instrument consists of a pair of circular metal diaphragms connected back to back to form an elastic capsule as shown in Fig. 8.8*a*. One pressure is applied to the inside of the capsule which is surrounded on the outside by the other pressure. The deflection of the capsule is found to have a linear relationship with the pressure provided the movement is not excessive. The diaphragms used in a capsule gauge are generally dished, flat or corrugated in shape (Fig. 8.8*b*).

A deflection many times greater than that of a single capsule can be obtained when a series of corrugated metal diaphragms each containing a central hole are joined in a leak-proof manner at the innner or outer edges as shown in Fig. 8.9. If $P_1 > P_2$ the diaphragm stack is compressed. The number of diaphragms used may vary from two (the capsule) up to twenty or so.

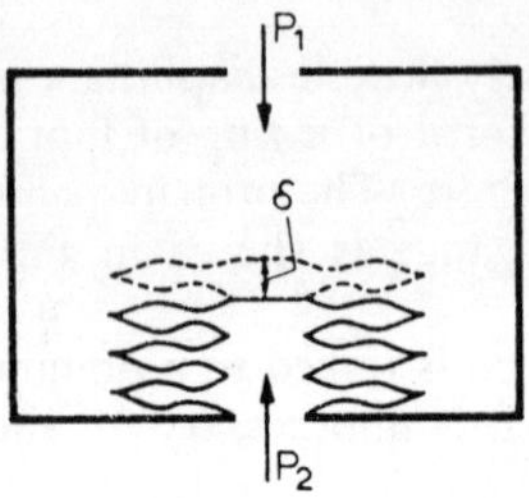

Fig. 8.9 DIAPHRAGM STACK
The stack is compressed an amount δ under the action of a differential pressure $P_1 > P_2$

The *metallic bellows* is similar in shape to the diaphragm stack but its method of construction is entirely different. Initially a thin walled tube is taken and formed, by hydraulic presses, into the corrugated shape shown in Fig. 8.10. Bellows can be produced in quite large diameters, up to about 300 mm. The larger sizes, however, are used mainly on control valves and not for pressure-measuring instruments.

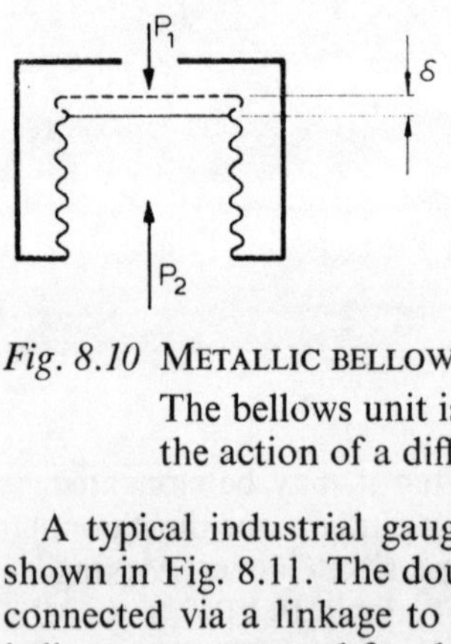

Fig. 8.10 METALLIC BELLOWS UNIT
The bellows unit is deflected an amount δ under the action of a differential pressure $P_1 > P_2$

A typical industrial gauge used for differential pressure measurement is shown in Fig. 8.11. The double bellows arrangement on the left-hand side is connected via a linkage to a recording pen or pointer while the right-hand bellows acts as a seal for the range adjusting spring.

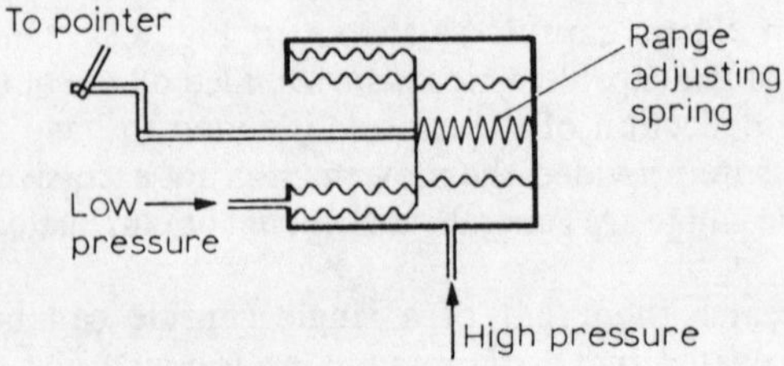

Fig. 8.11 INDUSTRIAL BELLOWS GAUGE USED TO MEASURE DIFFERENTIAL PRESSURE

The materials most commonly used for the manufacture of diaphragms are steel, phosphor bronze, nickel silver and beryllium copper. Bellows materials have largely consisted of 80–20 brass (80 per cent copper and 20 per cent zinc), phosphor bronze, stainless steel and beryllium copper.

The Ring Balance

This apparatus consists of a hollow or annular circular ring divided into two sections by a partition B and a seal fluid (Fig. 8.12). The sections on either side of the partition are connected to the two pressure sources. The ring is pivoted on a knife edge at its centre and when a pressure difference is applied across the partition a rotating moment is set up. The ring starts to rotate in a direction

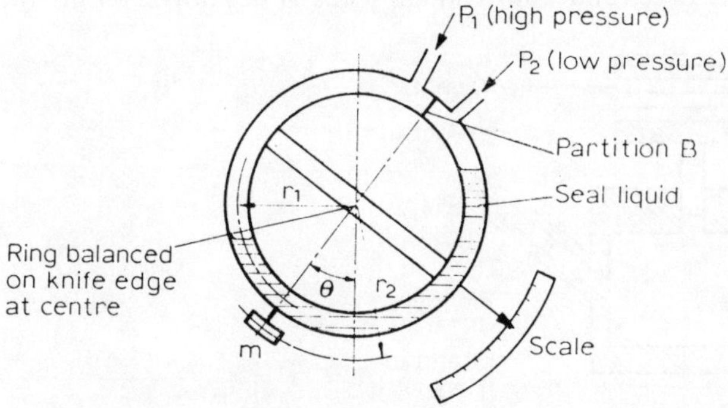

Fig. 8.12 THE RING BALANCE

away from the higher pressure until the opposing moment, created by the mass at the foot of the ring, balances the rotating moment. Now

Rotating moment $= (P_1 - P_2)\ Ar_1$

where $P_1 - P_2 =$ differential pressure
$A =$ cross-sectional area of the tube
$r_1 =$ mean radius of the ring.

Also

Restoring moment $= mgr_2 \sin \theta$

where $m =$ mass
$r_2 =$ radius of its point of application
$\theta =$ angle of rotation.
$g =$ acceleration due to gravity.

Therefore

$$(P_1 - P_2)Ar_1 = mgr_2 \sin \theta$$

or

$$P_1 - P_2 = \frac{mgr_2}{Ar_1} \sin \theta$$

Therefore the differential pressure applied to the ring balance is proportional to the angle of rotation, i.e. the angle of rotation is a measure of differential pressure. Typical ranges for this instrument would be within the limit 0–300 mm water.

8.3 Gauge Pressure Measurement

The *deadweight free-piston gauge* has been used for the precise determination of steady pressures for nearly eighty years. The gauge (Fig. 8.13) consists of an accurately machined piston of known weight which is inserted into a close-fitting cylinder, both of known cross-sectional area. A number of masses of known weight are first loaded on one end of the free piston and fluid pressure is then applied to the other end until enough force is developed to lift the

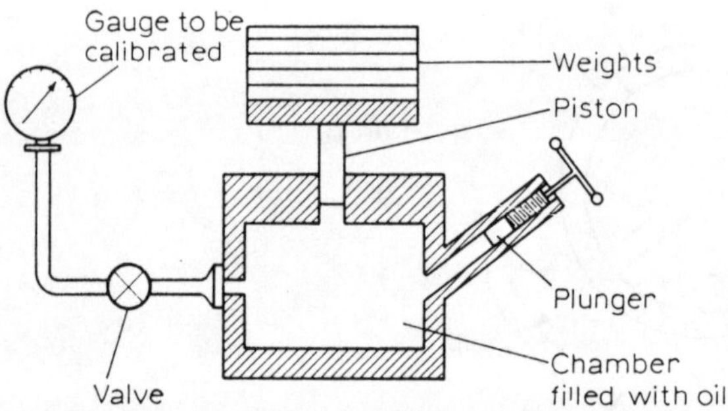

Fig. 8.13 THE DEADWEIGHT PRESSURE TESTER

piston–weight combination. When the piston is floating freely within the cylinder the piston is in equilibrium with the system pressure. Therefore

P (dead weight pressure) $= F_e/A_e$

where F_e is the equivalent force of the piston–weight combination, and A_e is the equivalent area of the piston–cylinder combination.

The film of fluid leaking from the system through the piston–cylinder clearance provides lubrication and the piston is also rotated to further reduce the friction.

Perhaps the most commonly used gauge for the measurement of gauge pressure is the *Bourdon tube gauge*. The gauge consists of a metal tube of approximately elliptical cross-section which is formed into a C-shape by special machines. The tube is fixed at one end, which is open to accept the applied pressure, but free at the other end, which is closed. When the pressure within the tube is increased the cross-section of the tube tends to become round and the entire tube tends to straighten out. Thus the free end of the tube will move, and this motion will be proportional to the pressure difference between the inside and the outside of the tube. The motion of the free end may be magnified and transmitted to an indication pointer by a lever and rack-and-pinion arrangement (see Fig. 8.14). The reference pressure in the case containing the Bourdon tube is usually atmospheric, and therefore the pointer indicates gauge pressure.

Two alternative forms of the Bourdon tube are the helical type and the flat spiral. The principle of operation for both these gauges is the same as for the C-type, i.e. one end of the tube is closed and sealed, but left free to move, the other end being fixed. If a pressure is now applied to the inside of the tube

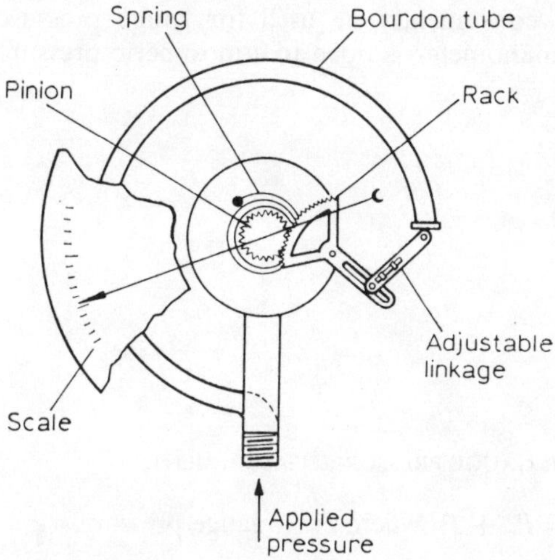

Fig. 8.14 BOURDON TUBE PRESSURE GAUGE

greater than that outside then the free end will deflect and this movement can be transferred to an indicating pointer as before.

Bourdon tubes are normally used for measuring pressures up to 500 MPa. The materials most commonly used in this country are phosphor bronze or

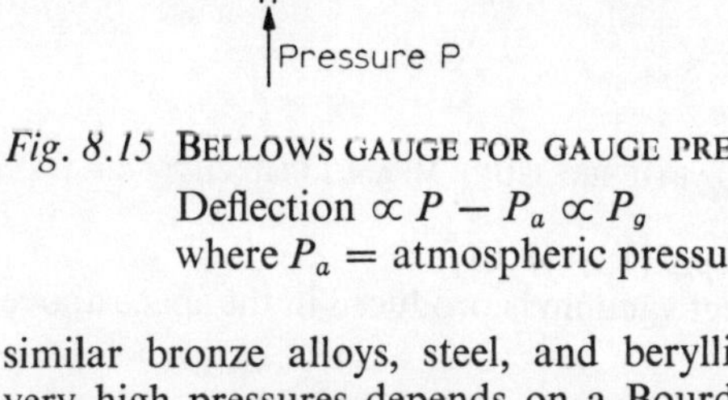

Fig. 8.15 BELLOWS GAUGE FOR GAUGE PRESSURE MEASUREMENT
Deflection $\propto P - P_a \propto P_g$
where P_a = atmospheric pressure and P_g = gauge pressure

similar bronze alloys, steel, and beryllium copper. The measurement of very high pressures depends on a Bourdon tube constructed from a metal with a high elastic limit. Therefore phosphor bronze which has a fairly low value is ruled out for high pressure applications and a tough steel or beryllium copper is used.

Single diaphragms, diaphragm stacks and bellows may also be used for gauge pressure measurement. The basic arrangements are the same as for the measurement of differential pressure but in this case one of the pressures is arranged to be atmospheric. Fig. 8.15 illustrates an industrial bellows type gauge pressure instrument.

The U-tube manometer may conveniently be used for gauge pressure measurement if one limb of the manometer is open to atmospheric pressure as shown in Fig. 8.16.

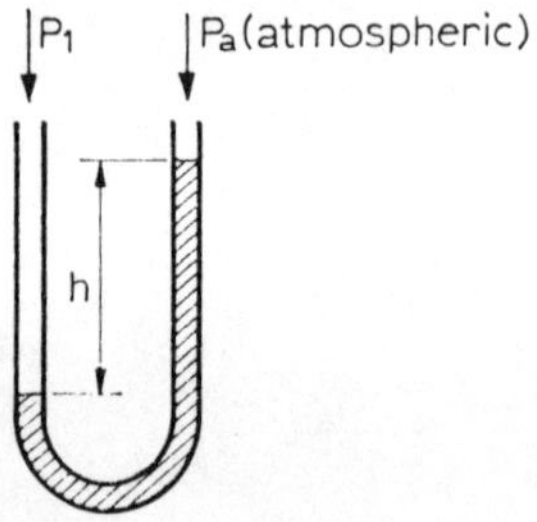

Fig. 8.16 U-TUBE MANOMETER FOR GAUGE PRESSURE MEASUREMENT

Now $P_1 - P_a = \rho gh$ and $P_1 = P_g + P_a$ where P_g = gauge pressure. Therefore

$$P_g = \rho gh \qquad \text{or} \qquad h = P_g/\rho g$$

Thus h is a measurement of the gauge pressure P_g.

8.4 Absolute Pressure Measurement

The *cistern manometer* discussed in section 8.2 may be modified by closing the end of the tube as shown in Fig. 8.17.

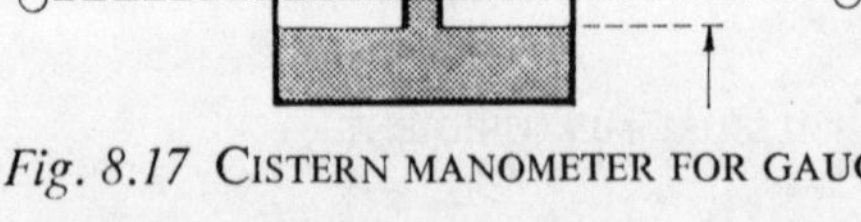

Fig. 8.17 CISTERN MANOMETER FOR GAUGE PRESSURE MEASUREMENT

It may be shown that $P_1 - P_2 = \rho gh(1 + A_2/A_1)$.

If P_2 is now made zero, i.e. a perfect vacuum is produced in the space above the liquid, then

$$P_1 = \rho gh(1 + A_2/A_1)$$

and therefore the height of the liquid column is proportional to the absolute pressure P_1. A particular case is the barometer where P_1 is equal to the atmospheric pressure P_a. The most common cistern barometer in general use is the the Fortin type, in which the height of the mercury surface in the cistern can be adjusted (Fig. 8.18).

The cistern in this case is a leather bag supported in a bakelite housing. The datum-adjusting screw is turned until the mercury in the cistern just

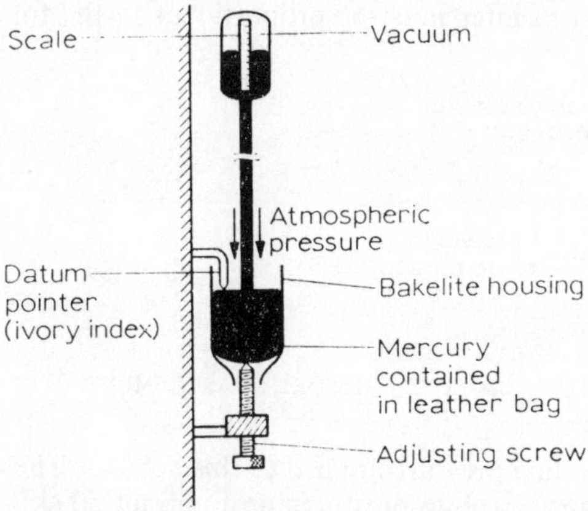

Fig. 8.18 THE FORTIN BAROMETER

makes contact with the ivory index, when the mercury surface is aligned with zero on the instrument scale. The indicated height of the mercury column in the glass tube is then determined by lining up the lower edge of the sighting ring with the top of the meniscus in the tube. The vernier scale is then read to give the indicated mercury height at the barometer temperature t.

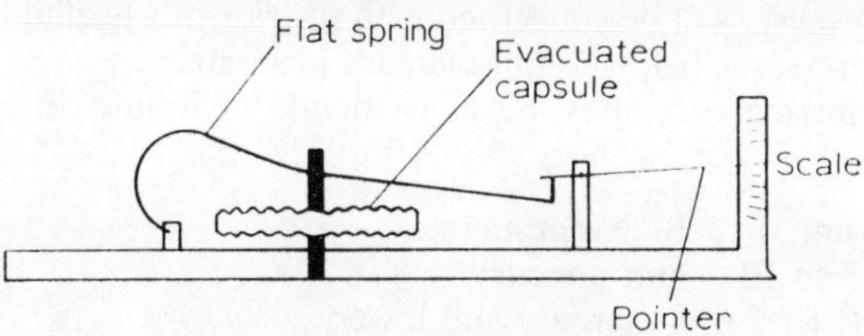

Fig. 8.19 THE ANEROID BAROMETER

A barometer using a bellows capsule or diaphragm measuring element is called an *aneroid barometer*. The capsule is evacuated until it contains a high vacuum and then sealed. The force on the bellows due to atmospheric changes is opposed by a strong spring and the motion of the bellows as the atmospheric pressure changes is transmitted through a mechanical magnification system to a pointer, which indicates the atmospheric pressure in millimetres of mercury on a scale. A typical aneroid barometer is shown in Fig. 8.19. The aneroid barometer is calibrated using the Fortin mercury barometer as a standard. Multiplication is so high that movement friction is important and therefore the gauge should be tapped to reduce friction for a true reading.

Diaphragm stacks or *bellows instruments* can be used to measure absolute pressure as shown in Fig. 8.20. If one bellows is evacuated, then by suitable matching, the effects on the two bellows due to atmospheric pressure and ambient temperature are equal over a reasonable range. As these effects are in opposition they will not influence the instrument pen or pointer. With the external atmospheric effects balanced out and negligible pressure in one bellows, the deflection of the pen or pointer must be proportional to the total

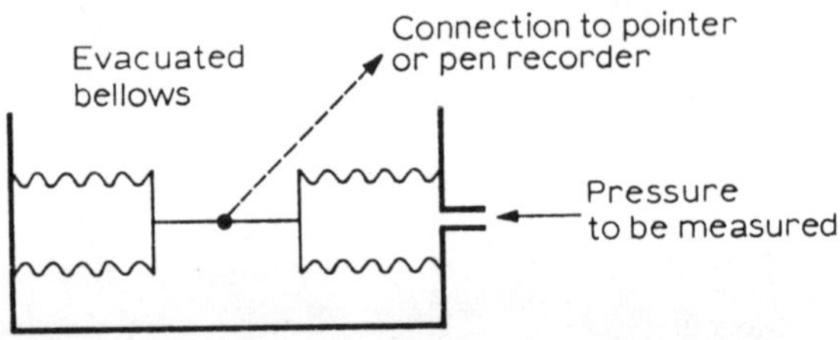

Fig. 8.20 BELLOWS INSTRUMENT FOR ABSOLUTE PRESSURE MEASUREMENT

pressure in the other, i.e. the absolute pressure applied to this bellows. These gauges may be used for measuring absolute pressures up to about 500 kPa.

8.5 High Vacuum Measurement

Some of the instruments previously described can be used for the measurement of vacuum conditions. For example Bourdon tube instruments may be obtained which measure pressures down to 20 mm mercury and diaphragm instruments may be obtained which measure pressures as low as 1 mm mercury. This section, however, will be concerned with instruments available for the measurement of pressures below 1 mm mercury absolute.

The region below 1 mm mercury may be conveniently split into three sections as follows:

(*a*) Medium high vacuum: 1 to 10^{-3} mm mercury
(*b*) High vacuum: 10^{-3} to 10^{-7} mm mercury
(*c*) Ultra high vacuum: 10^{-7} mm mercury and lower.

The Macleod Gauge

The basic elements of the Macleod Gauge are shown in Fig. 8.21. The capillary A is sealed at the top and the limbs B and C are connected to the vacuum system. A and B are the same diameter and are capillary tubes while C is of wider bore and helps to decrease errors due to capillary.

Initially the moveable reservoir is lowered until the mercury column drops below the level of the opening W. At this stage all the three limbs are connected to the pressure to be measured. The reservoir is then raised until the mercury fills the bulb and rises in the capillary A such that the gas in A is compressed. In practice the mercury in B is always raised to the same level as the top of the capillary A, which is the zero on the scale. The gas in limb A is compressed according to Boyle's Law.

If V is the volume of the bulb and limb A above the level W, p the pressure of the gas in the system connected to B and C, v the volume of the gas in the sealed limb after compression, and p_1 the pressure of the gas in this limb after compression, then

$$pV = p_1 v \qquad \text{where } p \text{ and } p_1 \text{ are measured in mmHg}$$

Also

$$v = ah \qquad \text{and} \qquad h = p_1 - p$$

where a is the cross-sectional area of the limb A and h is the difference in levels between A and B

Substituting

$$p = \frac{ah^2}{V - ah}$$

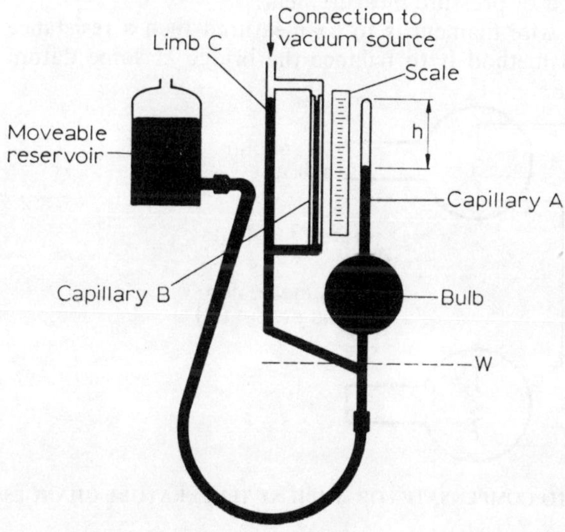

Fig. 8.21 THE MACLEOD GAUGE

For the measurement of low pressures the value of V should be large compared with a. The ratio V/a is termed the compression ratio and in practice its upper limit is restricted due to the fact that if a is made too small the mercury tends to stick in the capillary tube and if V is made too large an excessive weight of mercury may be involved.

The range of the Macleod Gauge is from 10 to 10^{-4} mm mercury and it is frequently used as a calibration device for other high vacuum measuring equipment. A serious disadvantage of the Macleod Gauge is that if the gas whose pressure is being measured contains any condensable vapours then Boyle's Law is not obeyed and errors may occur.

The Pirani Gauge

The Pirani Gauge (Fig. 8.22) operates on the principle that if a heated wire is placed in a chamber of gas the thermal conductivity of the gas is pressure-dependent and therefore the transfer of energy from the wire to the gas is

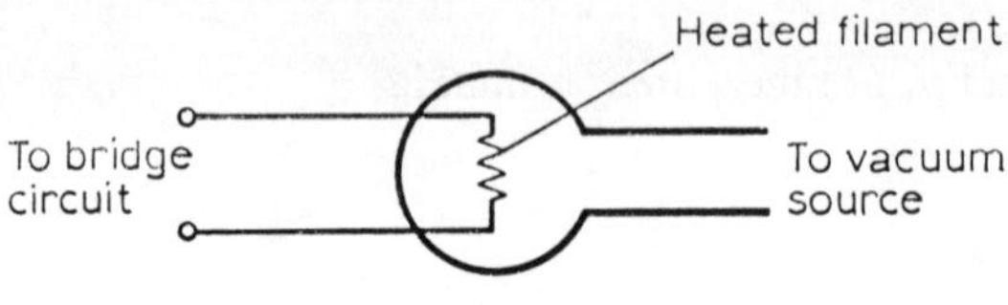

Fig. 8.22 PIRANI GAUGE

proportional to the gas pressure. If the supply of heating energy to the wire is kept constant and the pressure of the gas is varied then the temperature of the wire will alter and is therefore a method of pressure measurement. Alternatively if the temperature of the wire alters so does the resistance thus providing a second method of pressure measurement.

If the resistance of the wire filament is to be measured then a resistance bridge is used. The usual method is to balance the bridge at some datum

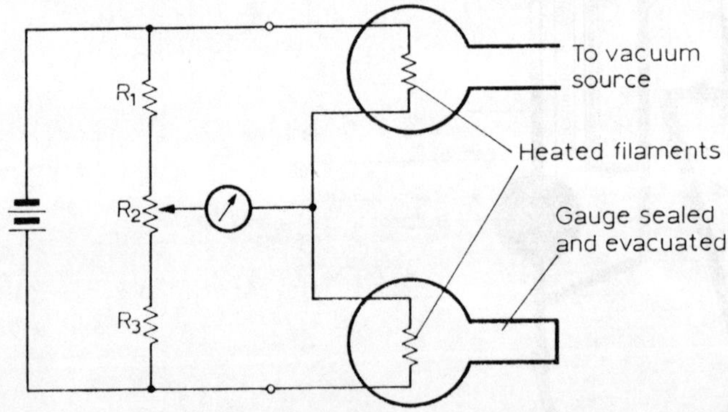

Fig. 8.23 ARRANGEMENT TO COMPENSATE FOR AMBIENT TEMPERATURE CHANGES

pressure and use the out-of-balance currents at all other pressures as a measure of the relative pressures.

The heat loss from the filament is also a function of the ambient temperature and compensation for this effect may be achieved in practice by connecting two gauges in series as shown in Fig. 8.23. The measurement gauge is first evacuated and both it and the sealed gauge are exposed to the same environment conditions. The bridge circuit is then adjusted through the resistor R_2 to give a null condition. When the measurement gauge is exposed to the test pressure conditions the deflection of the bridge from the null reading will be independent of changes in ambient temperature.

The range of the Pirani gauge is from 10^{-4} mm to 5 mm of mercury approximately.

Thermionic or Hot Cathode Ionization Gauge

The gauge consists of a thermionic triode assembly contained in a glass envelope which may be connected to the system whose pressure is to be measured. A typical assembly is shown in Fig. 8.24. The anode is at a negative potential and the grid is at a positive potential with respect to the filament. The electrons emitted from the filament collide with gas molecules and ionize them; the positive ions then travel to the negatively charged anode. An ionization current will now travel round the anode circuit, this current being a measure of the abolute gas pressure in the gauge. Ultimately electrons will

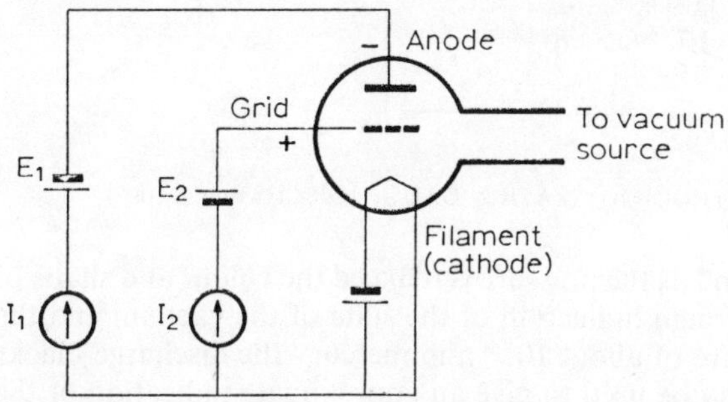

Fig. 8.24 THERMIONIC OR HOT CATHODE IONIZATION GAUGE

reach the positively charged grid and form an electron current round the grid circuit. The following approximate relation holds:

$$p = kc\, I_1/I_2$$

where p = absolute pressure
c = a probability factor
k = a constant depending on the potentials used in the gauge, and the geometry of the electrodes
I_1 = ionization current (anode circuit current)
I_2 = electronic current (grid circuit current).

Provided therefore that the potentials and the electronic current are not altered the absolute gas pressure is proportional to the ionization current.

In the triode arrangement shown in Fig. 8.24 the filament is either pure tungsten, thoriated tungsten or platinum alloy coated with barium and strontium oxides. The grid is normally in the form of a cylindrical molybdenum wire helix surrounding the filament. The anode is usually an open-ended nickel cylinder surrounding the grid but in one design it consists of flat nickel plates parallel to the flattened sides of the helix.

A disadvantage of the hot ionization gauge is the tendency to produce x-rays when ionizing electrons strike the grid. The x-rays cause a secondary emission at the anode resulting in a current in the anode circuit in the same direction as that due to the ions. Because of this disadvantage the lower limit of the gauge is usually restricted to 10^{-8} mm mercury.

The Cold Cathode Ionization Gauge

In its most simple form this gauge consists of a discharge tube comprising a glass tube, with two sealed-in electrodes connected to the vacuum system (Fig. 8.25). When a high voltage is applied to the electrodes a discharge

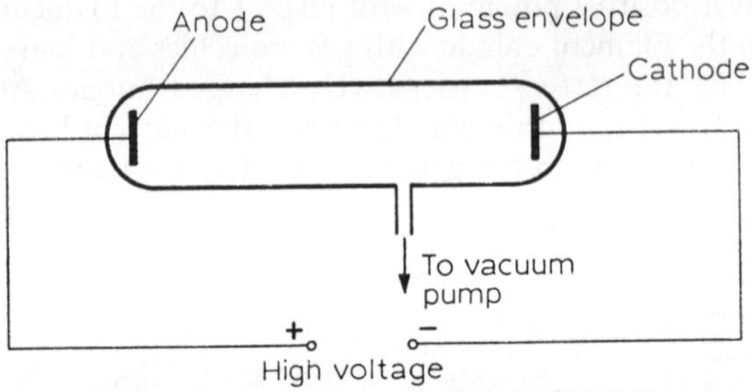

Fig. 8.25 SIMPLE COLD CATHODE IONIZATION GAUGE (DISCHARGE TUBE)

occurs at a low pressure and as the pressure is reduced the colour and shape of the discharge can give a rough indication of the state of the vacuum and the gases present. At a pressure of about 10^{-2} mm mercury the discharge blacks out and this condition may be used to give an approximate indication of the vacuum present.

A more sophisticated development is the *Penning gauge* in which two electrodes, an anode and a cathode, are contained in a glass envelope (Fig. 8.26). In the simplest design the cathode consists of two parallel flat plates

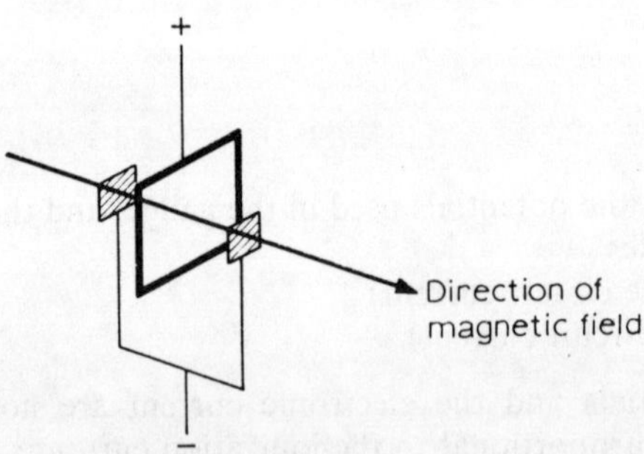

Fig. 8.26 TYPICAL ARRANGEMENT OF THE ANODES AND CATHODES FOR THE PENNING GAUGE

whereas the anode consists of a parallel loop of wire between them. The gauge is placed between the poles of an electromagnet so that the magnetic field is at right angles to the plates. Electrons emitted from the cathode move towards the loop anode in helical paths. The path length is increased by a large amount and the probability of collisions between electrons and molecules of gas to form ions is far greater than for the triode type of ionization gauge. As a result a relatively large discharge current is obtained for a moderately high voltage.

A typical calibration curve of pressure versus meter reading is shown in Fig. 8.27. It can be seen that the graph is linear for the lower pressures but above about 10^{-4} mm mercury non-linearities occur.

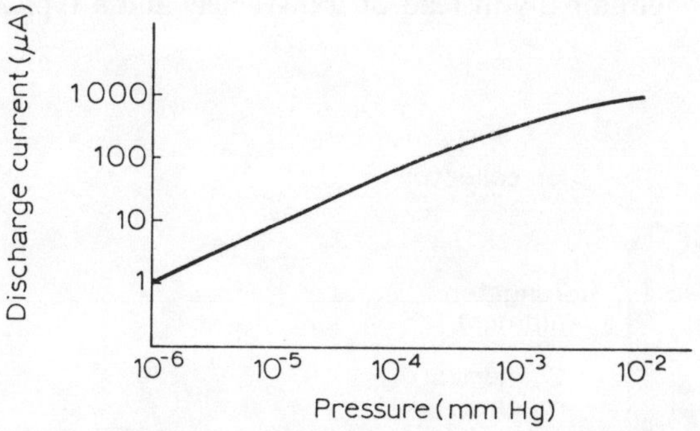

Fig. 8.27 TYPICAL CALIBRATION CURVE FOR THE PENNING GAUGE

The Cold Cathode Magnetron Gauge

A diagram of a typical gauge of this type is shown in Fig. 8.28. The cathode consists of two circular metal discs, the centres of which are joined together by a thin cylinder. The cylinder acts as an electron source. The anode is a metal cylinder placed co-axially to the thin cylinder joining the discs. The

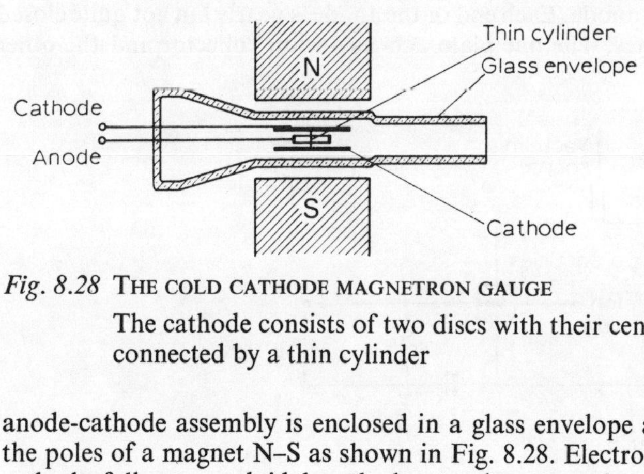

Fig. 8.28 THE COLD CATHODE MAGNETRON GAUGE
The cathode consists of two discs with their centres connected by a thin cylinder

anode-cathode assembly is enclosed in a glass envelope and placed between the poles of a magnet N–S as shown in Fig. 8.28. Electrons emitted from the cathode follow a cycloidal path due to the magnetic and electrical fields. They collide many times with gas molecules thus producing ions which travel straight to the anode. A current is thus set up and follows a linear relationship with pressure in the range $10^{-4} - 5 \times 10^{-1}$: mm mercury.

$$I = Kp^{1 \cdot 7}$$

where I = ionization current

K = a constant $\quad p$ = absolute pressure.

The Hot Cathode Magnetron Gauge

The object of the hot cathode magnetron gauge is to increase the path length of the electrons and hence increase the probability of producing ions. The magnetic field is applied longitudinally instead of transversely and a typical

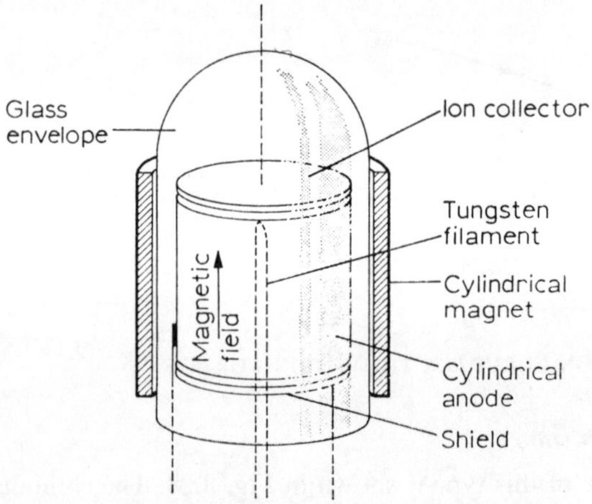

Fig. 8.29 THE HOT CATHODE MAGNETRON GAUGE

gauge is shown in Fig. 8.29. The tungsten filament is surrounded by a cylindrical molybdenum anode. Each end of the anode is nearly but not quite closed by two circular plates. The one plate acts as an ion collector and the other as a shield.

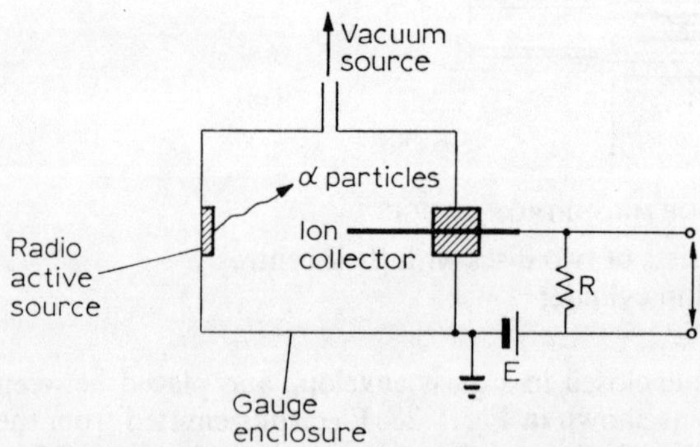

Fig. 8.30 THE ALPHA PARTICLE IONIZATION GAUGE (ALPHATRON)

The assembly is enclosed in a glass envelope and a cylindrical magnet slides over the envelope. The magnetic flux density is about 25×10^{-3} tesla and the normal operating voltages are -45 V on the ion collector and -10 V on the shield. The ion current has a linear relationship with absolute pressure between the limits 4×10^{-14} and 5×10^{-8} mm mercury.

The Alpha Particle Ionization Gauge (Alphatron)

Alpha particles emitted by a weak radioactive source collide with gas molecules to produce ions. The gauge contains a collector plate and grid assembly and positive ions are attracted towards the grid. An ionization current is created which is proportional to the gas pressure within the gauge. A schematic

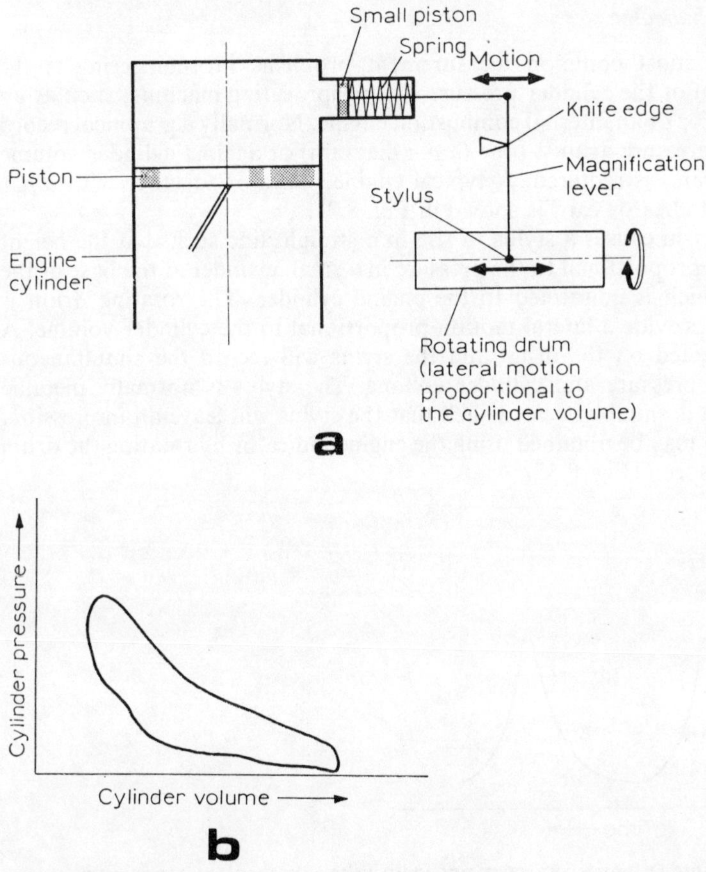

Fig. 8.31 THE MECHANICAL ENGINE INDICATOR

diagram is shown in Fig. 8.30. The ionic current is of the order of 2×10^{-10} A/mm mercury absolute pressure. At the lower pressures therefore the ionic current will be very small resulting in the need for complex d.c. amplification. A limit is placed on the lower end of the range therefore and the gauge is normally used within the range 10 mm to 10^{-3} mm mercury.

8.6 The Measurement of Fluctuating Pressure

When a pressure is changing slowly it is usually possible to follow this variation by taking periodic readings. As the rate of pressure variation increases, however, the difficulty of making these measurements will increase and the usefulness of the normal static pressure gauge diminishes. As the rate of pressure variation approaches the lowest natural frequency of vibration of a static pressure gauge, readings become impossible and some form of high-speed recording instrument must be used.

The Engine Indicator

One of the most common measurement problems in engineering is the measurement of the cylinder pressure in a reciprocating machine, such as an air compressor or an internal combustion engine. Normally a graphical record of cylinder pressure against time (a *p–t* diagram) or against cylinder volume (a *p–v* diagram) is required. A typical engine indicator together with a *p–v* diagram or indicator card is shown in Fig. 8.31.

A mechanism causes a stylus to rise in a straight line such that the height of the rise is proportional to the pressure in a small cylinder at the base of the indicator which is connected to the engine cylinder. The rotating drum is designed to provide a lateral motion proportional to the cylinder volume. A card is mounted on the drum and the stylus will record the simultaneous variation of pressure and cylinder volume. The stylus is normally metallic and the card is suitably treated such that the stylus will leave an impression. A *p–t* record may be obtained using the engine indicator by rotating the drum at constant speed (Fig. 8.32).

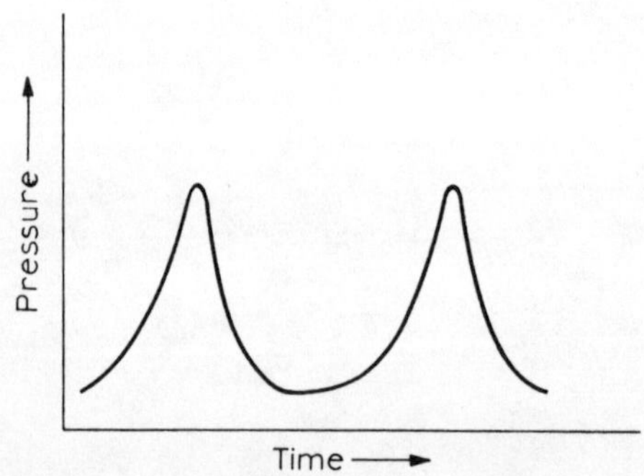

Fig. 8.32 TYPICAL PRESSURE/TIME RELATIONSHIP FOR ENGINE INDICATOR

The engine indicator described above has two major disadvantages:

(1) The serious problem of mechanical inertia.
(2) The effective volume of an engine cylinder will be changed when the engine indicator is connected. With small engines this increase in cylinder volume may be so great that *p–t* diagrams using a mechanical indicator are almost worthless.

Generally speaking electrical pressure transducers are used for recording *p–t* relationships at high speed. Electrical transducers may be divided into active and passive transducers. An active transducer is one which generates its own electrical output as a function of the mechanical displacement whereas a passive transducer requires an auxiliary electrical input which it modifies as a function of the mechanical displacement, the modified electrical signal being the output from the passive transducer.

Active Electrical Pressure Transducers

The only active electrical pressure transducer in common use is the piezoelectric transducer. This transducer operates on the principle discussed in Chapter 6 (pp. 174–75). Piezoelectric pressure pickups have the crystal

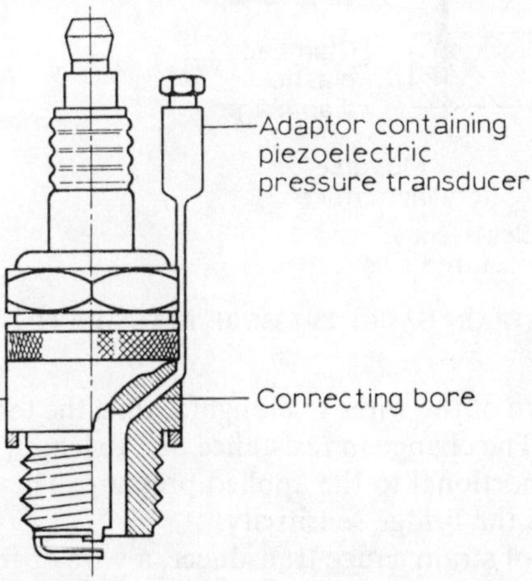

Fig. 8.33 PIEZOELECTRIC PRESSURE TRANSDUCER

geometry arranged to give maximum piezoelectric response in a desired direction with little or no response in other directions. Fig. 8.33 shows a piezoelectric pressure transducer of the Kistler type where the piezoelectric crystal is contained in a spark plug adaptor. The adaptor may be fitted to a standard type of spark plug and thus the combustion pressures of petrol engines may be easily measured.

Passive Electrical Pressure Transducers

Passive electrical pressure transducers may be conveniently subdivided under the following headings:

(1) Variable resistance types
(2) Capacitance types
(3) Variable inductance types.

Variable resistance transducers operate on the principle that the electrical resistance of a wire varies as the length of the wire is varied. In the unbonded strain gauge transducer four wires are connected between a fixed frame and a movable armature as shown in Fig. 8.34. The wires are located to the frame and movable armature by electrically insulated pins. The wires are installed under an initial tension and form the active legs of a conventional bridge circuit. When pressure is applied to the elastic element the movable armature

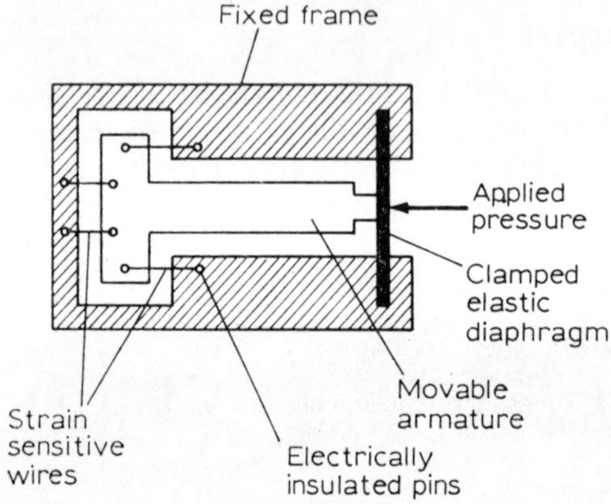

Fig. 8.34 UNBONDED STRAIN GAUGE PRESSURE TRANSDUCER

is displaced causing two of the wires to elongate while the tension in the other two wires is reduced. The change in resistance due to length changes causes a bridge unbalance proportional to the applied pressure. The use of four wires considerably increases the bridge sensitivity.

In the bonded type of strain gauge transducer, a wire or foil strain gauge is fastened by a suitable cement to a flexible plate that takes the load of the

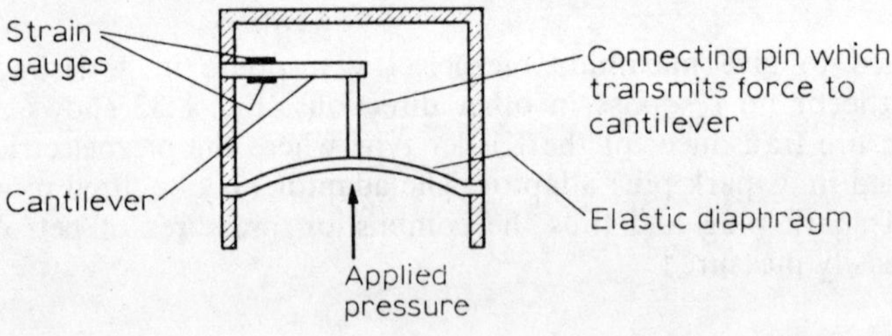

Fig. 8.35 BONDED STRAIN GAUGE PRESSURE TRANSDUCER

elastic element. A typical bonded strain gauge device is shown in Fig. 8.35. Normally two strain gauge elements are connected to the bridge measuring circuit so as to achieve temperature compensation (see Chapter 5).

Other variable resistance type pressure transducers operate on the principle of movable contacts such as those found in potentiometers or slide-wire rheostats. In a typical arrangement the elastic element, say a helical bourdon tube, is connected to a wire wound potentiometer. As pressure is applied to the open end of the bourdon tube the tube unwinds and causes the wiper, which is connected to the closed end of the tube, to move over the potentiometer thus varying the resistance of the measuring circuit (Fig. 8.36).

The *variable capacitance* type pressure transducer usually uses a metal diaphragm as the elastic element which serves as one plate of a capacitor. If

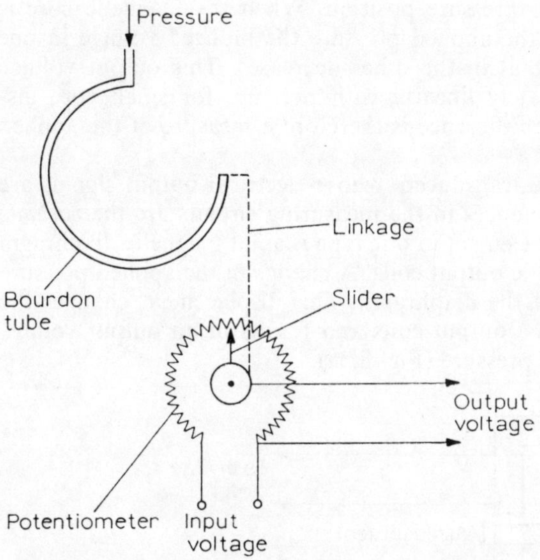

Fig. 8.36 POTENTIOMETRIC PRESSURE TRANSDUCER

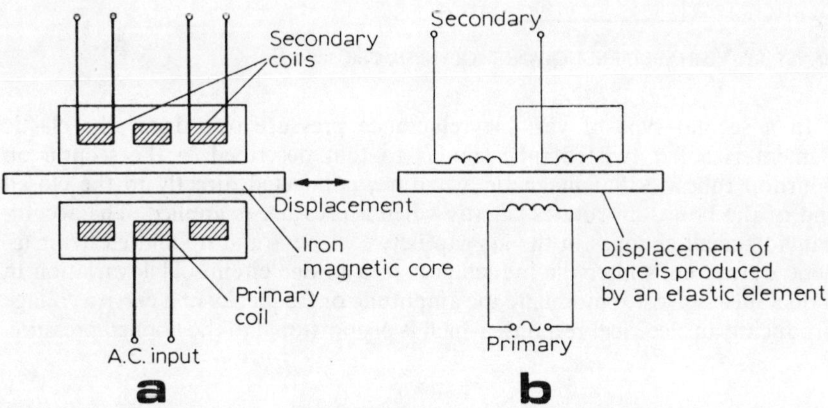

Fig. 8.37 THE LINEAR VARIABLE DIFFERENTIAL TRANSFORMER

the pressure is varied the diaphragm moves with respect to a fixed plate and the resulting change in capacitance can be measured by means of a suitable bridge circuit.

An important *variable inductance* type pressure transducer is the linear variable differential transformer type shown in Fig. 8.37. The electrical element is made up of three coils mounted in a common frame. A magnetic core centred within the coils is displaced by means of an elastic element. The centre coil is the primary winding of the transformer and the two outside coils form the secondaries. When the core is centred the induced voltages in these two outer coils are equal and 180° out of phase, this situation being taken to represent the zero pressure position. When the magnetic core is displaced by the action of the applied pressure the induced voltage in one secondary increases while that in the other decreases. This output voltage difference varies approximately linearly with pressure for small core displacements and this voltage difference is therefore a measure of the applied pressure.

Another class of pressure transducers whose electrical output signals are derived from variable inductances in the measuring circuits are the *variable reluctance* types. The elastic element in one type is a flat magnetic diaphragm located between two magnetic output coils. A change in the applied pressure produces a displacement of the diaphragm. This displacement changes the inductance ratio between the output coils and results in an output voltage proportional to the applied pressure (Fig. 8.38).

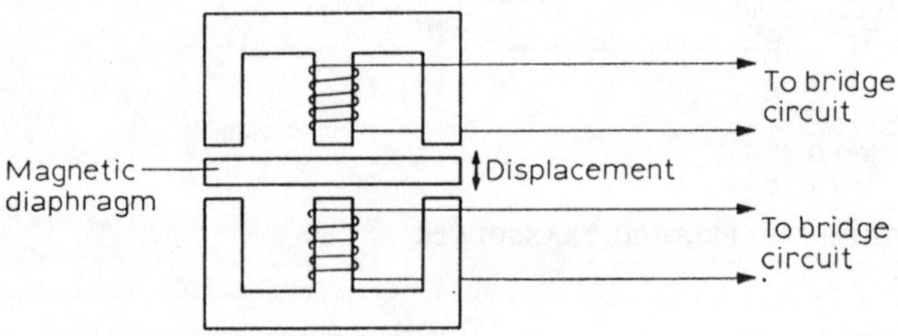

Fig. 8.38 VARIABLE RELUCTANCE TRANSDUCER

In a second type of variable reluctance pressure transducer the elastic element is a flat twisted tube similar to that described in the section on bourdon tubes. A flat magnetic armature, connected directly to the closed end of the bourdon, rotates slightly when a pressure is applied. The accompanying small changes in the air gap between the armature and electromagnetic output coils alter the inductances in a bridge circuit. This variation in inductance is used to modulate the amplitude or frequency of a carrier voltage producing an electrical response which is proportional to the applied pressure.

9 Measurement of Fluid Flow

In practice the most widely used flowmeter is the constriction-type of meter which depends on the measurement of differential pressure when fluid flows through a constriction. Before discussing this type of meter it is necessary to revise fluid flow theory.

9.1 Streamline or Laminar Flow

If the flow along a straight, smooth-walled pipe of uniform cross-section is considered and the paths of all the particles of fluid are parallel to the walls

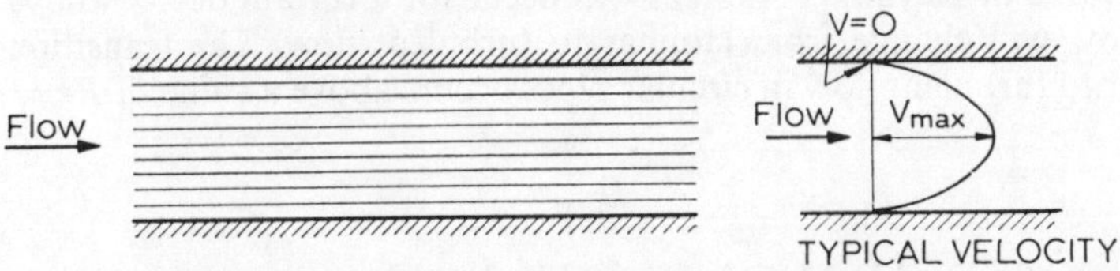

Fig. 9.1 STREAMLINE OR LAMINAR FLOW IN PIPES

of the pipe, then the flow is described as streamlined or laminar flow. All particles must have the same streamwise direction, but not necessarily the same magnitude of velocity. This type of flow occurs generally in smooth pipes when the velocity of flow is low and also in liquids having a high viscosity.

9.2 Turbulent Flow

If the flow velocity is now increased, some particles of fluid will have both streamwise and transverse velocity components and the flow pattern will

Fig. 9.2 TURBULENT FLOW IN PIPES

change. This velocity across the stream may produce violent swirls and eddies if the velocity of flow is high enough and such flow is called turbulent flow.

9.3 Reynolds Number

The flow through a certain device may be estimated by comparing it with the flow through a geometrically similar device provided the flow patterns are similar in both. In these circumstances the systems are said to be dynamically similar and this will occur provided the Reynolds Number is the same in both devices. The Reynolds Number *Re* is a dimensionless index and is defined by the equation:

$$Re = dV\rho/\eta$$

where d is the diameter at the plane of minimum cross-sectional area or "throat" of the device.
- V is the mean velocity at the throat
- ρ is the density of the fluid flowing
- η is the absolute viscosity of the fluid.

A critical value of Reynolds Number will occur for a certain device above which the flow will change from laminar to turbulent flow. The transition from laminar to turbulent flow in circular pipes occurs above a value of $Re =$ 2000.

9.4 The Energy of a Fluid in Motion: Bernoulli's Equation

Bernoulli's theorem states that the total energy of each particle of a fluid in motion remains constant provided no energy enters or leaves the system at any point, i.e.

$$H = Z + \frac{P}{\rho g} + \frac{V^2}{2g} = k$$

where H is the total pressure head (energy/unit weight, i.e. J/N)
- Z is the potential head or the energy the fluid has by virtue of its height above some fixed datum (J/N)
- $P/\rho g$ is the static pressure head or the energy the fluid has by virtue of its pressure (J/N)
- $V^2/2g$ is the velocity head or energy due to the kinetic energy of the fluid in motion (J/N)
- k is a constant.

Consider the laminar flow of a fluid through a smooth pipe as shown in Fig. 9.3.

The conditions at sections 1–1 and 2–2 can be represented by the following symbols:

	At section 1–1	*At section* 2–2	*Units*
Area	A_1	A_2	m^2
Mean velocity	V_1	V_2	m/s
Pressure	P_1	P_2	Pa
Density	ρ_1	ρ_2	kg/m^3
Height of section centroid above datum	Z_1	Z_2	m

Applying Bernoulli's theorem to the fluid,

$$Z_1 + \frac{V_1^2}{2g} + \frac{P_1}{\rho_1 g} = Z_2 + \frac{V_2^2}{2g} + \frac{P_2}{\rho_2 g}$$

The equation may be applied to the steady flow of incompressible fluids in pipes. It is also sometimes referred to as the Steady Flow Energy Equation

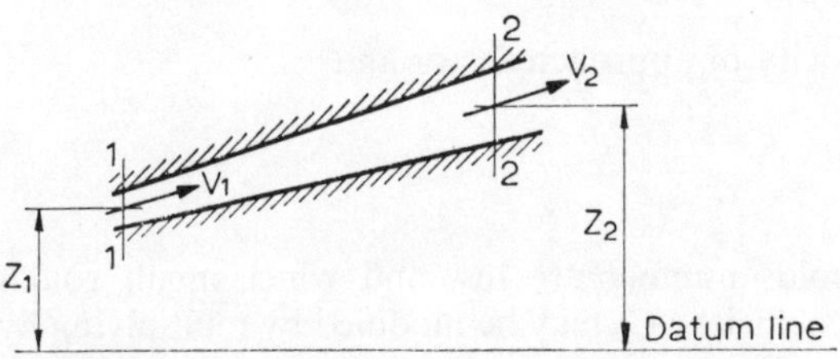

Fig. 9.3

provided there is no interchange of heat and work energy. In the case of incompressible flow (i.e. most liquids) the density remains constant. Therefore in the above equation, $\rho_1 = \rho_2$.

9.5 Constriction Flowmeters

The velocity of the flowing stream increases as it passes into the constriction and reaches a maximum in the plane of minimum cross-section of the flowing stream.

Because of the increase of velocity of the flowing stream at the plane of minimum cross-section, the kinetic energy will increase. Therefore if the total energy is to remain constant there must be a corresponding decrease in potential energy and a decrease in the pressure of the fluid.

Downstream of the plane of minimum cross-section the pressure on the pipe wall will increase as the area of the flowing stream increases and the velocity of the fluid falls to its original value. The pressure, however, does not quite regain its original value and the difference is known as the net pressure loss. The net pressure loss is due to the dissipation of energy as heat in the damping of turbulent eddies by internal friction.

Consider the streamline flow of an incompressible fluid through a constriction device as shown in Fig. 9.4.

If V_1, the velocity upstream of the constriction, is small compared with the velocity at the constriction V_2, then V_1 may be ignored. Applying Bernoulli's equation it may be shown that

$$Q = C_d A_2 \sqrt{(2gh)}$$

where Q is the volume flow rate (m^3/s)
C_d is the discharge coefficient

= actual volume flow rate/theoretical volume flow rate

A_2 is the area of minimum cross-section of the device
h is the differential pressure head measured by a U-tube manometer.

When the "velocity of approach" V_1 cannot be ignored compared with V_2 the volume rate may be calculated using the formula

$$Q = C_d E A_2 \sqrt{(2gh)}$$

where E is known as the velocity of approach factor and

$$E = \sqrt{\frac{1}{[1 - (A_2/A_1)^2]}}$$

When the values of Reynolds number are low and when small, rough pipes are used the discharge coefficient C_d may be modified by multiplying by

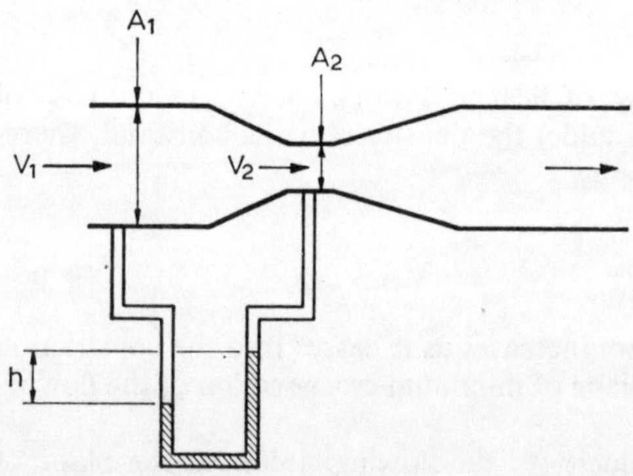

Fig. 9.4 FLOW OF FLUID THROUGH A CONSTRICTION DEVICE
A_1 Cross-sectional area upstream of the constriction
A_2 Cross-sectional area of the constriction

a correction factor z. The value of z depends on the area ratio, the Reynolds number, and the size and roughness of the pipe. Values of C_d and z for various constriction devices are tabulated in B.S. 1042, Part I 1964.

The above equation for volume flow can only be applied to incompressible fluids. When, however, the flow is compressible, as in the case of a gas, the equation is further modified by introducing an expansion factor ε.

9.6 Critical Flow for Compressible Fluids

A simple convergent tube, i.e. a nozzle, may be used as a flow-limiting device because when the flow rate is sufficiently high the pressure differential between the throat and the upstream pressure becomes quite large, and eventually sonic flow conditions may be achieved at the minimum flow area (throat). The flow is said to be "choked" under these conditions and the flow rate takes on its maximum value for the given inlet conditions. If the flow of an ideal gas is considered and isentropic flow is assumed the critical pressure ratio for this choked condition is

$$(P_2/P_1)_{crit} = \left(\frac{2}{\gamma + 1}\right)^{\gamma/\gamma+1}$$

where (P_2/P_1) is the ratio of the critical pressure at the throat to the upstream nozzle pressure

and $\gamma = c_p/c_v$

where c_p is the specific heat of the gas at constant pressure

c_v is the specific heat of the gas at constant volume.

9.7 Types of Constriction Device

The Orifice Plate

An orifice plate consists of a plate having an axial hole with a square edge on the upstream side and a bevel on the downstream side. The orifice plate is usually inserted between pipe flanges as shown in Fig. 9.5 and the pressure tappings may be in one of the following positions (Fig. 9.6):

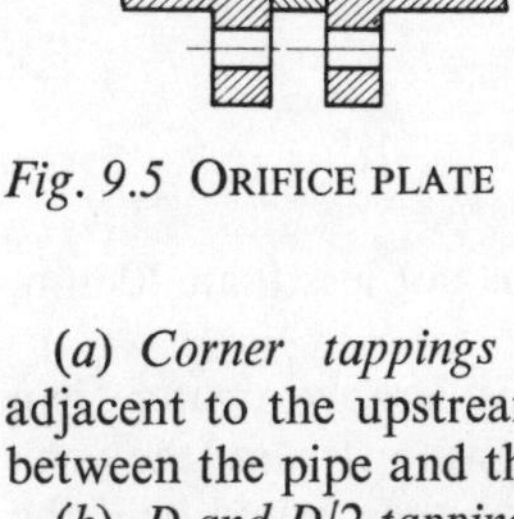

Fig. 9.5 ORIFICE PLATE

(*a*) *Corner tappings* The pressure tappings are located immediately adjacent to the upstream and downstream faces of the plate in the corners between the pipe and the plate.

(*b*) *D and D/2 tappings* The tappings are located at a distance of one pipe diameter upstream and one half of one pipe diameter downstream of the plate.

(*c*) *Flange tappings* The tappings are located in the pipe flanges at a distance of 25 mm from the upstream and downstream faces of the orifice plate.

Two further orifice plates used to meter viscous liquids are the conical entrance orifice plate and the quarter-circle plate. The former consists of a plate having an axial hole with a bevelled edge on the upstream side and a

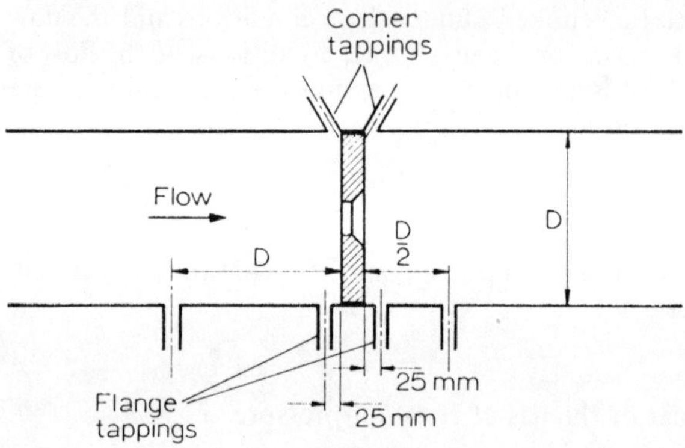

Fig. 9.6 Position of pressure tappings for an orifice plate

square edge on the downstream side. The pressure tappings are the same as for an orifice plate with corner tappings (Fig. 9.7*a*).

The quarter-circle plate consists of a plate having an axial hole with a rounded edge on the upstream side and a bevelled edge on the downstream side (Fig. 9.7*b*). The pressure tappings are the same for an orifice plate

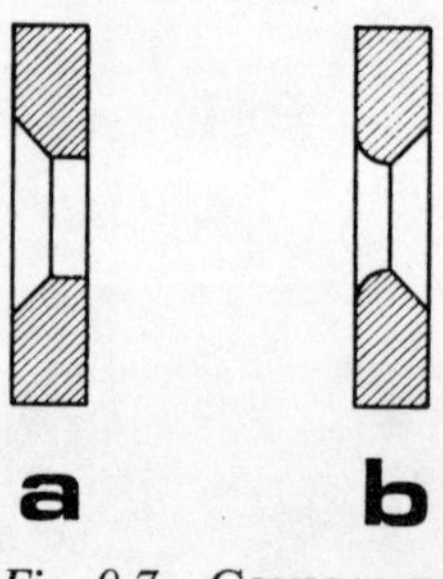

Fig. 9.7a Conical entrance orifice plate
b Quarter-circle orifice plate

with corner tappings or, provided the pipe diameter is not less than 40 mm, with flange tappings.

Since the cross-sectional area of the orifice is considerably greater than that at the plane of minimum cross-section of the flowing stream (the vena contracta) the values of C_d for orifices are smaller than they would be if the cross-sectional area of the vena contracta was used to calculate the flow rate. Typical values of C_d for orifice plates lie within the range 0·62–0·65.

Eccentric and Segmental Orifice Plates

When a fluid with suspended solid is encountered it is desirable to use a segment-type or eccentric orifice plate. The segment type (Fig. 9.8) has the circumference of the open part coincident with the pipe. Passage of solid

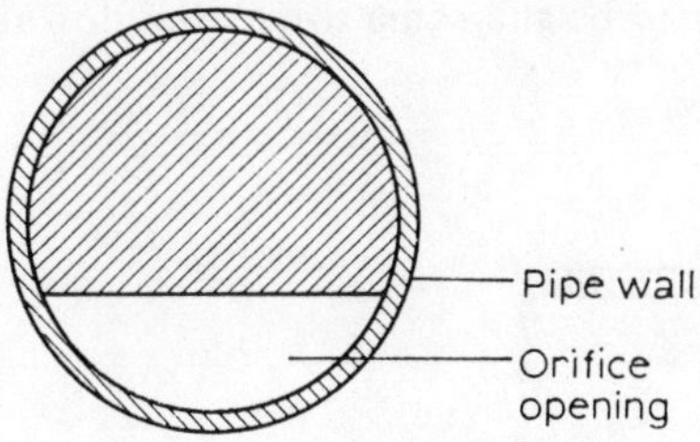

Fig. 9.8 THE SEGMENTAL ORIFICE PLATE

material is not interferred with and there is no build-up of solid matter against the upstream face of the orifice plate as would occur with the normal concentric orifice.

The eccentric orifice plate is similar to the segmental type in that the lower part of the orifice opening is flush with the lower part of the pipe (Fig. 9.9).

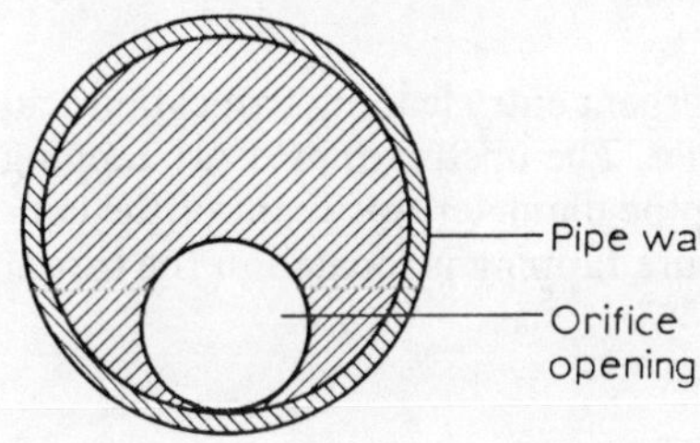

Fig. 9.9 THE ECCENTRIC ORIFICE PLATE

The Nozzle and the Venturi Nozzle

The standard flow nozzle (Fig. 9.10) consists of a bell-mouthed convergent entry leading to a short cylindrical throat which projects into the downstream

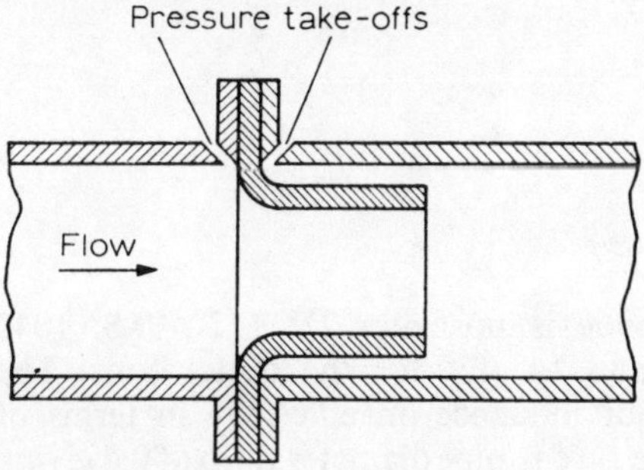

Fig. 9.10 THE FLOW NOZZLE

pipe. The pressure tappings are located immediately adjacent to the device in the corners and between the device and the pipe.

A refinement of the flow nozzle is the venturi nozzle which consists of a nozzle followed by a venturi outlet (Fig. 9.11). The upstream tapping is located as for a nozzle but the downstream tapping is located in the throat. The exit diameter of the venturi nozzle may be the same size as the downstream pipe or rather less (truncated).

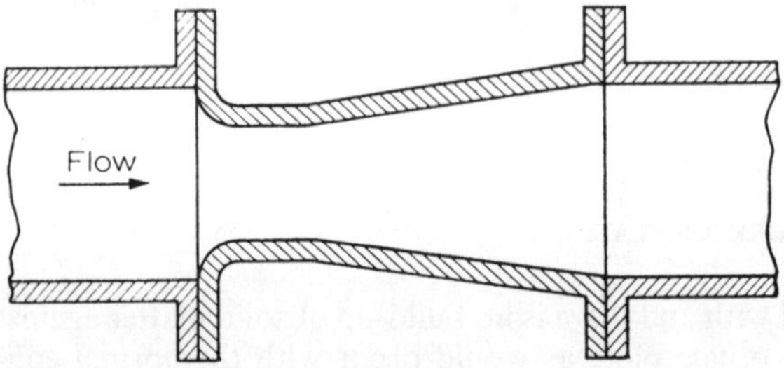

Fig. 9.11 THE VENTURI NOZZLE

The Venturi Tube

The venturi tube consists of a conical convergent entry leading to a cylindrical throat followed by a conical divergent outlet. The upstream pressure tapping is located at a distance of one half of one pipe diameter upstream of the convergent entry, while the downstream pressure tapping is located in the throat (Fig. 9.12*a*).

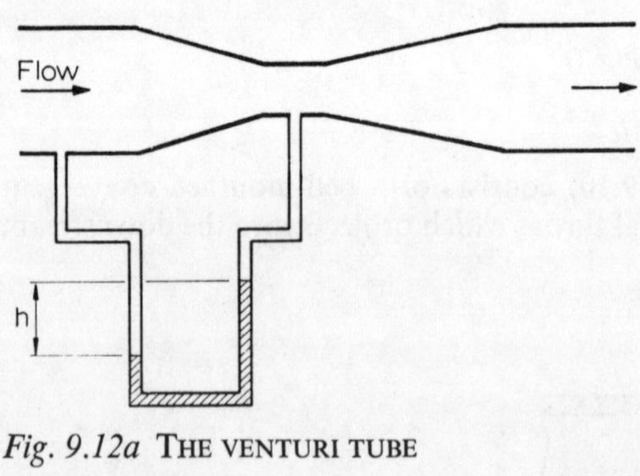

Fig. 9.12a THE VENTURI TUBE

The conical angle for the inlet cone is normally 21° ± 2°. B.S. 1042 specifies two conical angles, 5°–7° and 14°–15° for the outlet cone. The design of the outlet cone has a major influence on efficiency in terms of pressure loss. As an example consider a throat/pipe diameter ratio of 0·6 giving an area ratio of 0·36. Using a 5°–7° outlet cone the net pressure loss is about 11% of the differential pressure between inlet and throat. With the 14°–15° cone this loss is about 18%. These figures can be compared with about 65% for an orifice plate having the same area ratio.

The value of C_d for a venturi will approach unity, a typical value being 0·97.

The Dall Tube

The Dall tube is similar in principle to a venturi tube. It consists of two truncated cones separated by a narrow throat slot of length between 0·03*d* and 0·1*d* where *d* is the throat diameter (Fig. 9.12*b*).

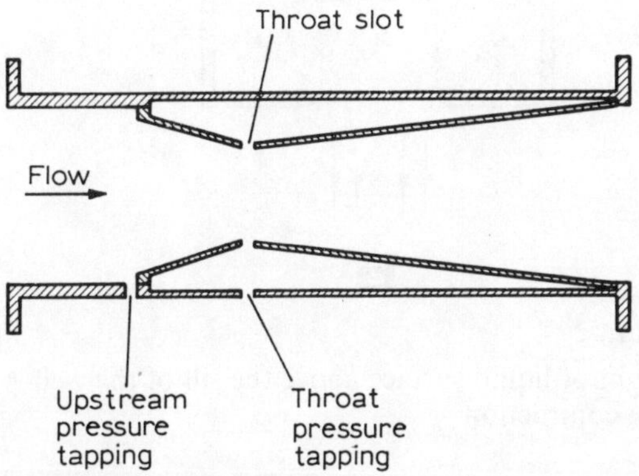

Fig. 9.12b THE DALL TUBE

The inlet cone has an included angle between 40° and 50° and the outlet cone an angle between 12° and 17°. The diameter of the inlet cone is less than that of the pipe producing a sudden step.

This step produces an impact pressure in addition to the static pressure at the step. The upstream pressure connection is made just before the step whereas the other connection is made at the throat. The Dall tube has the advantage of being much shorter than the normal venturi and the pressure loss may be as low as 5% of the differential pressure, whereas the pressure loss with a normal venturi would be between two and three times this value.

The Weir

Discharges over weirs are extensively used to determine the mean flow velocity for open channel flows of liquids, when the channel dimensions and the depth of the liquid are known. The mean velocity and hence the volume flow may be found by taking only one measurement, i.e. the height of liquid *h* over the sill of the weir (Fig. 9.13*a*).

Weirs are normally classified by their shape and the two most common are:

(a) The rectangular notch weir

A diagram of a typical rectangular notch is shown in Fig. 9.13. It can be shown that for a rectangular notch

$$Q = \tfrac{2}{3} C_d B \sqrt{(2gh^3)}$$

where Q is the volume flow rate (m^3/s)
C_d is the discharge coefficient of the weir
B is the breadth of the weir notch (m)
h is the height of the surface of the liquid above the sill of the weir (m).

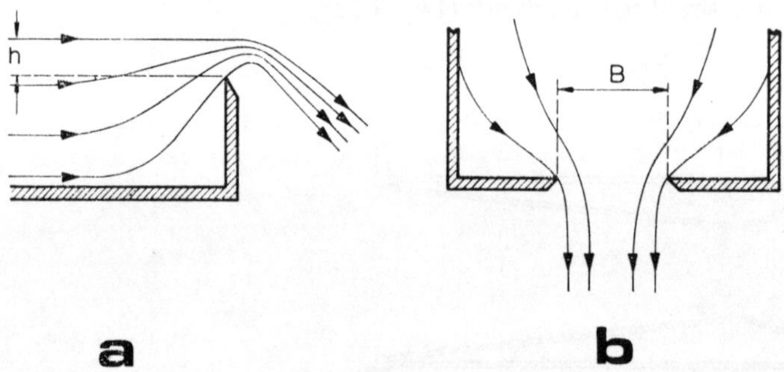

Fig. 9.13 RECTANGULAR NOTCH
a View showing height of liquid surface above the sill of the weir
b View showing side contraction

(b) The triangular or V-notch weir

A diagram of a typical triangular notch is shown in Fig. 9.14. It can be shown that the volume flow rate for a triangular notch is given by the following equation:

$$Q = \frac{8}{15} C_d \tan \tfrac{1}{2}\theta \sqrt{(2gh^5)}$$

where θ is the angle of the triangular notch.

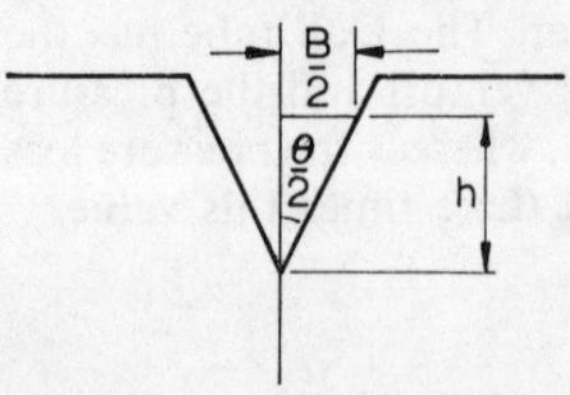

Fig. 9.14 TRIANGULAR NOTCH

9.8 The Selection of Constriction Devices

Weirs are used for open channel flow measurement, particularly when the fluid is flowing in large quantities. If the flow rate is sufficiently high and high orders of accuracy are not required, it may be convenient to measure the flow rate by passing the fluid over one of the standard weirs described above and applying the appropriate formula. V-notch weirs can be used to measure smaller flow rates than rectangular notch weirs, however, because the same flow rate gives a larger head of liquid over the weir than for a rectangular notch.

When measuring closed channel flow, the choice of device usually rests between an orifice plate, a nozzle or a venturi nozzle, and a venturi tube. The remainder of this section will therefore discuss factors which may influence the choice of one of these devices for a particular application.

A factor of prime importance is the initial cost of the device and in practice orifice plates are the most widely used due largely to their low initial cost and simplicity of construction. Orifice plates may also be easily installed by insertion in an existing flanged joint in the pipeline. Nozzles are very short and require only a small installation space whereas venturi tubes and venturi nozzles require an installation length of between two and six pipe diameters.

When initial cost is considered it is very important also to consider running costs on large installations due to "permanent head loss" or net pressure loss. The net pressure loss will be appreciable for orifices and nozzles where the fluid issues from the orifice or throat in the form of a jet, but very much smaller for venturi nozzles and venturi tubes where the fluid fills the whole cross-sectional area of the tube throughout its length. The loss depends on the angle of the divergent outlet and is greater for the shorter forms of venturi nozzle and venturi tube.

All the devices may be used for metering liquids, gases, and vapours, but only the conical entrance and quarter-circle orifice plates are suitable for metering viscous fluids flowing in small pipes (as small as 25 mm) at low Reynolds numbers (as low as 250 for conical entrance and 1000 for quarter-circle). Venturi tubes may be used for viscous liquids at Reynolds numbers down to 5000 but only in pipes greater than 50 mm internal diameter.

Except in the case of viscous fluids the lower limit of flow rate that can be measured is smallest for orifice plates with corner or D and $\frac{1}{2}D$ tappings. It is several times larger for nozzles and venturi tubes and several times larger still for venturi nozzles and orifice plates with flange tappings. There is no upper limit to the rate of flow which can be measured except in the case of devices designed for metering viscous fluids, the upper limit of Reynolds number for these devices being 200 000 for conical entrance and 100 000 for quarter-circle orifice plates.

The devices so far described are not capable of metering pulsating flow. However, pulsations in the system may be damped by providing a sufficient combination of volumetric capacity and throttling between the measuring device and the source of pulsations, which may be either upstream or downstream of the device. The volumetric capacity is provided by the volume of the pipeline and any receivers used, while throttling is provided by the frictional resistance of the pipeline, the pressure loss in bends and fittings, and the pressure loss in the metering device itself.

All the devices require a certain minimum length of straight pipeline upstream of the devices, this minimum length being longer the larger the area ratio of the device. The minimum length is therefore shorter for nozzles, venturi nozzles and venturi tubes than for orifice plates, assuming the same rate of flow in the same size of pipe at the same pressure difference. The minimum length may also be shorter if multiple pressure tappings are used instead of single tappings. The minimum length will normally be within the range 10 to 100 pipe diameters depending on the area ratio and the layout of the upstream pipeline.

The lower limit of pipe size through which fluid flow may be measured varies with the surface finish of the upstream pipe and the measuring device used. Provided, however, the internal finish of the upstream pipe is sufficiently smooth orifice plates may be used in pipes as small as 25 mm internal diameter, except where flange tappings are used when the lower limit is 50 mm. Nozzles, venturi nozzles and venturi tubes may be used in pipes down to 50 mm internal diameter. When the surface is less smooth the lower limit of pipe diameter is greater and tables of permissible lower limit of pipe size for varying stages of upstream pipe roughness are included in B.S. 1042, Part 1 1964.

If flow having a pressure difference greater than the critical is to be measured, all the devices are unsuitable except for the flow nozzle.

9.9 Fluid Velocity Meters

Fluid velocity meters may be classified as follows:

(1) Instruments which measure the local velocity at a point in the channel or duct through which the fluid is flowing.
(2) Instruments which measure the mean velocity.

9.10 Measurement of Local Velocity

The two most important instruments used for measuring local fluid velocity are:

(*a*) the pitot static tube
(*b*) the hot wire anemometer.

Both of these instruments are used mainly for the measurement of gas velocities, usually air.

The Pitot Static Tube

The pitot static tube combines total and static pressure tappings in one probe in order to find the local flow velocity. However, by traversing the stream and finding the velocity at several points it is possible to obtain the average stream velocity and hence, by multiplying by the cross-sectional area, the quantity of fluid flowing.

The instrument (Fig. 9.15) consists of two concentric tubes, the outer tube having a series of holes drilled at right angles to the direction of flow. The inner tube when placed in the fluid stream transmits the sum of static and dynamic pressures while the annular space between the tubes transmits the static pressure only. If the two pressures are connected across a U-tube manometer the difference between the total pressure head and the static pressure head can be measured. This pressure difference will be equal to the velocity head or pressure head produced by the loss of kinetic energy.

Now if the velocity of the fluid stream is V_1 the pressure head produced by the loss of kinetic energy of the fluid is $V_1^2/2g$.

Therefore $h = V_1^2/2g$ where h is the head, in metres, of the fluid flowing, measured by the manometer.

Therefore $V_1 = \sqrt{(2gh)}$.

The whole of the stream flowing on to the end of the pitot tube may not be brought to rest, as some may be deflected around the edge. The value of V_1

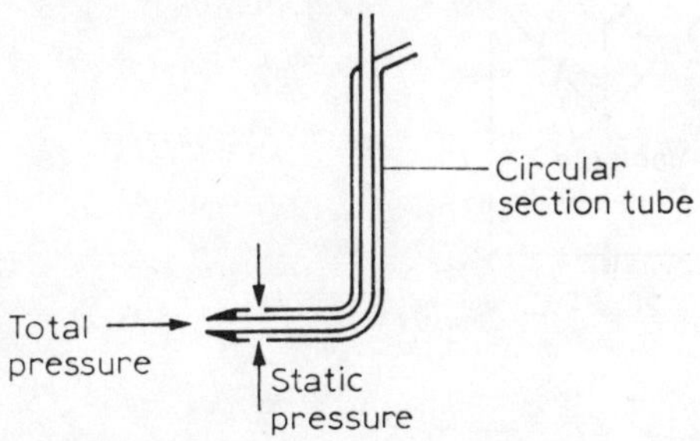

Fig. 9.15 PITOT STATIC TUBE FOR LOCAL VELOCITY MEASUREMENT

may therefore differ slightly from the true velocity and in order to compensate for this variation a coefficient C is introduced. The value of this coefficient will depend on the design of the tube and for a correctly designed tube this coefficient is unity.

The Hot Wire Anemometer

A very fine electrical resistance wire will be cooled when placed in an airstream. The rate of cooling will be a function of the temperature difference between the air and the object and will also be affected by the velocity of the air.

The wire, usually less than 0·025 mm diameter and approximately 25 mm long, is heated electrically as part of a "Wheatstone bridge circuit." It is cooled by the fluid flowing over it and the rate of cooling is proportional to the fluid velocity.

The hot wire anemometer is used in two ways:

(*a*) The wire is kept at a constant temperature by adjusting the current flowing through it.

(*b*) The voltage change in the Wheatstone bridge is recorded as the resistance of the wire varies.

The former is a null reading method and requires constant adjustment whereas the latter requires no resetting after an initial calibration curve of voltage against velocity has been plotted.

A typical hot wiré anemometer and bridge circuit is shown in Fig. 9.16 together with a calibration curve.

The wire used is usually made of platinum or platinum alloy but tungsten may be used when additional strength is required.

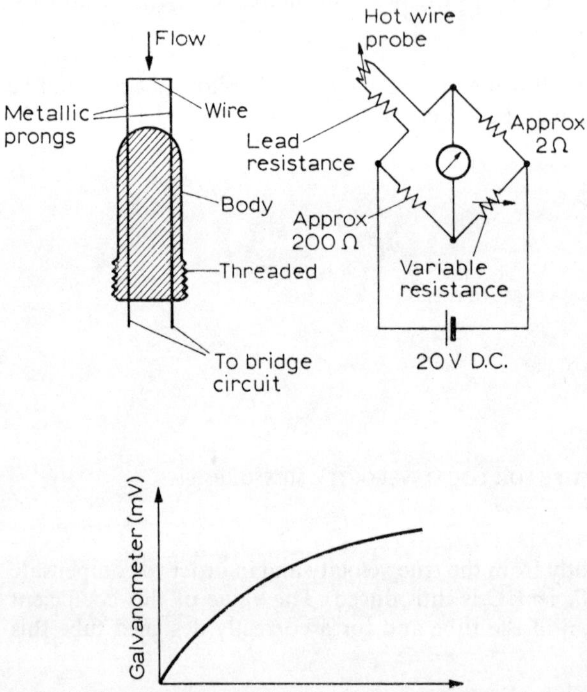

Fig. 9.16 HOT WIRE ANEMOMETER WITH TYPICAL BRIDGE CIRCUIT AND CALIBRATION CURVE

9.11 The Selection of a Local Velocity Measuring Instrument

Using a pitot tube it is possible to measure velocities as small as 1 m/s to an accuracy of about 0·5%. Above speeds of about 60 m/s in air, corrections must be made to the pitot tube reading to allow for compressibility effects, but even with supersonic flow fairly accurate results can be obtained by correcting the reading for pressure changes caused by a normal shock wave positioned at the entry to the total pressure tube.

The most common application of a pitot static tube is its use in aircraft when it is usually coupled to a Bourdon tube gauge calibrated in air-speed units (m/s) instead of pressure units.

The hot wire anemometer is widely used for measuring low flows as small as 0·03 m/s in turbulent regions. It can, however, also be used for measuring supersonic flows.

Generally the hot wire anemometer is a more expensive instrument than the pitot static tube. It is possible, however, where high accuracy and response is not essential to construct a hot wire probe for operation in air velocities as great as 60 m/s. The probe is constructed using a relatively thin piece of copper wire soldered to two darning needles which are mounted on a plastic stem. The ratio of the prong diameter to the wire diameter should be at least 10:1 in order to minimize the changes of resistance of the prongs as they are heated and cooled. This instrument is relatively inexpensive and can compete in accuracy and response with a pitot static tube.

9.12 Mean Velocity Meters

The most common devices for measuring mean flow velocity are:

(*a*) the cup anemometer
(*b*) the vane anemometer
(*c*) the current meter
(*d*) the turbine meter
(*e*) the reed anemometer.

The *cup anemometer* (Fig. 9.17) consists of three or four hemispherical hollow cups mounted in a horizontal plane with their lips vertical. As the flow

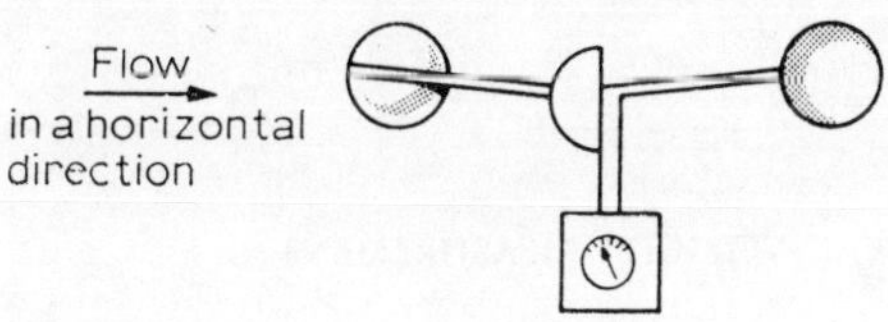

Fig. 9.17 Cup anemometer for mean velocity measurement

passes over the cups more drag force is experienced by the cup with its open end facing into the stream than the cups with their rounded faces towards the stream. The cup assembly is then caused to spin around its spindle which drives a geared dial and registers a reading approximately proportional to the flow velocity. The dial is calibrated in metres and a stop watch is used to calculate the velocity.

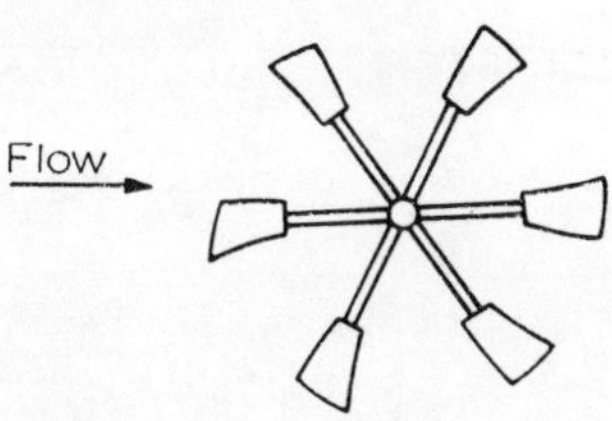

Fig. 9.18 Vane anemometer for mean velocity measurement

The *vane anemometer* (Fig. 9.18) consists of a small windmill which spins at different rates when subjected to varying velocities of flow. It is important that this instrument is correctly aligned to the flow direction if accurate readings are to be obtained.

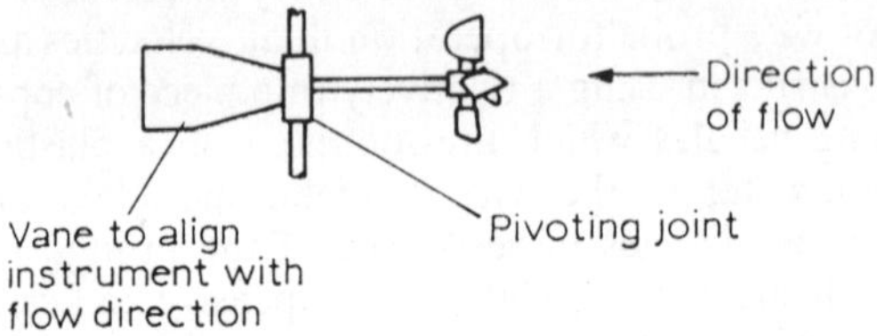

Fig. 9.19 THE CURRENT METER FOR MEAN VELOCITY MEASUREMENT

The *current meter* (Fig. 9.19) consists of a rotating vane and aligns itself to the direction of flow by means of a pivoting joint. The speed of rotation of the blades can be recorded electrically by means of a tachogenerator.

The *turbine meter* (Fig. 9.20) is similar in operation to a cup anemometer and depends on the application of an eccentric force applied by partly immersing the bladed rotor in the fluid flow.

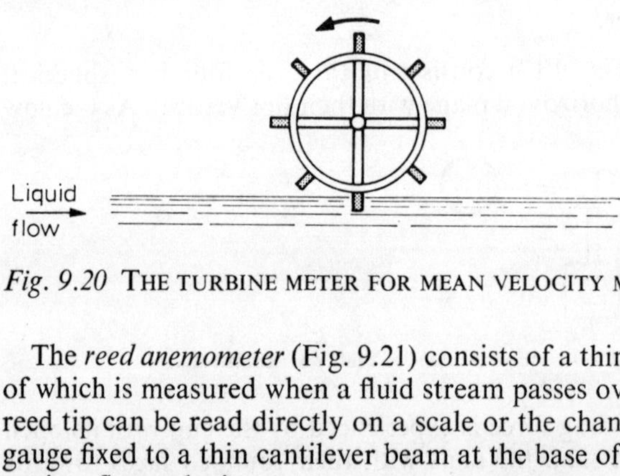

Fig. 9.20 THE TURBINE METER FOR MEAN VELOCITY MEASUREMENT

The *reed anemometer* (Fig. 9.21) consists of a thin metal reed, the deflection of which is measured when a fluid stream passes over it. The deflection of the reed tip can be read directly on a scale or the change in resistance of a strain gauge fixed to a thin cantilever beam at the base of the reed can be calibrated against flow velocity.

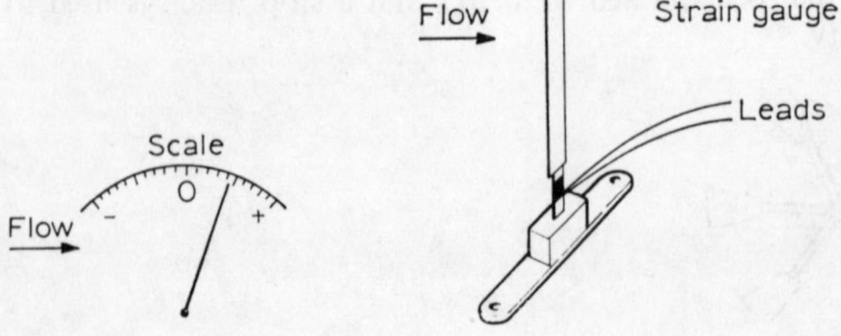

Fig. 9.21 REED ANEMOMETER FOR MEAN VELOCITY MEASUREMENT

9.13 The Selection of Mean Velocity Meters

The cup anemometer and vane anemometer are normally used for measuring the velocity of flow of gases. They may be selected to measure velocities in the ranges 0–15 m/s. The major application of the cup anemometer is the measurement of wind velocities in meteorology.

The current meter and turbine meter are normally used for measuring the flow velocity of liquids while the reed anemometer may be used for either liquids or gases.

9.14 Variable Area Flowmeters

If the pressure differential across a particular installation is constant and the flow pattern is unchanged, then the rate of flow through the orifice will depend on the orifice area and the "velocity of approach factor." That is

$$Q = C_d A_2 E \sqrt{(2gh)}$$

where A_2 is the cross-sectional area of the orifice.

If the pressure differential is held constant by adjusting the orifice area then the area of the orifice at any particular instant is effectively a measure of the rate of flow through the orifice. This is the principle of operation of variable area flowmeters of which the following are typical:

(*a*) the gate meter
(*b*) the orifice and plug meter
(*c*) the rotameter.

The Gate Area Meter

A typical gate flowmeter is shown in Fig. 9.22. The gate may be adjusted manually or by an automatically controlled electric motor to maintain a

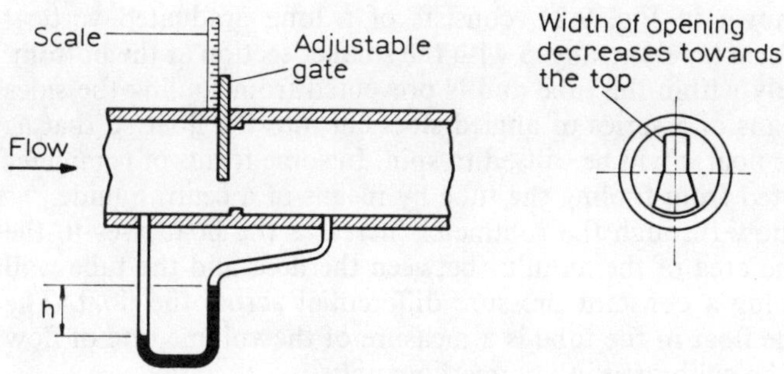

Fig. 9.22 VARIABLE AREA GATE METER

constant pressure drop across the orifice. The position of the gate is indicated by a scale which may be calibrated in terms of flow rate. It should be noted that the relationship between the flow rate and the area A_2 will not be strictly linear due to the fact that the velocity of approach factor will increase with an increase in A_2, i.e. the flow rate will increase more rapidly than the area A_2. Therefore, if the movement of the gate is to be directly proportional to the flow rate then the width of the orifice A_2 must decrease at the top.

The Orifice and Plug Area Meter

This meter is shown in Fig. 9.23. The tapered plug is of such a form that the area of the annular space between the orifice and the plug is proportional to

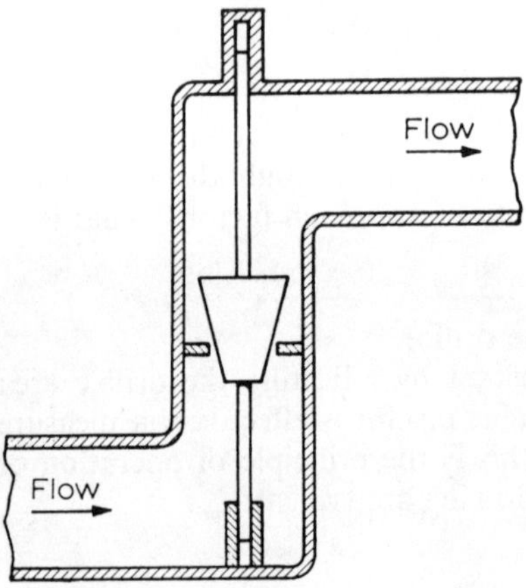

Fig. 9.23 ORIFICE AND PLUG METER (VARIABLE AREA)

the lift of the plug. Therefore the height by which the plug rises when a fluid flows past it is a measure of the rate of flow.

The Rotameter

The rotameter, shown in Fig. 9.24, consists of a long graduated vertical tube having a uniform taper, arranged with the smaller section at the bottom. A float moves freely within the tube and is prevented from fouling the sides of the tube by means of a series of angled slots cut into the float so that as fluid flows past the float it will be caused to spin. In some forms of rotameter the float is prevented from fouling the tube by means of a central guide.

As the rate of flow through the rotameter increases the float rises in the tube, increasing the area of the annulus between the float and the tube wall and thus maintaining a constant pressure differential across the float. The displacement of the float in the tube is a measure of the volume rate of flow and the tube may be calibrated in volume flow units.

The rotameter is by far the most widely used of the variable area meters, but the orifice and plug meter is used quite extensively to measure the flow of water into district water main pipes.

Rotameters may be used to measure the flow of both liquids and gases. It is important, however, that the temperature remains constant otherwise the viscosity and the density of the fluid being metered may change and temperature fluctuations can produce considerable errors, particularly when gases are being metered. A further source of error is the tendency of the float to oscillate which makes accurate scale reading difficult.

Glass tube rotameters may be selected to measure the flow rate of liquids ranging from very low flows of 30–300 ml/s up to flow rates as high as 0·5–5 litres/s. If air is the metered fluid, however, the equivalent range of flow is from about 0·2–2·0 ml/s to about 4–40 litres/s. When liquids are metered the smaller size glass tubes can be used up to a pressure of about 3·4 kPa

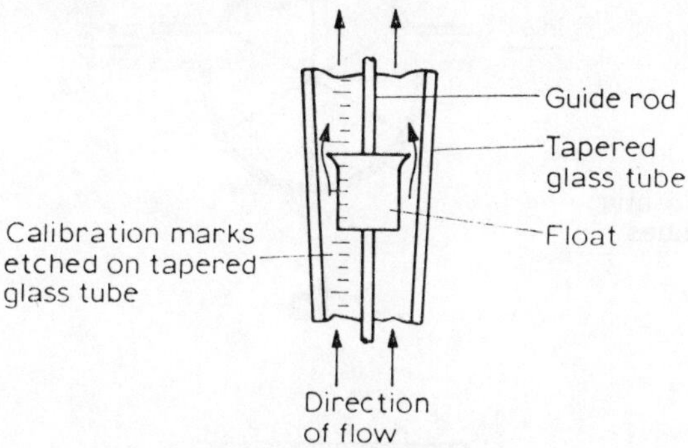

Fig. 9.24 VARIABLE AREA ROTAMETER

whereas the larger tubes can only be used up to about 0·8 kPa. When gases are metered the upper pressure limits are much lower.

Metal tube rotameters are used for measuring very high flow rates and may be used up to a pressure of about 200 kPa, a temperature of about 300°C, and a flow rate as large as 120 litres/s.

9.15 Quantity Flowmeters

Quantity meters measure the total quantity of fluid which has flowed in a certain time. If the quantity of fluid measured in a certain time is divided by that time, then the average rate of flow of the fluid can be determined. Quantity meters may be divided into two classes:

(1) Those which indicate the weight of fluid which has flowed.
(2) Those which indicate the volume which has flowed.

Weighing Meters

When liquid flows in sufficiently small quantities a weighing tank may be used to obtain the weight of liquid discharged in a certain time. Provided the density of the liquid is known, the volume flow rate may then be calculated.

Volumetric Meters

A large number of positive displacement meters are available for metering the flow of liquids where the highest degree of accuracy is required. The principle of operation is that as liquid flows through the meter it moves a measuring element which seals off the measuring chamber into a number of compartments of definite volume. As the measuring element moves these compartments are successively filled and emptied. Thus for one cycle of the measuring element a known quantity of liquid passes from the inlet to the outlet of the meter.

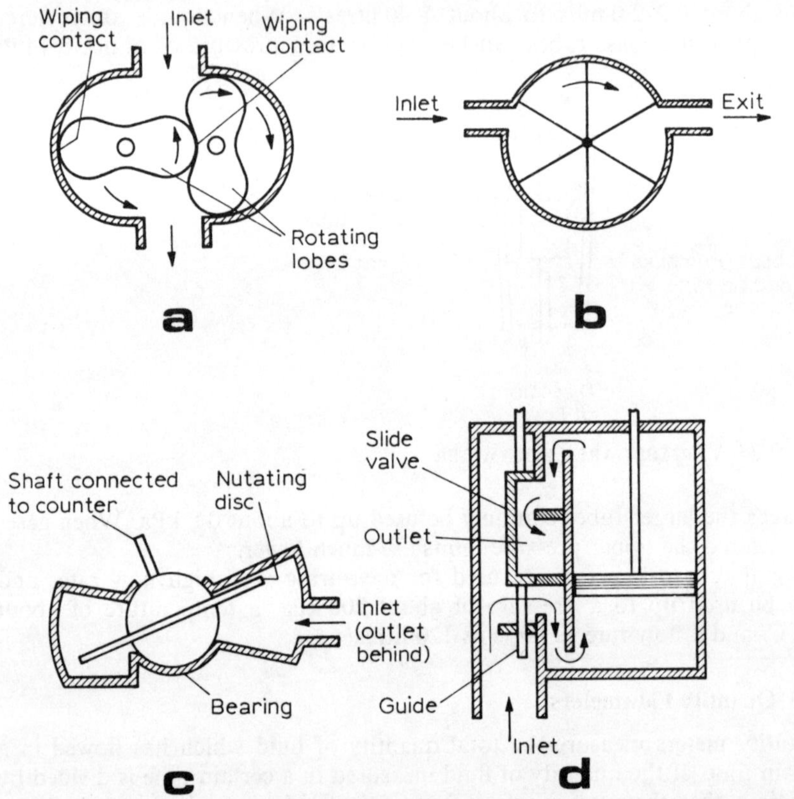

Fig. 9.25 VOLUMETRIC METERS FOR LIQUIDS
a Rotating lobe meter *b* Rotating vane meter
c Nutating disc meter *d* Reciprocating piston meter

The number of cycles of the measuring element may then be indicated by means of a pointer moving over a dial. Some typical liquid positive displacement flow meters are:

(*a*) the rotating lobe meter
(*b*) the rotating vane meter.
(*c*) the nutating disc meter
(*d*) the reciprocating piston meter

These meters are shown diagrammatically in Fig. 9.25.

Positive displacement meters are extremely useful for measuring pulsating flow as their accuracy is not affected by pulsations. Also accurate measurements may be made with liquids of higher viscosity than with many other types of flowmeter. Due to their high accuracy these meters are frequently used for metering oil and water for accounting purposes. If the temperature, density and viscosity of the metered liquid vary from the values at the calibration conditions, sources of error will occur. It is therefore usual to calibrate these meters "in situ" when they are to be used for metering expensive liquids for accounting purposes.

A number of volumetric meters are used for the metering of gases and the most common are:

(*a*) the bellows dry gas meter
(*b*) the water-displacement gas meter
(*c*) the rotating lobe gas meter.

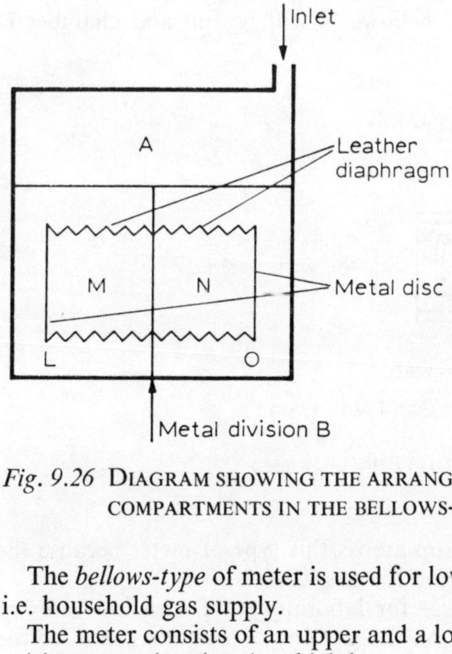

Fig. 9.26 DIAGRAM SHOWING THE ARRANGEMENT OF THE MEASURING COMPARTMENTS IN THE BELLOWS-TYPE GAS METER

The *bellows-type* of meter is used for low rate metering of low pressure gas, i.e. household gas supply.

The meter consists of an upper and a lower section, the upper section comprising a gas chamber A, which has a constant supply of gas from the inlet. A simple valve arrangement allows the gas supply to pass at predetermined intervals into the four measuring chambers below. The lower section is divided vertically by means of a metal division B to each side of which is attached a bellows comprising a leather diaphragm and a metal disc. The lower part of the case is therefore split into four separate measuring compartments as shown in Fig. 9.26.

The movement of the bellows is used to operate the valves which regulate the gas supply into the four measuring chambers and the quantity of gas delivered is measured by the number of times the measuring chambers are filled and emptied. This is achieved by counting the number of horizontal movements of the two diaphragms.

The method of operation of the instrument is as follows:

(*a*) Gas enters the chamber O and compresses the belows N (which is full of gas), thus causing gas in N to exhaust through an exit port.
(*b*) When N is half exhausted a valve opens permitting gas to enter bellows M. Hence the bellows M expands and exhausts gas in chamber L to the outlet.
(*c*) When bellows N is completely exhausted and chamber O is full of gas, a valve opens N to the gas supply thus causing the bellows to expand and the gas in O to be exhausted. The operations are arranged so that when N is exhausted and O is full, bellows M is half full and chamber L half exhausted.
(*d*) When bellows M is full and chamber L exhausted, a valve allows gas to enter L thus causing M to exhaust. At the stage when M is half exhausted and L is half full, bellows N will be full and chamber O exhausted.

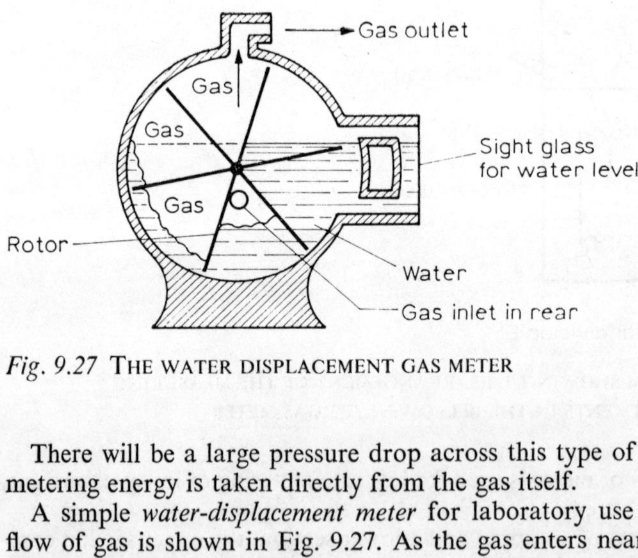

Fig. 9.27 THE WATER DISPLACEMENT GAS METER

There will be a large pressure drop across this type of meter because the metering energy is taken directly from the gas itself.

A simple *water-displacement meter* for laboratory use with low rates of flow of gas is shown in Fig. 9.27. As the gas enters near the centre of the instrument it flows into a segment of the rotor causing the rotor to "float" on the water surface. The rotor will be caused to rotate by the gas flow and gas is discharged at the top outlet. Metering consists of counting the revolutions as each rotor segment will discharge a known quantity of gas. At small flow rates the pressure drop across the meter will be low, but will increase rapidly with increased flow rates. These meters are normally designed for gas flow rates of 6–90 litres/hour.

The *rotating lobe meter* is similar to that used for liquids. These meters are normally used up to a pressure of about 70 Pa but models are available up to 340 Pa.

9.16 Electromagnetic Flowmeters

If a conducting fluid is allowed to flow with velocity U along a pipe of internal diameter d and an electromagnetic system directs a magnetic field B across a section of the pipe so that it acts at right angles to the direction of motion of the fluid, an induced voltage V is produced (Fig. 9.28).

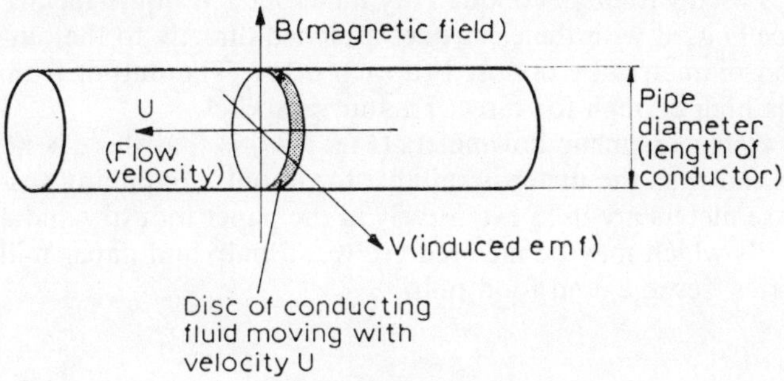

Fig. 9.28 PRINCIPLE OF THE ELECTROMAGNETIC FLOWMETER

By Faraday's Law of Induction, when an electrical conductor moves through a magnetic field in a direction at right angles both to the magnetic field and its length, an e.m.f. is generated of value

$$V = BlU \text{ volts}$$

where B is the magnetic flux density (tesla, i.e. W/m^2)
l is the length of the conductor (m)
U is the velocity of the conductor (m/s)

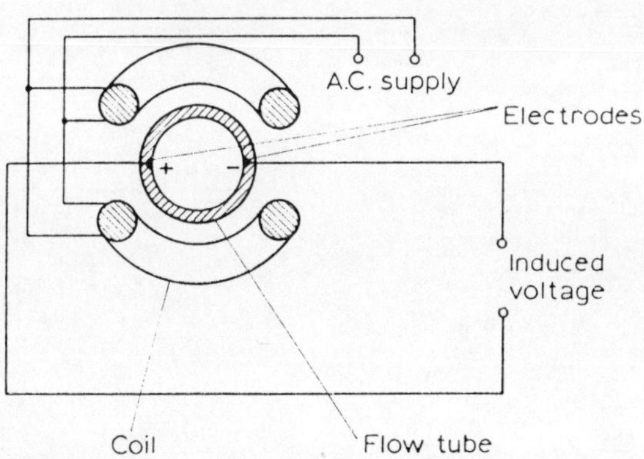

Fig. 9.29 ELECTROMAGNETIC FLOWMETER

The length of the conductor is proportional to the diameter of the flow tube and the velocity is proportional to the mean flow velocity. If the magnetic flux density is constant the induced voltage will therefore be proportional to the mean flow velocity and the value of the induced voltage will thus represent a direct indication of mean flow velocity.

In practice two types of meter are commonly used. The first has a non-conducting pipe liner to prevent short-circuiting the e.m.f. and is used for fluids with low conductivities such as water. The electrodes are mounted flush with the non-conducting liner such that they make contact with the fluid. The output voltage from this type of meter is low and will normally require amplification.

The second type is used with high conductivity fluids such as liquid metals. A stainless steel pipe is used with the electrodes attached directly to the outside of the pipe, and diametrically opposed to each other. The output from this type of meter is high enough for direct readout purposes.

An advantage of electromagnetic flowmeters (Fig. 9.29) is that there is no obstruction to flow and thus the meter is suitable for liquids containing suspended matter. These meters are used extensively in the paper industry and a few examples of fluids which may be metered are wood pulp and paper mill stocks, cement slurries, sewage, and food pulp.

10 Viscometry

Viscosity is the "stickiness" of a fluid and arises from the cohesion and interaction between the fluid molecules. It is a measure of the fluid resistance to shear when the fluid is in motion, either by a body moving through a fluid or by a fluid moving around a body.

Pressure and temperature changes affect viscosity but whilst a change in pressure affects viscosity to a negligible extent only, temperature changes cause considerable variations. It should be noted that the viscosity of liquids decreases with increasing temperature whereas the viscosity of gases increases.

The presence of viscosity represents the reluctance of a fluid to motion and objects moving in viscous fluids will soon come to rest unless power is applied to maintain the motion. In practice when a body is required to move through a viscous fluid a large proportion of the available power is required to overcome viscous effects. It is therefore most important to know the magnitude of the drag forces acting on a body and, as these forces are proportional to the fluid viscosity, instruments have been designed to measure this property.

10.1 Newton's Law of Viscosity

It has been shown experimentally that the flow velocity is not uniform across the section of a conduit and varies from point to point. The velocity distribution may be represented by profiles as shown in Fig. 10.1.

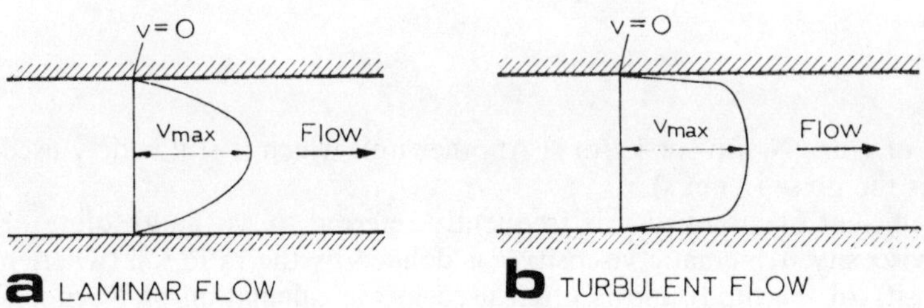

Fig. 10.1 TYPICAL VELOCITY PROFILE

It is convenient to assume that fluid layers slide relative to adjacent layers in the direction of flow and that the relative velocity between adjacent layers varies across the flow. Consider two layers at an infinitesimal distance δy apart as shown in Fig. 10.2.

Over a distance δy the velocity changes by δv and thus the velocity gradient dv/dy (where $dv/dy = \lim_{\delta y \to 0} \delta v/\delta y$) at a point P on the velocity profile is the tangent of the angle θ as shown in Fig. 10.2. A Newtonian fluid is one in which the shear stress between adjacent layers is proportional to the velocity gradient, i.e.

$$\tau \propto dv/dy$$

or

$$\tau = \eta \frac{dv}{dy} \qquad \text{(i)}$$

where τ is the shear stress between layers
dv/dy is the velocity gradient
η is a constant known as the *coefficient of viscosity.*

From equation (i) it can be seen that

$$\eta = \tau \frac{dy}{dv} \qquad \text{(ii)}$$

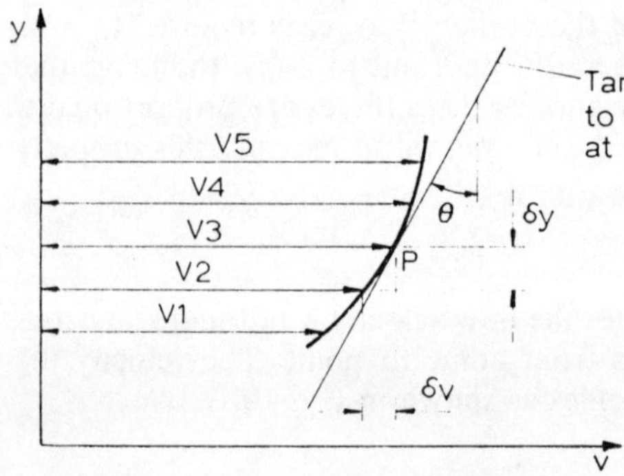

Fig. 10.2

The units of η are N s/m^2 or kg/m s. Another unit which is still widely used however is the poise (g/cm s).

The coefficient of viscosity η is frequently referred to as the absolute or dynamic viscosity. Kinematic viscosity ν is defined by the ratio η/ρ (where ρ is the density of the fluid) and is often used in the calculation of Reynolds Number (see Chapter 9). The units of kinematic viscosity are m^2/s or the stokes (cm^2/s).

10.2 Efflux Viscometers

Certain instruments have been developed in the oil industry for the measurement of viscosity and a number of these are now in general use. These instruments are known as short-tube or orifice viscometers and the principal types are: Saybolt Universal and Saybolt Furol (U.S.A.), Redwood No. 1 and No. 2 (Great Britain), and Engler (Germany and other continental countries).

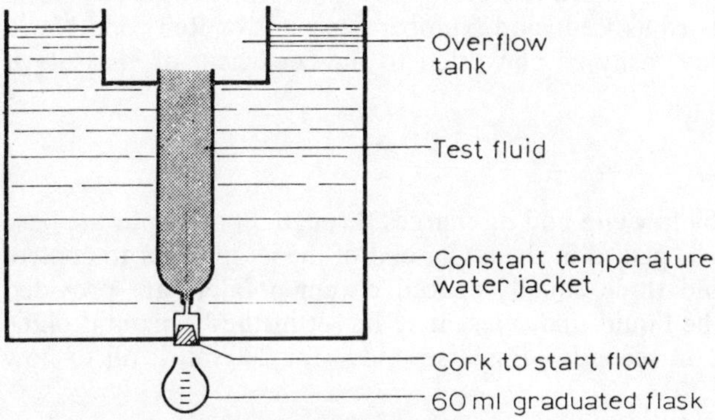

Fig. 10.3 THE SAYBOLT VISCOMETER

The Saybolt Viscometer

The time taken in seconds to discharge 60 ml through the outlet tube is measured and used as a scale of viscosity for comparison. The results are usually expressed in terms of Saybolt Universal seconds but may be readily converted to the absolute or dynamic viscosity if required. As the coefficient of viscosity depends on the temperature of the fluid under test, care must be taken to ensure that the fluid remains at a constant temperature throughout the test. A water jacket is used for this purpose (Fig. 10.3).

The U.S. Bureau of Standards recommends the use of the Saybolt Furol Viscometer for oils in which the time of flow in a Saybolt Universal Viscometer is inconveniently large. This viscometer is identical to the Universal apparatus with the exception of the outlet tube which is the same length but has a larger diameter.

The Redwood Viscometer

The Redwood viscometer is similar in construction to the Saybolt instrument and is shown in Fig. 10.4. The time is measured to collect 50 ml of fluid in the flask and the viscosity may be expressed in Redwood seconds which is

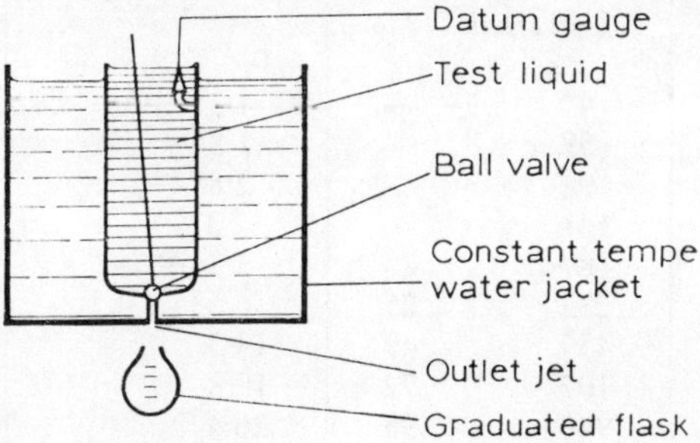

Fig. 10.4 THE REDWOOD VISCOMETER

proportional to viscosity. Two jet sizes are available and are denoted by the numbers 1 and 2. The number 2 jet has a larger diameter than the number 1 and is normally used for the more viscous liquids. The results of the Redwood test are usually expressed as Redwood Number I seconds or Redwood Number 2 seconds but may easily be converted to the coefficient of viscosity if required.

The Engler Viscometer

The oil is held in a shallow cup and discharges through an orifice in the base of the cup. A stopper is placed in the upper end of the jet in order to control the flow of liquid and three equally spaced datum pointers are provided so that the level of the liquid under test may be set in the horizontal plane (Fig. 10.5). The time in seconds—Englerseconds—for 200 ml of oil to flow

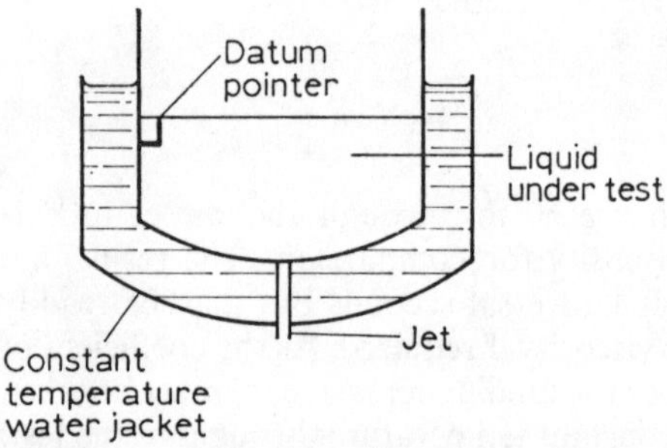

Fig. 10.5 THE ENGLER VISCOMETER

through the orifice gives a measure of the viscosity of the oil. The results obtained using this instrument are often expressed in "Engler degrees" which is the ratio of the time of flow of 200 ml of oil at the temperature of the test to the time for the same volume of water at 20°C.

TABLE 10.1

$m^2/s \times 10^4$	REDWOOD SEC		SAYBOLT SEC		ENGLER DEG
	No. 1	*No. 2*	*Universal*	*Furol*	
0·05	38	—	42	—	1·3
0·1	52	—	59	—	1·84
0·2	86	—	98	—	2·88
0·3	125	—	141	—	4·07
0·4	165	—	187	23	5·34
0·5	205	23	233	25	6·63
1·0	400	41	459	49	13·3
1·5	600	61	707	72	19·8
2·0	809	82	946	95	26·4

Table 10.1 shows viscosity conversions between kinematic viscosity in m^2/s, Redwood seconds, Saybolt seconds, and Engler degrees.

10.3 Torque or Rotational Viscometers

A thin film of the liquid to be tested is enclosed between two vertical concentric cylinders. With low viscosity liquids the inner cylinder is usually

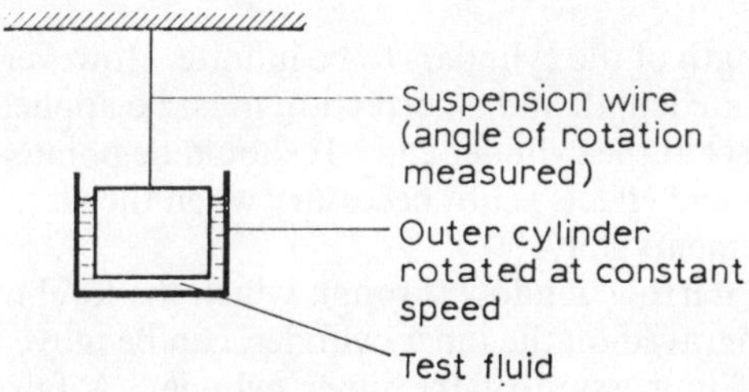

Fig. 10.6 THE TORQUE OR ROTATIONAL VISCOMETER

suspended by a torsion wire or spring and the outer cylinder is rotated at constant speed, usually by means of an electric motor. The torque produced can be readily measured by the deflection of the inner cylinder as shown in Fig. 10.6.

For measurements on liquids of high viscosity which give a larger retarding couple the method usually adopted is to apply a known torque to the inner

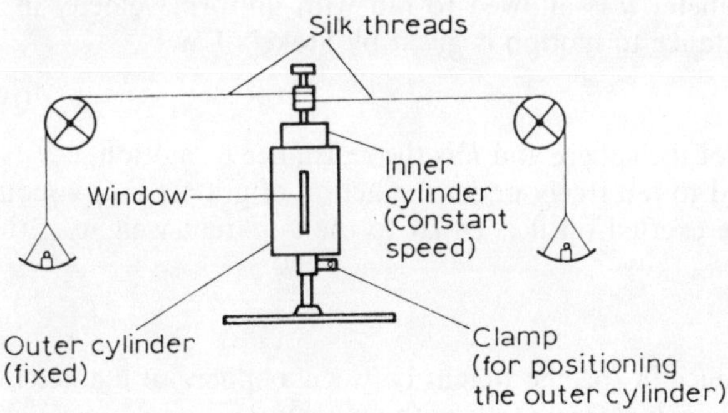

Fig. 10.7 SEARLE'S VISCOMETER

cylinder by a system of weight and pulleys. The rate of rotation of the cylinder when the motion has become uniform is then observed. Searle's viscometer operates on this principle, i.e. the inner cylinder rotates and the outer cylinder is fixed. The inner cylinder (Fig. 10.7) is carried on a spindle pivoted at top and bottom, the lower pivot acting as a support for the outer cylinder.

Two equal masses carried by silk threads passing over ball-bearing pulleys and wound round a drum fixed to the spindle provide the known couple.

If the couple G maintains a constant speed equal to one revolution in t seconds the viscosity can be found from the equation:

$$\eta = \frac{Gt}{8\pi^2 h} \times \frac{(b^2 - a^2)}{a^2 b^2} \qquad \text{(iii)}$$

where a is the radius of the inner cylinder
b is the radius of the outer cylinder
h is the immersed length of the inner cylinder.

The above formula assumes the length of the cylinders to be infinite. However, in practice both cylinders are of finite length and a correction must be applied to compensate for the viscosity effect at the cylinder ends. It should be pointed out, however, that a correction for end effects is not necessary when the viscometer is used for relative measurements only.

The outer cylinder, which has a narrow window through which the level of the liquid can be read on a scale engraved on the inner cylinder, can be moved up or down to vary the depth of immersion of the inner cylinder. A false bottom, consisting of a perforated plate loosely fitting the outer vessel, keeps conditions at the lower end almost constant for different heights of liquid. The couple required to produce a given rate of rotation will be proportional to $h + l$, where h is the depth of immersion of the inner cylinder and l is the correction to be applied for the end effect. It therefore follows that the correction l may be found from two sets of observations.

10.4 The Falling Sphere Viscometer

When a sphere of diameter d is allowed to fall with uniform velocity in a viscous liquid the resistance to motion is given by Stoke's Law:

$$F = 3\pi\eta ud \qquad \text{(iv)}$$

where u is the velocity of the sphere and F is the resistance to motion.

If a sphere is allowed to fall freely under the action of gravity in a viscous medium then the force exerted on it is equal to the apparent weight of the sphere:

$$F = \Delta w \tfrac{1}{6}\pi d^3 \qquad \text{(v)}$$

where Δw is the difference in specific weight between a sphere of diameter d and the fluid through which it freely falls with velocity u.

Equating (iv) and (v) gives

$$3\pi\eta ud = \Delta w \tfrac{1}{6}\pi d^3$$

or

$$\eta = \Delta w d^2/18u \qquad \text{(vi)}$$

The essence of the method is to allow a steel ball-bearing of known size to fall freely in a viscous liquid. The time taken for the sphere to pass with uniform velocity between two fixed points on the cylinder containing the test liquid is noted and the velocity calculated. The apparatus used is shown in

Fig. 10.8. For normal use a stop-watch may be used to record the time, and velocities up to 10^{-2} m/s can usually be recorded to within one per cent.

The observed velocity of fall in the cylinder of diameter D will differ slightly from the velocity of free fall u, because of the interference of the

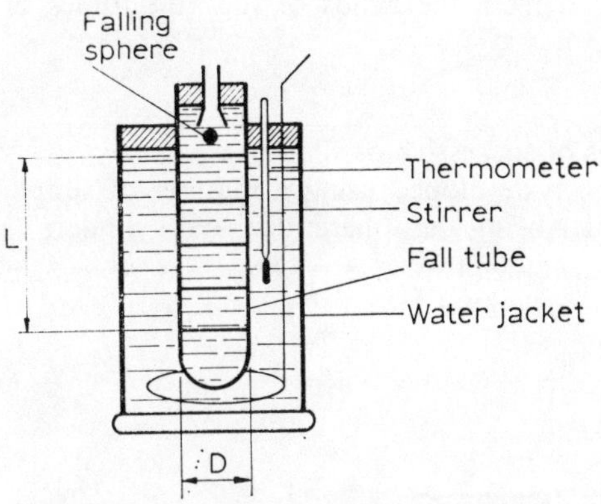

Fig. 10.8 THE FALLING SPHERE VISCOMETER

cylinder boundaries on the sphere. If v is the observed velocity in the cylinder of diameter D the value of u can be obtained from the following relationship:

$$\frac{u}{v} = 1 + \frac{9d}{4D} + \left(\frac{9d}{4D}\right)^2 \qquad \text{(vii)}$$

The fall of the sphere is often timed over the middle half of the tube but additional markings should be provided separating the timing distance into two or three equal parts so that the uniformity of the motion may be checked. Also as the sphere must be falling at constant speed between the marks an initial length is required so that the acceleration from rest has reached zero by the time the sphere reaches the first timing mark.

10.5 Oscillational Viscometers

Consider a torsional pendulum consisting of a bob having moment of inertia I about an axis of symmetry coincident with the suspension wire and oscillating in a viscous medium (Fig. 10.9). The dynamic viscosity can be determined from

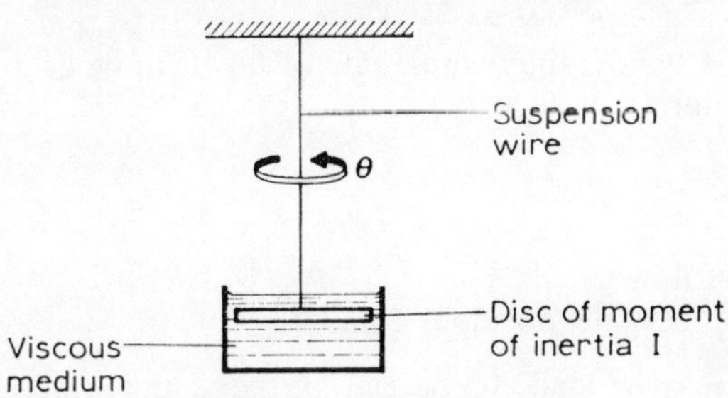

Fig. 10.9 TORSIONAL PENDULUM

measurements of the periodic time and the logarithmic decrement of the amplitude.

A typical curve showing the angular displacement of the bob from its equilibrium position as the time increases is shown in Fig. 10.10. The logarithmic decrement δ can be determined from the ratios of the amplitudes of successive oscillations:

$$\delta = \log_e a_1/a_2$$

where a_1 and a_2 are the amplitudes of successive oscillations.

Oscillational viscometers have been developed using a number of simple geometrical shapes, the most popular being the sphere, disc and cylinder.

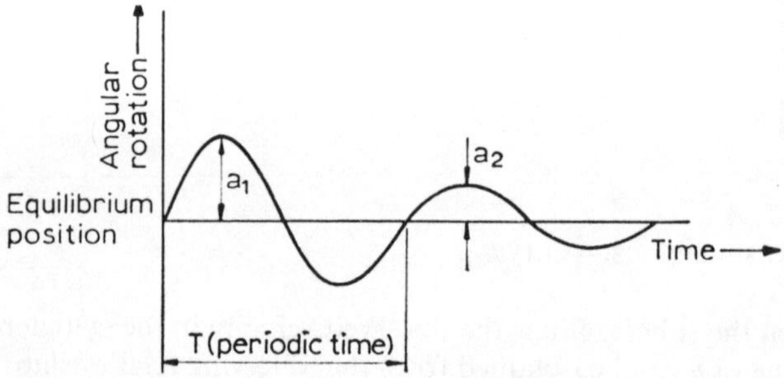

Fig. 10.10 OSCILLATIONS OF TORSIONAL PENDULUM IN VISCOUS MEDIUM

10.6 Capillary Tube Viscometers

Consider the tube cross-section shown in Fig. 10.11. If the Reynolds number (see Chapter 9) is below 1 000, laminar flow will exist in the tube and the familiar parabolic velocity profile will be experienced as shown. If the fluid is

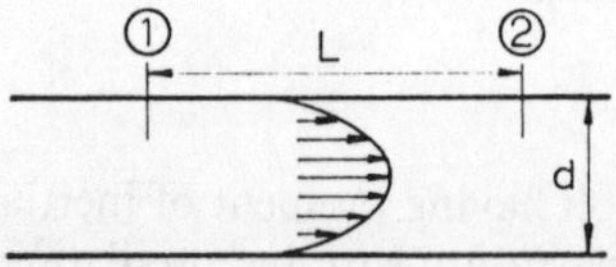

Fig. 10.11 LAMINAR FLOW THROUGH A CAPILLARY TUBE

incompressible and the flow is steady, the volume rate of flow can be determined by the Poiseuille equation:

$$Q = \frac{\Delta p \pi d^4}{128 \eta L} \qquad \text{(viii)}$$

where Q is the volume rate of flow
Δp is the pressure drop between positions 1 and 2.

A viscosity determination may be made by measuring the volume rate of flow and pressure drop for flow in such a tube. A small diameter capillary tube is used to reduce the size of the Reynolds number and ensure laminar flow.

Capillary viscometers can be divided into two classes:

(1) Instruments for measuring absolute viscosity directly from the dimensions of the apparatus and experimental data obtained.
(2) Viscometers used for determining relative viscosities by reference to certain standard liquids.

Greater accuracy is usually obtained with viscometers designed for relative measurements than that found in absolute instruments. This greater accuracy is due partly to simplification in design and also to the elimination of errors made in the measurement of dimensions. The precise measurement of absolute values is therefore a subject chiefly of historical interest. The apparatus illustrated in Fig. 10.12 was that used by Poiseuille.

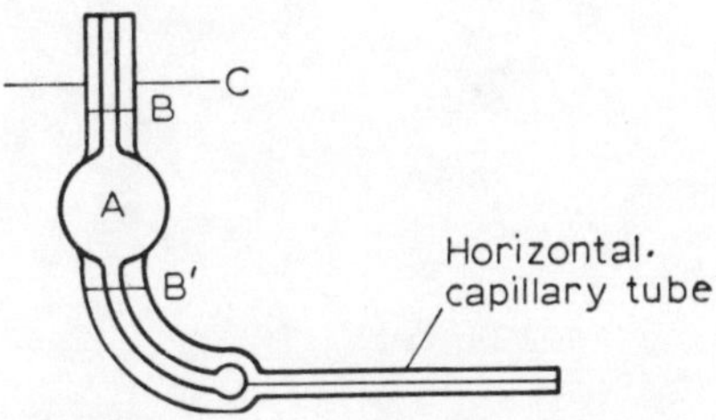

Fig. 10.12 POISEUILLE'S VISCOMETER

The bulb A and the horizontal capillary attached are filled with and immersed in the liquid under test. The upper tube is joined to a compressed air reservoir connected to some form of manometer. The liquid in bulb A is forced out through the capillary and the time for the level to fall from B to B′ noted. The effective pressure is that indicated by the manometer less a small head of liquid which varies during the experiment from BC to B′C. Provided the pressure in the compressed air reservoir is reasonably large the head to be subtracted may be taken as $\frac{1}{2}(BC + B'C)$.

Poiseuille's viscometer had many disadvantages some of which are listed as follows:

(*a*) The exact length of the capillary could not easily be measured.
(*b*) Owing to the shape of bulb A differences in the rate of drainage down the walls led to errors in the volume delivered for different rates of flow or with different liquids.
(*c*) Solid impurities in the fluid were almost certain to enter the capillary.

Many capillary viscometers have been designed to overcome these difficulties. The basic principle of all of these instruments however is the same as the Poiseuille viscometer already discussed.

Viscometers for Relative Measurements

Relative viscometers can be divided into two classes:

(1) those using an externally applied pressure,
(2) those relying solely on the hydrostatic head of liquid in the viscometer to produce flow.

Instruments using an Externally Applied Pressure

Absolute viscometers can readily be used for relative measurements, and because of this, construction is simplified. A typical device used by Ubbelohde is shown in Fig. 10.13. The viscometer is filled by suction until the

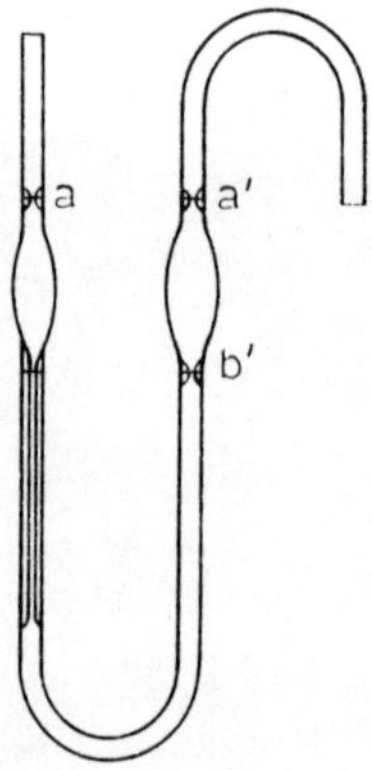

Fig. 10.13 UBBELOHDE'S VISCOMETER

liquid stands at a and b′. The liquid is then forced through the capillary by air pressure, the time taken for the liquid to move between the marks b′ and a′ being noted. If the kinetic energy term is neglected the viscosity of the test liquid can be calculated from the relationship

$$\eta = \eta' Pt/P't' \tag{ix}$$

where P is the air pressure when the liquid under test is forced through the capillary
t is the time taken for the liquid under test to pass between the marks b′ and a′
P' and t' are the pressure and time for a standard liquid of known viscosity η'.

Gravity Viscometers

If the jet of a simple pipette is replaced by a capillary tube and the liquid under test allowed to flow out under its own hydrostatic head a measure of the kinematic viscosity can be obtained by timing the meniscus between two marks. A typical pipette viscometer used by Barr is shown in Fig. 10.14.

The pipette PT is enclosed in an outer vessel UW. The test liquid is sucked up into the pipette as far as the mark Q; the pipette is then placed in the outer vessel as shown and allowed to drain. When a reading is required the liquid is sucked up to a level above R and the fall from R to S timed.

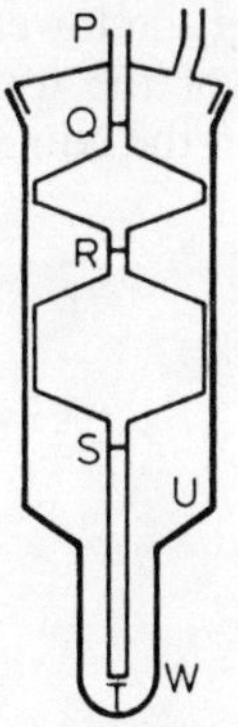

Fig. 10.14 BARR'S PIPETTE VISCOMETER

Many relative measurements made in viscometry have been carried out with instruments based on the Ostwald Viscometer shown in Fig. 10.15. The apparatus is charged with a constant volume from a pipette and the liquid is then drawn into the left-hand limb above the mark A. The liquid is then allowed

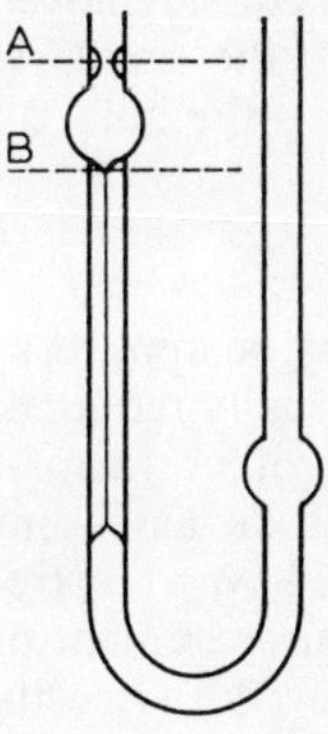

Fig. 10.15 THE OSTWALD VISCOMETER

to flow back and the time taken to fall from A to B measured. The absolute viscosity η of a test liquid can be calculated using the relationship

$$\eta = \eta' \rho t / \rho' t' \qquad \text{(x)}$$

where ρ is the density of the test liquid
t is the time taken for the test liquid to fall between A and B
ρ' and t' are the density and time for a standard liquid of known viscosity η'.

Suspended Level Viscometers

Ubbelohde introduced a modification of his viscometer to include a suspended level. The lower end of the capillary is widened into a spherical shoulder as shown in Fig. 10.16. By introducing an air vent the liquid is induced to flow only down the walls of the bulb D thus automatically fixing the lower liquid level at the lower end of the capillary. A small sample of the test fluid is introduced into tube 1, tube 3 is blocked with the finger, and the liquid

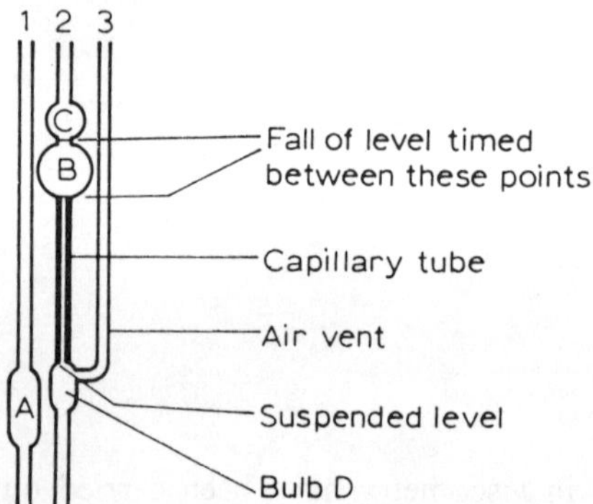

Fig. 10.16 THE SUSPENDED-LEVEL VISCOMETER

drawn into tube 2 to a point above the upper mark. The finger is then removed from tube 3 and the liquid is allowed to flow freely through the capillary. The fall of the upper level is then timed between the marks on the bulb B.

10.7 The Viscosity of Gases

Many methods used for measuring the viscosity of liquids may be applied to gases and possibly vapours. A modification to the theory is usually required, however, to allow for the compressibility of the gas and other factors. Capillary viscometers are used extensively for measurements on gases but as gases are much less viscous than liquids the capillary must have an extremely fine bore. Because of the fineness of the bore, accurate measurement of the tube radius is made more difficult but this difficulty is removed if viscosity comparisons only are to be made.

Because a gas is compressible the density will decrease along the tube and therefore it is not simply the volume but the mass traversing any cross-section which is constant. If ρ is the density and Q the volume passing any cross-section per second then the product ρQ is constant. Since the density is proportional to the pressure the product pQ will also be constant, where p is the pressure of the gas at a particular cross-section. The absolute viscosity can then be calculated from the formula

$$p_0 Q_0 = p_1 Q_1 = \frac{(p_0^2 - p_1^2)\pi d^2}{256\eta L} \qquad \text{(xi)}$$

where p_0 and p_1 are the pressures at the tube entrance and outlet respectively
Q_0 is the volume of liquid entering the tube per second
Q_1 is the volume of liquid flowing out per second
L is the length of the tube.

A typical capillary viscometer used for measuring the viscosity of gases is the apparatus used by Schultze (Fig. 10.17). In this apparatus the pressure difference ($p_0 - p_1$) is maintained at a constant value by regulating the supply or withdrawal of liquid. Gas contained in the bulbs a and b is forced through the capillary at constant pressure by raising the mercury c up the scale d.

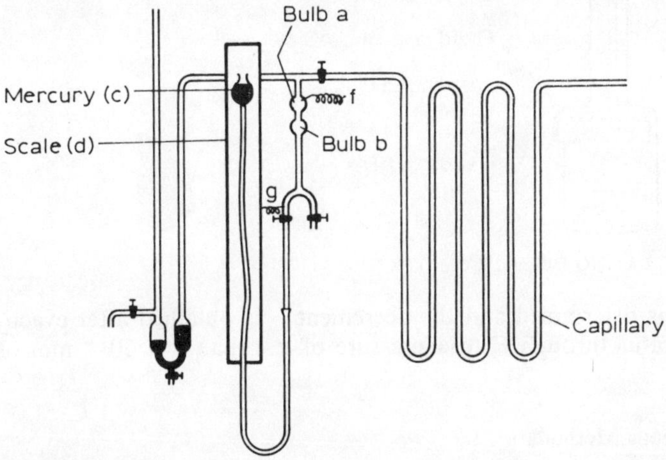

Fig. 10.17 SCHULTZE'S VISCOMETER

The volume of gas is recorded electrically as the mercury meniscus passes the contacts f and g.

Viscometers of the concentric cylinder type have frequently been used for measuring the viscosity of gases. The apparatus is usually of the form in which the inner of two concentric cylinders is suspended while the outer one rotates at a steady speed. Viscometers of this type have already been discussed in connection with liquids (section 10.3).

The oscillating disc viscometer has been used extensively for gases and vapours. In this instrument, which has already been discussed in connection with liquids, observations are made of the damping imposed on an oscillating disc attached to a suspension wire. Fig. 10.18 shows the suspended disc A oscillating midway between two fixed discs of slightly larger diameter, B_1 and B_2.

The fixed discs are held apart by three pegs and are supported by a rod D, which passes up the tube and is sealed at the top. The suspension wire is fused to a small projection from the top of this rod. The dynamic viscosity may be calculated from the formula:

$$\eta = c\left(\frac{\delta}{T} - \frac{\delta_0}{T_0}\right) \qquad \text{(xii)}$$

where δ and T are the logarithmic decrement and periodic time respectively at the temperature and pressure of the test and δ_0 and T_0 the corresponding values in vacuo; c is a constant for the apparatus and is normally found by calibrating with a gas of known viscosity. The logarithmic decrement for

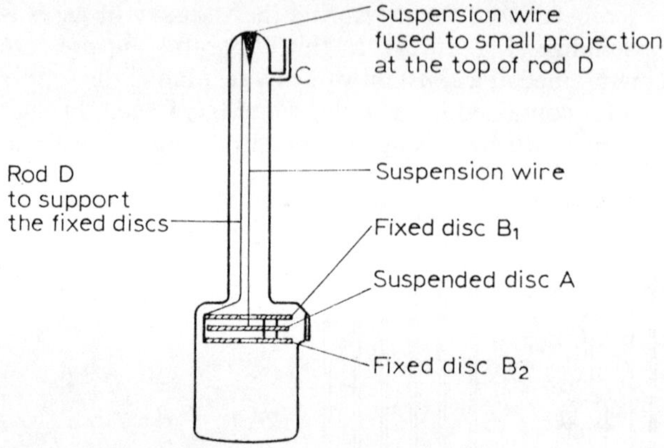

Fig. 10.18 OSCILLATING DISC VISCOMETER

dry air at 17°C is determined and the decrement δ_0 is obtained after evacuating the apparatus through C to a pressure of less than 1×10^{-4} mm of mercury.

10.8 Miscellaneous Methods

Oscillation of a Liquid in a U-tube

The liquid contained in a U-tube of uniform bore is displaced by applying a pressure to the one side and then suddenly releasing it (Fig. 10.19). If a_1 and

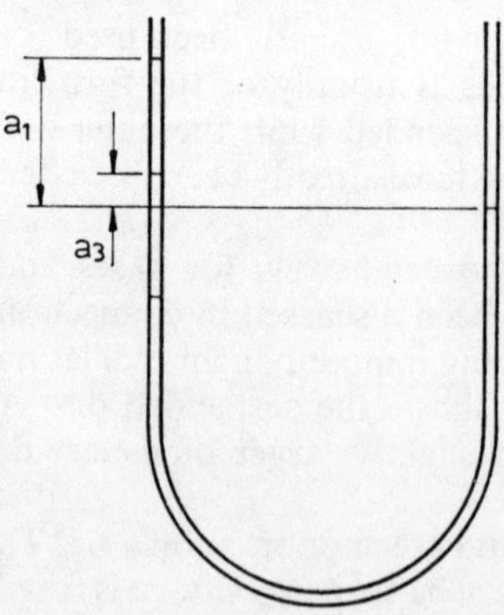

Fig. 10.19 OSCILLATION OF LIQUID IN A U-TUBE

a_3 represent successive amplitudes on the same side of the equilibrium level when $a_1/a_3 > 5$ the damping is large and the logarithmic decrement δ is constant and independent of the initial amplitude. The dynamic viscosity can now be calculated from the following formulae:

$$\eta = \tfrac{1}{4}\delta r^2 \rho / T \qquad \text{(xiii)}$$

and

$$T^2 = \frac{2\pi^2 l}{g}\left(1 + \frac{\delta^2}{4\pi^2}\right) \qquad \text{(xiv)}$$

where l is the total length of the liquid when measured along the axis of the U-tube of radius r, ρ is the density of the liquid, and T is the periodic time of oscillation.

Consistent results can only be obtained using a viscometer of this type when $a_1/a_3 > 5$, using initial amplitudes of up to 160 mm.

The Rolling Sphere Viscometer

A simple method for determining the viscosity of a liquid is to measure the speed with which a sphere will roll down an inclined cylindrical tube filled with liquid. As the ball is always in contact with the tube at one point visual observation is possible in a glass cylinder even with opaque liquids. The range of the instrument may be easily extended by using a ball of different diameter.

The Falling Cylinder System

In the falling sphere viscometer discussed earlier the clearance between the sphere and the container is large. This apparatus however operates on the principle of a cylindrical metal bulb falling in a vertical metal tube of only slightly greater diameter. The apparatus used by Bridgman is shown in Fig. 10.20.

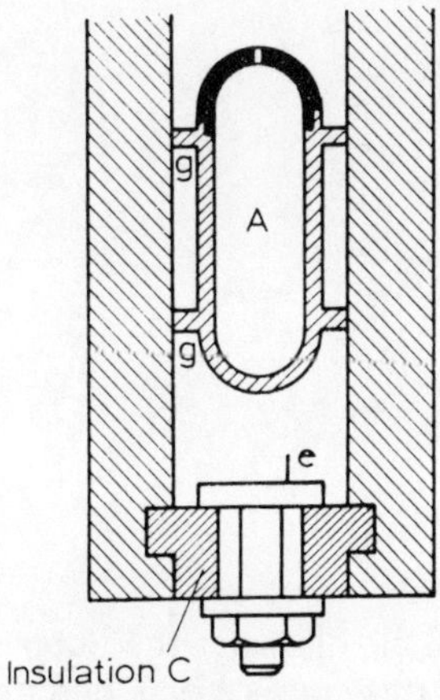

Fig. 10.20 THE FALLING CYLINDER VISCOMETER

The steel bulb A is made in two parts so that masses may be placed inside to accelerate the fall if necessary. The falling bulb forces liquid from the bottom to the top of the tube through the annulus and a coaxial fall is ensured by having three small projections at each end of the tube. These guides (gg in the Figure) at each end make electrical contact with the walls of the tube and the circuit is completed when the bulb touches the electrode e, insulated from the outer tube by C. The measured time of fall is proportional to the viscosity of the liquid and the coefficient of viscosity can be calculated from the formula:

$$\eta = (\sigma - \rho)\frac{(a-b)^3}{3b}\frac{t}{s} \qquad \text{(xv)}$$

where σ is the density of the bulb, ρ is the density of the liquid, a and b are the radii of the tube and bulb respectively, and t is the time taken for the bulb to fall a distance s.

The above formula does not allow for the change in density of a liquid due to compression and it is usual to calibrate such an instrument with a liquid of known viscosity. Then for a fixed distance of fall the viscosity may be calculated from the formula:

$$\eta = kt(\sigma - \rho) \qquad \text{(xvi)}$$

where k is the constant for the instrument.

The Air Bubble Viscometer

Various industries have made viscosity comparisons by observing the rate of rise of an air bubble in a vertical tube containing the test liquid. Provided the

Fig. 10.21 THE AIR BUBBLE VISCOMETER

bore of the tube is only slightly larger than the bubble diameter the bubble will rise slowly. In its most simple form this apparatus consists of a burette tube closed by taps at each end. The tap at the lower end is arranged so that a bubble of constant volume can be released into the vertical tube (Fig. 10.21).

A slower rate of rise may be obtained by inclining tubes at a small angle to the horizontal. A purely comparative system may be obtained using a series of inclined tubes filled to the same level with oils having viscosities increasing in constant steps. A similar tube is then filled to the same level with the sample and the time taken for an air bubble to rise in the sample is compared with the standards. The range within which the viscosity of the sample must be is then quickly determined.

The Direct-reading Viscometer

The principle of operation of the apparatus described by Rhodin depends on the viscous drag exerted on the walls of a movable body placed in a channel

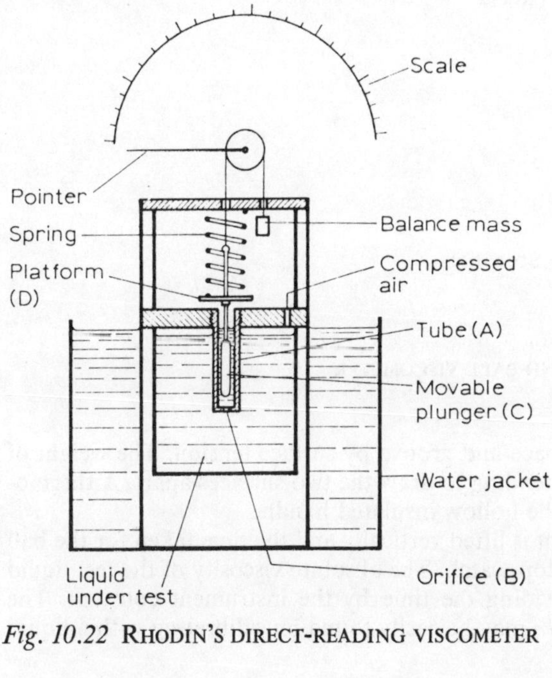

Fig. 10.22 RHODIN'S DIRECT-READING VISCOMETER

through which the fluid under test flows at a constant velocity. The body is capable of movement in the direction of flow against the action of a spring or mass.

The instrument is shown in Fig. 10.22. Liquid is forced up the tube A via the aperture B by means of compressed air which is maintained at constant pressure by a relief valve. The viscous drag on the plunger C is balanced by compression of the spring; plunger movement is indicated on the scale by means of the pointer. The balance-mass may be adjusted so that its weight equals the apparent weight of the plunger immersed in the liquid. To compensate for variations in specific gravity masses may be placed on the platform D.

The Michell Cup and Ball Viscometer

The apparatus (Fig. 10.23) consists of a cup attached to a handle and a steel ball of the same curvature as the cup. Contact between the surfaces of the ball and cup is prevented by three small projections on the surface of the cup. The projections are accurately ground to project about 0·01 mm so that a space of this thickness remains between the ball and the cup. The interspace and a groove formed round the edge of the cup are filled with liquid

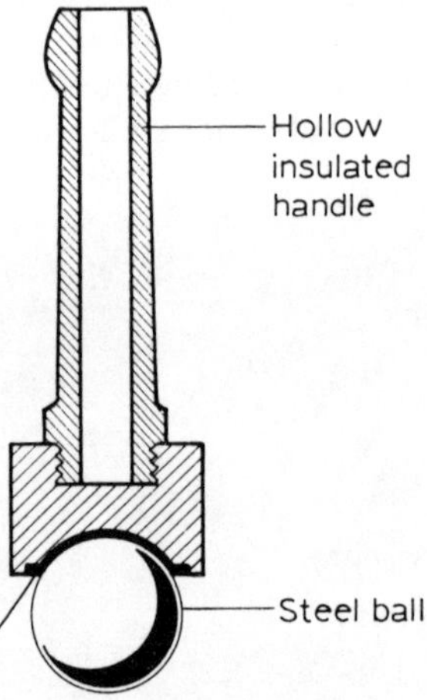

Fig. 10.23 MICHELL CUP AND BALL VISCOMETER

which is held in the interspace and groove by surface tension. The weight of the ball provides the force tending to draw the two surfaces apart. A thermometer may be inserted in the hollow insulated handle.

In practice the instrument is lifted vertically and the time taken for the ball to fall off is noted with a stop-watch. The absolute viscosity of the test liquid may be determined by dividing the time by the instrument constant. The instrument constant $k = t/\eta$ may be easily found by calibrating with a liquid of known viscosity.

10.9 Viscosity Index

The Viscosity Index of an oil is an indication of the relative rate of change of the viscosity with temperature change and is used extensively in the oil industry.

The Viscosity Index can be calculated using the following relationship:

$$\mathrm{VI} = \frac{100(L - U)}{L - H} \qquad \text{(xvii)}$$

where $U =$ the kinematic viscosity at 100°F of the sample.

$L =$ the kinematic viscosity at 100°F of an oil of zero viscosity index having the same viscosity at 210°F as the sample.

$H =$ the kinematic viscosity at 100°F of an oil of 100 viscosity index having the same viscosity at 210°F as the sample.

The values of L and $(L - H)$ are tabulated in the Institution of Petroleum Standards I.P. 73/53. Therefore if the kinematic viscosity of an oil is known at both 100°F and 210°F the viscosity index can be easily calculated.

As the temperature rises a fluid tends to become less viscous and the rate at which the fluid will thin out depends on the viscosity index. As an example consider two fluids which have the same viscosity at 100°F but one fluid has a viscosity index of 80 whereas the other has a viscosity index of 70. An increase in temperature will cause the fluid of viscosity index 70 to thin out faster than the fluid of viscosity index 80.

11 Measurement of Temperature

Measuring the temperature of a body depends upon the establishment of thermodynamic equilibrium between the body and the device used to sense the temperature, e.g. a thermometer bulb or thermocouple wires. In practice this condition is rarely attained since it is difficult to establish complete instantaneous equilibrium and great care must be exercised in choosing a method suited to the problem so that satisfactory conditions for temperature measurement are obtained. Temperature sensors possess thermal characteristics dependent largely upon their size and shape and the materials from which they are made, and these characteristics, for instance high heat capacity, affect their ability to conduct, radiate or reflect heat from their surroundings. The introduction of a temperature sensor into a body tends to modify the temperature conditions at that point, and the body will be modified by any holes drilled, or by insertion of protective pockets to accommodate temperature sensors.

In most cases the sensor is connected to a recording instrument by means of an intermediate system, such as a bourdon tube or electrical conductors, along which the signal is carried. The intermediate system and the recorder may be subject to temperature or other changes and compensating devices are usually incorporated to reduce or eliminate errors.

Heat is transferred from place to place by conduction, convection and radiation, and it is necessary to take these into account when choosing a temperature measurement method.

11.1 Conduction

Heat flow along a conductor is analogous to flow of electricity. Heat flows along a conductor provided that the conductor is subject to a temperature gradient. Some substances are good conductors, e.g. metals; others are poor conductors, e.g. glass, plastics, oxides. The temperature gradient across a poor conductor is steep; thus poor conductors are good insulators. To a rough approximation the conductivities of metals are of the order of one thousand times those of other solids and liquids except mercury, which is a liquid metal, and ten thousand times as great as gases.

It is desirable that heat be conducted as rapidly as possible to the temperature sensor which should be directly immersed in the heated medium or, if this is not possible, a protective pocket should be provided and the sensor immersed in a good conductor such as mercury. This technique will reduce time lag.

11.2 Convection

Liquids are poor thermal conductors, and gases for most practical purposes can be regarded as non-conductors. When liquids and gases transfer heat by moving from one place to another, this is called convection. Convection may be either "natural" or forced. The domestic hot water radiator is an example of natural convection; it transfers most of its heat by warming the surrounding air which expands, rises and is replaced by cooler air. On the other hand, the electric fan heater transfers most of its heat by mechanically directing a stream of air over a hot element; this is forced convection.

11.3 Radiation

Radiant heat (infra red radiation), like light, is considered to take the form of an electromagnetic wave; it travels at the same speed as light (300 000 km/s) in vacuum and air; it can be focused in the same way and is reflected, transmitted, absorbed and radiated by materials. Its essential differences from light are (*a*) it is not visible, (*b*) it has a longer wave-length ranging from 0·75 μm to about 100 μm, (*c*) when it falls on the skin it produces the sensation of warmth.

All substances radiate heat at all temperatures above absolute zero; thus if an attempt is made to measure the temperature of a hot gas surrounded by cooler walls using a thermocouple it is found that the temperature recorded is lower than that of the gas. The couple and the walls exchange radiation; thus the thermocouple grows cooler by emitting more radiation than it receives until it reaches radiation equilibrium with the cooler walls. They will then continue to transfer radiation equally to each other.

11.4 Temperature Scales

Many substances change their state from solid to liquid and liquid to gas at a constant temperature. The effect of barometric pressure on the solid/liquid change of state is negligible, but the liquid/gas change is affected by barometric pressure, an increase in pressure raising the boiling point and a decrease reducing it.

The positions of the freezing and boiling point graduations on a mercury thermometer are established by placing the thermometer in melting ice and then in steam from boiling water. In both cases distilled water is used, and for the boiling point allowance must be made for any deviation from a barometric pressure of 760 mm of mercury. These fixed points, defined as 0°C and 100°C, are often called the ice point and the steam point.

Mercury in glass thermometers depend on change of volume of the mercury with change of temperature. Other types of thermometers, based on other properties such as change of electrical resistance or change of pressure, give their own temperature scales and these agree exactly with each other only at the fixed points where their readings are defined as 0°C and 100°C or some other accepted equilibrium temperature. The scale established by a platinum resistance thermometer differs from the mercury in glass scale except at the fixed points, because the platinum resistance response does not follow the same law as that of mercury in glass. A comparison of two such scales is shown in Table 11.1.

TABLE 11.1

Mercury in glass (°C)	0	50	100	200	300
Platinum resistance (°C)	0	50·25	100	197	291

11.5 The Absolute Thermodynamic Temperature Scale

The absolute thermodynamic temperature scale was originally established in 1854. The term "thermodynamic" means that the temperature scale conforms to the laws of thermodynamics and is independent of any thermometric substance.

The Second Law of Thermodynamics applied to a "perfect" gas states that:

$$\text{Pressure} \times \text{Volume} = \text{a constant} \times \text{Absolute Temperature}$$

The scale may be established by various methods but the constant volume gas thermometer is the most widely used. The unit degree on the scale was originally defined as one hundredth part of the temperature interval between the freezing point and boiling point of pure water. This temperature interval was called the fundamental interval. The present thermodynamic scale is called the kelvin scale and its sole defining point is the triple point of water 273·16 K (0·01°C). The present kelvin degree does not differ from the degree originally established in 1854. The ice point corresponds to 273·15 on the kelvin scale and is thus zero on the Celsius scale.

No gases perfectly obey the gas laws but at a pressure of about one atmosphere their behaviour approaches the ideal and their temperature scales correspond. The departure of a gas from perfect behaviour can be observed and corrections made. The constant volume gas thermometer which consists essentially of a metal bulb filled with hydrogen or nitrogen is used to measure increase in pressure which, because temperature change is directly proportional to pressure change, provides a means whereby thermodynamic temperature may be established accurately.

The establishment of the thermodynamic scale by gas thermometry is a difficult technique requiring great skill, and in view of this the International Practical Temperature Scale was established which conforms closely to the absolute scale using instruments for fixed point determination which conveniently and accurately reproduce temperatures within the scale range.

11.6 The International Practical Temperature Scale (IPTS) 1968

The primary fixed points on the IPTS are shown in Table 11.2. They are determined at specified equilibrium states between phases of pure substances and in most cases at one standard atmosphere (101 325 Pa).

The methods used to establish the scale are shown at Table 11.3, and intermediate values between the points are established by the use of a formula appropriate to the interval under consideration. A large number of secondary points are also defined in the IPTS which allow intermediate temperatures on the primary scale to be determined and also provide fixed points for instrument calibration beyond 1337·58 K. Some of the secondary points are shown at Table 11.4.

TABLE 11.2 *The Defining Fixed Points of the IPTS* (1968)

Equilibrium state	*Kelvin scale* T K	*Celsius scale* t°C
Equilibrium between the solid, liquid and vapour phases of equilibrium hydrogen (triple point of equilibrium hydrogen)	13·81	−259·34
Equilibrium between the liquid and vapour phases of equilibrium hydrogen at a pressure of 33 330·6 Pa (25/76 standard atmosphere)	17·042	−256·108
Boiling point of equilibrium hydrogen	20·28	−252·87
Boiling point of neon	27·102	−246·048
Triple point of oxygen	54·361	−218·789
Boiling point of oxygen	90·188	−182·962
Triple point of water	273·16	0·01
Boiling point of water	373·15	100
Freezing point of zinc	692·73	419·58
Freezing point of silver	1235·08	961·93
Freezing point of gold	1337·58	1064·43

The English translation of the IPTS 1968 appeared in *Metrologia*, vol, 5, no, 2, April 1969.

TABLE 11.3 *Methods Used to Establish the IPTS*

Range	*Method*	*Output measured*
13·8 K to the antimony point 630·74°C	Platinum resistance thermometer	Electrical resistance
630·74°C to the gold point 1064·43°C	P1/P1 + 10% Rh thermocouple	Electromotive force
Above 1337·58 K (1064·43°C) the IPTS is defined by Plancks Law of Radiation		

TABLE 11.4

Substance[1] (*solid/liquid equilibrium*)	*Temp.* *T* K	*Temp.* *t* °C
Mercury	234·288	−38·862
Water (ice point)	273·15	0
Bismuth	544·592	271·442
Lead	600·652	327·502
Antimony	903·89	630·74
Nickel	1 728	1 455
Platinum	2 045	1 772
Rhodium	2 236	1 963
Iridium	2 720	2 447
Tungsten	3 660	3 387

[1] The triple points and boiling points of a number of gases are also specified for the range from 13·956 K.

There is no IPTS from absolute zero to 13·81 K, but temperatures in the range 0 K to 20 K are measured by magnetic, vapour pressure and ultrasonic (noise) thermometers.

11.7 Temperature-measuring Instruments

Temperature-measuring instruments can be divided into two groups, electrical and non-electrical.

Non-electrical methods

1. Liquid, vapour pressure and gas thermometers
2. Bimetal strip thermometers
3. Refractory cones, paints and crayons

Electrical methods

4. Electrical resistance pyrometers
5. Thermocouple pyrometers
6. (i) Total radiation pyrometers
   (ii) Photoelectric pyrometers
   (iii) Optical pyrometers

Those listed under 6 measure emitted radiation in various ways. Non-electrical methods are dealt with in sections 11.8–11.13 and electrical methods from section 11.14.

11.8 Liquid in Glass Thermometers

Mercury is usually used in liquid in glass thermometers although other liquids, such as alcohol and pentane, which have lower freezing temperatures than mercury and do not cause contamination through breakage, are also used, see Table 11.5.

TABLE 11.5 *Liquids used in liquid in glass thermometers*

Liquid	*Range* (°C)	*Range* (K)
Mercury	− 35 to +510	238 to 683
Alcohol	− 80 to + 70	293 to 343
Toluene	− 80 to +100	293 to 373
Pentane	−200 to + 30	73 to 303
Creosote	− 5 to +200	268 to 573

Increase in temperature causes the liquid to expand and rise up the stem. When measuring temperatures above the boiling point of mercury (357°C at atmospheric pressure) the space above the liquid is filled with nitrogen under pressure, thus raising the boiling point and allowing temperatures up to 510°C to be measured (Fig. 11.1).

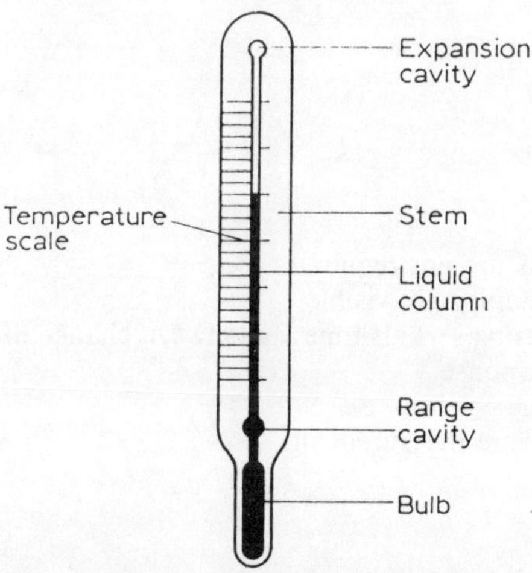

Fig. 11.1 LIQUID IN GLASS THERMOMETER

Range variation is provided for by incorporating a small cavity above the bulb which the mercury has to fill before rising up the stem. Liquid in glass thermometers have considerable heat capacity and do not respond quickly to rapid changes of temperature; and glass which has been heated and then cooled does not immediately return to its original volume, thus tending to produce a low reading, although for many purposes this discrepancy is of little importance.

Thermometers may be classified as "chemical" or "industrial", the former being used as standards in laboratories or for checking industrial plant, and are carefully and regularly calibrated against standard instruments when in use. Such calibrations may be carried out by the National Physical Laboratories or other independent authority, or by the manufacturer or user.

Industrial thermometers may be fitted into pipelines or other apparatus to give routine temperature indications. A protective sheath is often provided to reduce the risk of breakage; an extra long stem may be incorporated for dipping into hot liquids or the thermometer may be of the registering type to indicate maximum and minimum temperatures. Wherever possible thermometers are calibrated, standardized and used when totally immersed up to the readings but where necessary they may be standardized for partial immersion.

As with all measuring instruments, a tolerance must be applied to the accuracy of the readings. In the case of thermometers this is based on a percentage of the scale range but also depends upon how one degree division is divided. In general the smaller the scale length for one degree, and the greater the temperature range, the broader the accuracy tolerance. Special types having a small scale range, usually 5°C and a one degree interval 50 mm long, can give estimated values of the order of 0·001° to 0·002°C when used in conjunction with a low power telescope.

The essential characteristics of liquid in glass thermometers are summarized below.

Essential Characteristics

(1) Inexpensive.
(2) Simple.
(3) Easily portable.
(4) Fragile.
(5) Additional indication instruments are not required.
(6) Can be used only where stem readings are visible.
(7) Relatively high heat capacity causing serious time lag between change of temperature and thermometer response.
(8) Not suitable for distant reading.
(9) Not suitable for surface temperature measurements.

Application

SOLIDS The thermometer may be immersed in a mercury or aluminium powder filled hole in the solid to ensure rapid heat conduction. Steady temperature conditions in the area of measurement are necessary to ensure accurate readings and the hole volume should be small compared with the volume of the solid.

LIQUIDS Total immersion, i.e. immersion up to the column reading in the continually stirred liquid, is the most certain way to measure the mean liquid temperature but where part of the column is exposed corrections should be applied.

GASES Because gases are poor heat conductors, reading errors are caused by radiation exchange between the thermometer and the container walls which may be at a different temperature from the gas. Bright metal shields round the thermometer, forced gas flow or fanning are used to improve measurement accuracy. Vapour condensation must be avoided as the latent heat of condensation can cause reading errors.

11.9 Liquid in Metal Thermometers

Within the range −40°C to +650°C, and particularly where continuous indication is required at positions up to 65 m from the temperature point, liquid in steel thermometers offer many advantages. These instruments are of the bulb, capillary bourdon tube type, filled with liquid under pressure, and measuring change of volume of the liquid. The bulb has a high heat capacity and is subject to considerable lag but it has the advantage of being more robust than a glass thermometer. The most widely used liquids, with their corresponding ranges of application, are shown in Table 11.6.

TABLE 11.6 *Liquids used in liquid in metal thermometers*

Liquid	*Range* (°C)	*Range* (K)
Mercury	−39 to +650	234 to 923
Xylene	−40 to +400	233 to 673
Alcohol	−46 to +150	227 to 423
Ether	+20 to + 90	293 to 363

The bourdon tube may be spiral or helical and under the influence of increasing temperature the volume of liquid expands tending to straighten or unwind the tube; the movement of the free end is transmitted through a linkage to a pointer which moves over a scale.

The main advantages of the instruments are as follows:

(1) An approximately linear scale.
(2) Wide temperature range.
(3) Ample power available to operate the bourdon tube giving good pointer control.
(4) The liquid can be under high pressure thus reducing "head" errors.
(5) There are no inaccuracies due to barometric pressure variations.
(6) Long capillaries allow large differences in bulb and indicator levels.
(7) Remote indication of temperature; errors in the system may arise from:
   (*a*) Bourdon tube zero errors, e.g. if the tube temperature changes then the indicator will give incorrect readings; this may be compensated by means of a bimetal helix attached to the indicator and acting in opposition to the error.
   (*b*) Capillary tube errors due to change of temperature. This error may be reduced in one of the following ways: a micro-bore capillary ensuring that any volume change is insignificant compared with the temperature effect on the bulb, or a second capillary alongside the main capillary not attached to the bulb and operating a second bourdon tube mechanically connected to the main bourdon tube in opposition to the temperature error (Fig. 11.2).

(*c*) Thé cross-section of the capillary may be made into an annulus by inserting an "invar" wire along its whole length which reduces the volume of liquid in the capillary so that its cubical volume change when subjected to temperature change is exactly equal to the change in volume in the space between the wire and the capillary walls. Another method based on the same principle is the introduction of compensating chambers at intervals along the capillary tube. These chambers compensate for both the liquid in them and also for the liquid in the intervening portions of capillary tube (Fig. 11.3).

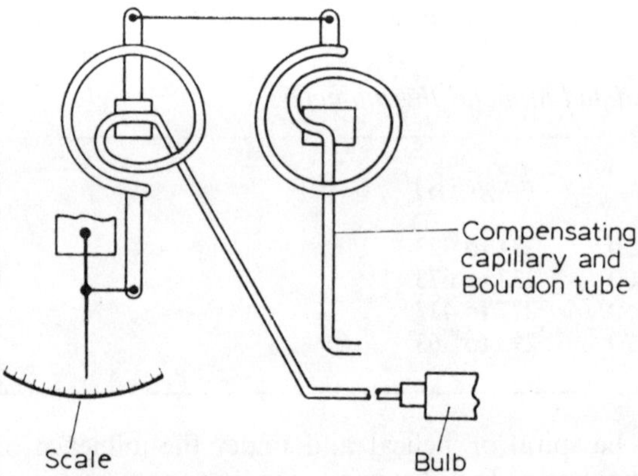

Fig. 11.2 COMPENSATED LIQUID IN METAL THERMOMETER

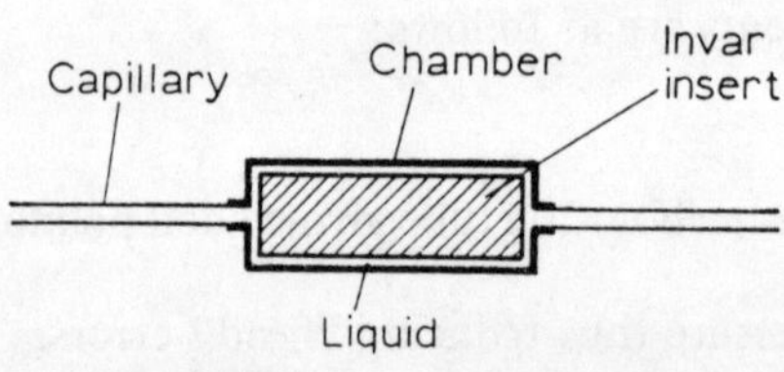

Fig. 11.3 CAPILLARY COMPENSATING CHAMBER

Accuracy This is of the order of $\pm\frac{1}{2}\%$ of the scale range up to 300°C and $\pm 1\%$ above this temperature.

Bulbs and Pockets

The bulb size of these instruments usually ranges from about 25 mm up to 375 mm in length and 12 to 18 mm in diameter depending on requirements, and to reduce time lag to a minimum the surface area to liquid volume ratio is made as large as possible. To register small temperature changes the volume of liquid in the bulb must be increased, thus small temperature ranges require longer bulb lengths. Increased bulb length may be achieved by using special bulb shapes to improve sensitivity (Fig. 11.4). The bulb may be directly immersed, in which case the instrument response is rather better than the mercury in glass thermometer which often requires a metal pocket for protection. When used in high pressure systems above 3·5 MPa it is usual to use a pocket which enables the bulb to be removed without emptying the system. Brass, titanium, monel and specially coated pockets are among the metals and methods used to afford corrosion protection.

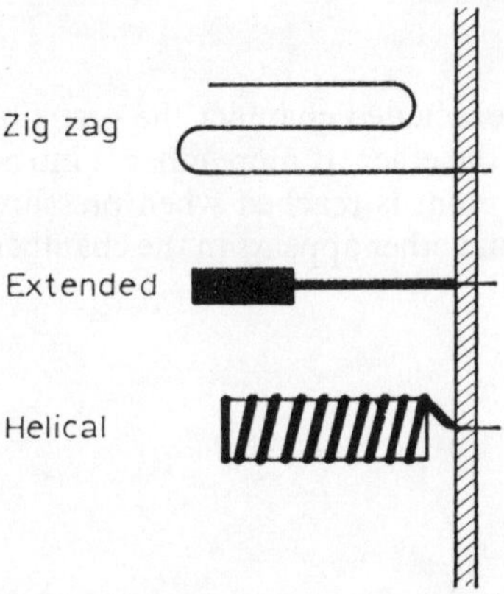

Fig. 11.4 SPECIAL BULB SHAPES

11.10 Constant Volume Gas Expansion Thermometers

In this bulb, capillary, bourdon tube thermometer used mainly for measuring liquid temperatures, change of temperature is sensed through change of pressure of an inert gas which fills the instrument, and the instrument thus acts as a pressure gauge. Temperatures in the range −130°C to 540°C may be measured, a given instrument range being not usually less than 50°C.

The main advantage of this instrument is its rapid response to temperature change as gas has a much lower heat capacity than liquids or solids.

Gases have a much higher coefficient of cubical expansion than liquids or solids, e.g. N_2 $3{\cdot}6744 \times 10^{-3}$ and Ne_2 $3{\cdot}6617 \times 10^{-3}$ as compared with Hg $0{\cdot}181 \times 10^{-3}$ and carbon steel $0{\cdot}033 \times 10^{-3}$, but the ratio of bulb size to capillary plus bourdon tube must be very large if ambient temperature effects on the latter are to be small enough not to affect the accuracy of the instrument.

Capillary Errors

(1) Capillary errors increase with bulb temperature.
(2) Capillary errors decrease as the differential range increases (maximum temperature minus minimum temperature).

Two methods are used to compensate for these errors:

(1) A large differential range,
(2) A second or auxiliary capillary as described in section 11.9.

By adequate compensation and careful design these thermometers offer the advantages of low temperature measurement, small bulb size compared, say, with mercury in steel, and long capillaries (up to 60 m).

Bourdon tube compensation is by means of a bimetal strip; there are no "head" errors, but changes of barometric pressure or altitude make zero reading adjustment necessary.

Accuracy is $\pm\frac{1}{2}\%$ of the range up to 320°C and above this range $\pm 1\%$.

11.11 Vapour Pressure Thermometers

If a vapour, e.g. ether, is introduced into an evacuated chamber the pressure in the chamber rises owing to the pressure of the ether. If more ether is introduced the pressure continues to rise until a point is reached when pressure change does not occur, a small quantity of liquid ether appears in the chamber

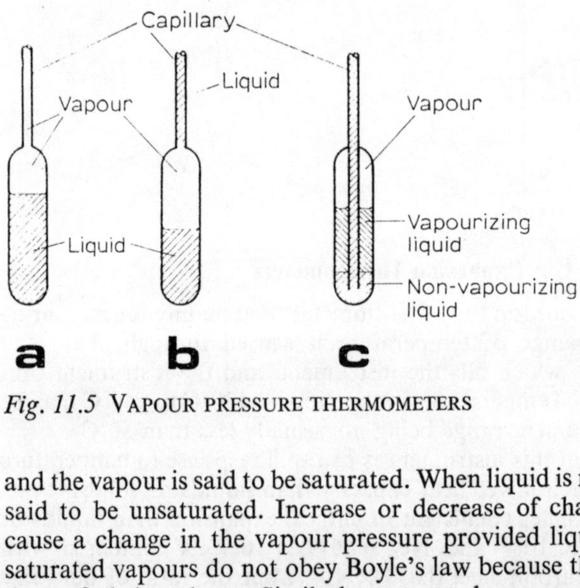

Fig. 11.5 VAPOUR PRESSURE THERMOMETERS

and the vapour is said to be saturated. When liquid is not present the vapour is said to be unsaturated. Increase or decrease of chamber volume does not cause a change in the vapour pressure provided liquid is still present; thus saturated vapours do not obey Boyle's law because their pressure is independent of their volume. Similarly, saturated vapours do not obey Charles's law because their pressure changes roughly exponentially with change in temperature.

The vapour pressure thermometer has a bourdon capillary and bulb partially filled with a volatile liquid such as methyl chloride or sulphur dioxide; the rest of the system is filled with vapour. An increase in temperature at the bulb causes further vaporization, the resulting pressure increase causing movement of the bourdon tube.

The condition described occurs in the instrument when the ambient temperature of the capillary and bourdon tube is higher than that of the bulb (Fig. 11.5*a*). When the ambient temperature of the capillary and bourdon is less than that of the bulb they are filled with condensed vapour, that is, liquid, and the pressure of the vapour in the bulb is transmitted through the liquid (11.5*b*). The rearrangement of the liquid/vapour system and certain other closely related factors involving lag in response and inaccuracy of reading makes vapour pressure thermometers unsuitable for cross-ambient temperature measurement. Differences in level between the bulb and the bourdon tube from that established at calibration introduce a "head" error, for although the pressure at the bulb may remain constant the pressure due to a capillary full of vapour will be less than that due to a similar column of liquid.

A thermometer which overcomes the cross-ambient difficulties has the bourdon, capillary and part of the bulb filled with a non-vaporizing liquid (11.5*c*), which prevents the volatile fluid from entering the capillary and bourdon tube, but these can be successfully used only when the bulb is in the upright position.

TABLE 11.7 *Liquids used in vapour pressure thermometers*

Liquid	*Boiling point* (°C)	*Critical*[1] *temperature* (°C)	*Range* (°C)
Methyl chloride	−23·7	143·1	0–50
Diethyl ether	34·5	193·8	60–160
Ethyl alcohol	78·5	243·1	90–170
Water	100	374·0	120–220
Toluene	110·5	320·6	150–250
Argon	−185·7	−122	(2)

[1] The temperature at which both liquid and vapour have equal densities.
[2] Used for measuring the temperature of liquid gases down to −253°C (20 K).

Vapour pressure thermometers have a lower initial cost than mercury in steel but because of their limitations they are unsuitable for universal application. Their main characteristics are as follows:

(1) A non-linear scale; this can be modified by a lever system or by choosing an instrument whose scale over the mid-section into which the temperature readings should fall is approximately linear.
(2) Scale range; this is of the order of 100°C or less and is accurate to ±1% of the differential range, e.g. diethyl ether 60°C–160°C.
(3) The time lag of these instruments is much less than for most other types.
(4) Non-ferrous metals can be used throughout.
(5) Distant reading 65 m.
(6) Temperatures between −50°C and 310°C are usually measured but the range for most commercial instruments lies within 0°C and 250°C.
(7) Normal ambient temperature change at the capillary and bourdon does not appreciably affect the accuracy of the dial reading and is virtually independent of the capillary length.
(8) Barometric altitude and head errors are corrected by resetting the pointer.

(9) Smaller bulb sizes can be used than for a mercury in steel thermometer of the same range.
(10) Where the introduction of mercury or other contaminants into the measured substance through breakage might be harmful, e.g. food, a harmless filler liquid can be chosen.

A selection of liquids used in vapour pressure thermometers is given in Table 11.7.

11.12 Bimetal Strip Thermometers

A bimetal strip consists of two different metals or alloys having different coefficients of expansion (e.g. invar and brass) firmly bonded together by welding. When the strip is heated or cooled, differential expansion or contraction occurs causing the strip to bend (Fig. 11.6). The metal having the higher coefficient of expansion is shown hatched on the diagram.

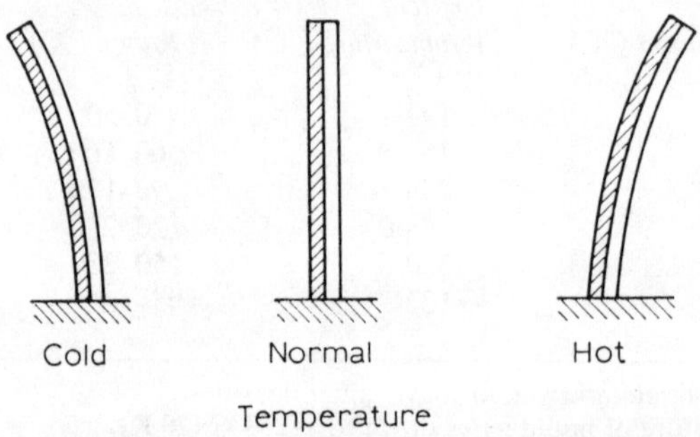

Fig. 11.6 BIMETAL STRIP

Many switching devices, such as those in oven thermostats, electric irons and car winker lamps, operate on this principle. They are also used as temperature sensors or as compensators for ambient temperature change in liquid, in metal, vapour pressure, gas expansion thermometers, aneroid barometers, and as balance wheel compensators in some watches.

The usual industrial form of bimetal thermometers (Fig. 11.7*a*) is a continuous strip wound into several compensating helices one inside the other; this has the advantage of compactness while providing the long length of strip required for adequate indicator movement, and also ensures that there is no lateral displacement of the free-floating transmission spindle which need not then be constrained by positive bearings.

Spiral bimetal strip thermometers (Fig. 11.7*b*), are widely used for measuring ambient and oven temperatures and clip-on types are available for checking the temperature of hot water pipes. Steam chambers, asphalt, tar macadam, dough and paint are among the many applications for these thermometers.

Essential Characteristics

(1) Inexpensive: often used instead of liquid in glass thermometers.
(2) Compact: the volume of metal used in the protective sheath is small and thus of low thermal capacity, reducing lag.
(3) Robust: good resistance to mechanical shock due to freely floating spindle.
(4) Close linearity throughout the temperature range.
(5) Range of application: low temperature −30°C to 200°C; high temperature, 0°C to 550°C.
(6) Accuracy: (low temperature) 1% of scale range; (high temperature) 2% of scale range.
(7) Stem diameters as small as 4 mm are available. Stem lengths range from 75 mm for liquids of low specific heat and thermal conductivity, e.g. water. Longer stems are required for oil, and for air and gas 250 mm may be required.

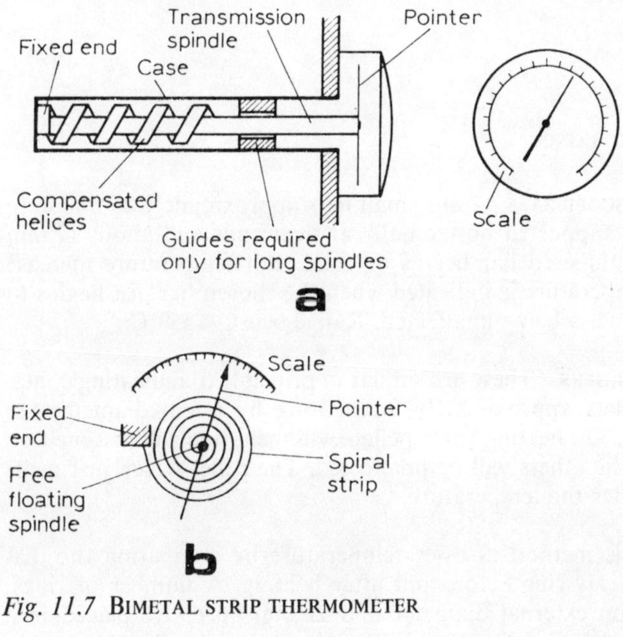

Fig. 11.7 BIMETAL STRIP THERMOMETER

In dry, clean non-corrosive gases stems may be perforated to improve response, and stainless iron (13% Cr) which has low conductivity may be used to reduce heat leakage.

11.13 Temperature Indication by Change of State of Solids

Temperature indication by change of state of solids falls into two groups:

(1) *Change of shape or size*. This group includes pyrometric cones (Fig. 11.8), thermoscope bars, heat recorders and rings.
(2) *Change of colour*. Paints and crayons.

PYROMETRIC CONES These are made of various mixtures of materials, china clay, talc, feldspar, quartz, etc., which are arranged in a numbered series to correspond to temperatures between 600°C and 2000°C. A minimum of three cones having various melting points are mounted on to a refractory plate. The required temperature is indicated when the selected cone reaches its end point, i.e. its tip just touches the plate (Fig. 11.8).

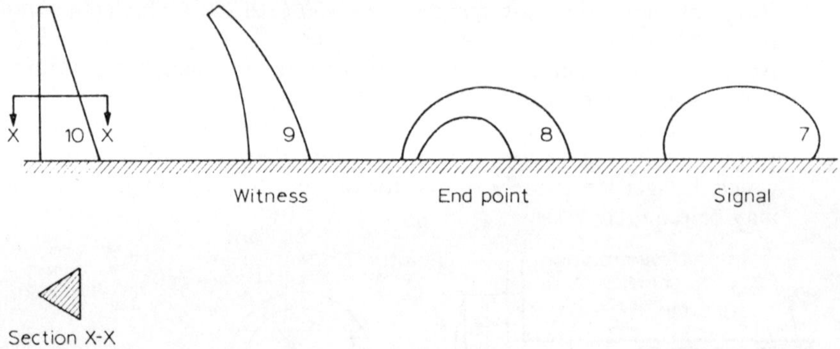

Fig. 11.8 PYROMETRIC CONES

HOLDCROFT THERMOSCOPE BARS Four small bars approximately 55 mm long by 6 mm square are supported horizontally at their ends and about 12 mm apart. The lowest numbered bar begins to bend with temperature increase and the required temperature is indicated when the chosen bar just begins to bend, the remaining bars being unaffected. Range 600°C–1550°C.

WATKINS HEAT RECORDERS These are similar in principle to indicating cones. Five cylindrical pellets approximately 9 mm long by 6 mm diameter are mounted in a holder. On heating some pellets will melt, one will just melt on its top surface and the others will be unaffected. The pellet which just melts on its surface indicates the temperature.

BULLERS RINGS This method assesses temperature by comparing the diameter of an unfired clay ring before and after heating. A number of rings, approximately 64 mm external diameter and 22 mm bore, are placed in a holder, several holders being distributed throughout the kiln. Rings are withdrawn during the heating cycle, cooled and measured. Firing is completed when a given contraction is obtained. Range 960°C–1400°C.

PAINTS AND CRAYONS These are useful for small-scale heating of steel parts, weldments, for stress relief and pre-heating, etc. The paint or crayon changes colour or appearance at a fixed temperature.

Cones, bars, recorders and rings are used in heat treatment kilns for ceramic wear, grinding wheels, bricks, refractories, electrical porcelain, earthenware, sanitary ware, tiles and china, and temperatures are measured to about ±10°C.

They are often used in conjunction with a pyrometer which will give the actual temperature throughout a firing cycle at a particular point in the furnace or kiln, and act as a check against the rate of heating which must not be too rapid if correctly fired products are to be obtained.

11.14 Thermoelectricity

In certain circumstances heat may be converted into electrical energy or electrical energy converted into heat. This thermoelectric reversible process should not be confused with the irreversible conversion of electrical energy

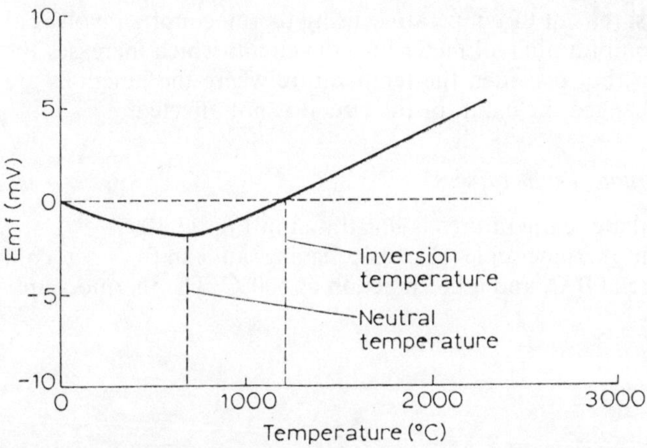

Fig. 11.9 VARIATION OF E.M.F. WITH TEMPERATURE (W–Mo thermocouple)

into heat (the Joule effect) that occurs when an electric current passes through a resistance causing an increase in temperature of the resistance.

Thermoelectric effects arise in two ways:

(1) A potential difference always exists between two dissimilar metals in contact with each other.
(2) A potential gradient exists even in a single metal if there is a temperature gradient.

Thus if two dissimilar metals are joined together to form a closed circuit (a thermocouple) and one junction is maintained at a different temperature from the other, a current will flow spontaneously through the circuit; this is called the Seebeck effect. The magnitude of the current and its direction depends upon the thermoelectric characteristics of the two metals concerned and the temperature of their junctions.

The variation of thermocouple e.m.f. with temperature is illustrated by the curve for the tungsten/molybdenum thermocouple (Fig. 11.9).

In this case when the cold junction temperature is constant at 0°C, an increase in temperature at the hot junction produces an e.m.f. increasing in the negative direction, and at about 800°C the e.m.f. is a maximum; this is known as the neutral temperature. Further increasing the temperature to about 1200°C reduces the e.m.f. to zero; this is known as the inversion temperature, beyond which the e.m.f. becomes positive. If a thermoelectric pyrometer is used to measure a range of temperature in which the neutral temperature for the couple falls, then two different temperatures will give the same e.m.f. value; thus a temperature range measured by a thermocouple should not include its neutral or inversion temperature.

The Law of Intermediate Metals

In practice the measurement of temperature using thermocouples involves at least the introduction of a millivoltmeter into the circuit which increases the number of junctions, but provided the temperature where the junctions are introduced is not changed the e.m.f. of the circuit is not affected.

The Law of Intermediate Temperatures

The law of intermediate temperatures is illustrated in Fig. 11.10.

Consider a given thermocouple (11.10*a*), having an e.m.f. E_1, a cold junction temperature at 0°C, and a hot junction at 500°C. The thermocouple

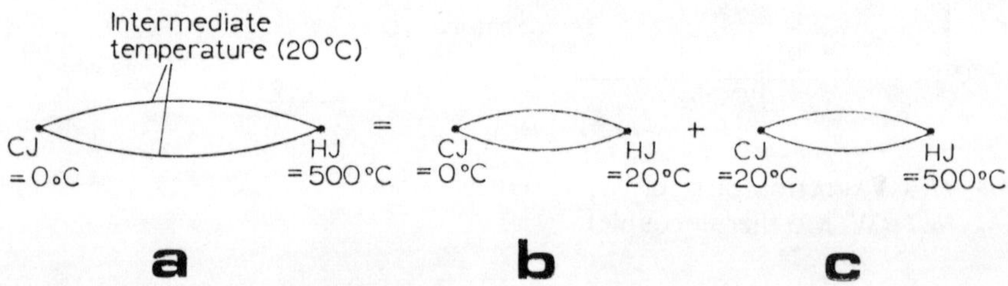

Fig. 11.10 LAW OF INTERMEDIATE TEMPERATURES

is the sum of two other similar couples, one having a cold junction at 0°C and a hot junction at 20°C (11.10*b*), the other having a cold junction at 20°C and a hot junction at 500°C (11.10*c*). The e.m.f.'s for these two couples are designated E_2 and E_3 respectively. Thus

$$E_1 = E_2 + E_3$$

This law is used when making corrections to thermocouple readings when the cold junction temperature is different from the temperature at which the couple was calibrated. Thus if a thermocouple calibrated at 0°C is used at a cold junction temperature of 17°C, the thermocouple will read correctly if the "lost" e.m.f. is added to the reduced e.m.f.; the "lost" e.m.f. values may be taken from tables of corrections for the couple concerned. In the case of base metal couples where the temperature/e.m.f. relationship is virtually linear the addition of the difference between the calibration and actual temperature to the scale reading is all that is required; for precious metal couples the correction is between 0·4 and 0·5 of this difference. Adjustment for ambient temperature variation can be made by disconnecting the cold junction and resetting the scale pointer to ambient temperature, the junction then being reconnected. Automatic compensation can also be incorporated by means of a bimetal strip which moves the scale pointer the required amount.

11.15 Thermoelectric Pyrometers

The essential features of thermoelectric pyrometers are:

(1) Two dissimilar conductors in the form of rod or wire electrically insulated except at the hot junction. The hot junction wires may be welded or soldered together or may be completely separated depending upon their application.
(2) A refractory sheath to afford protection from injurious furnace gases. An outer metal sheath may also be used to prevent mechanical damage.
(3) Cold junction temperature control.
(4) An instrument for measuring the e.m.f., either a millivoltmeter or for more accurate measurement a potentiometer system.
(5) Compensating leads to allow the measuring instrument to be sited at a considerable distance from the couple without the necessity for using expensive couple materials as extension leads. Compensating leads for precious metal thermocouples are usually pure copper and a suitable copper nickel alloy which have the same thermoelectric e.m.f. as the couple wires themselves, but within the temperature range of the cold junction. A ballast resistance is incorporated in the millivoltmeter to ensure that it has a large resistance compared with the couple circuit so that any temperature changes and thus any changes of resistance of the compensating leads will be too small to affect the accuracy of the system (Fig. 11.11).

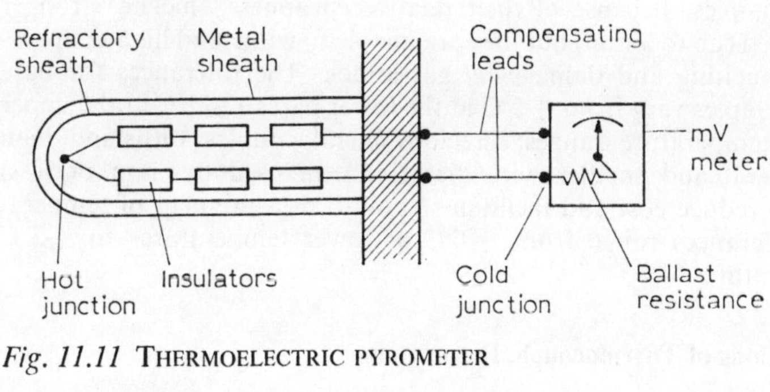

Fig. 11.11 THERMOELECTRIC PYROMETER

Cold Junction Temperature Control

This may be achieved by immersing the cold junction in a thermos flask filled with melting ice, but this is only suitable under laboratory conditions.

A modern method for both laboratory and industrial cold junction control is the Zeref/Zerac method, manufactured by De La Rue Ltd. (Fig. 11.12).

The cold junction is immersed in a copper cylinder containing double distilled water; the cylinder is closed at one end and fitted with a bellows at the other end. Heat is extracted by a compact thermoelectric solid state heat pump. As ice is formed the resulting expansion inflates the bellows which at a certain stage operates a microswitch and stops the pump; the ice begins to turn to water and the bellows deflates, eventually switching on the heat pump; thus the cold junction is kept at the true ice point of water. The Zerac ice point reference system can maintain 100 cold junctions at the reference temperature.

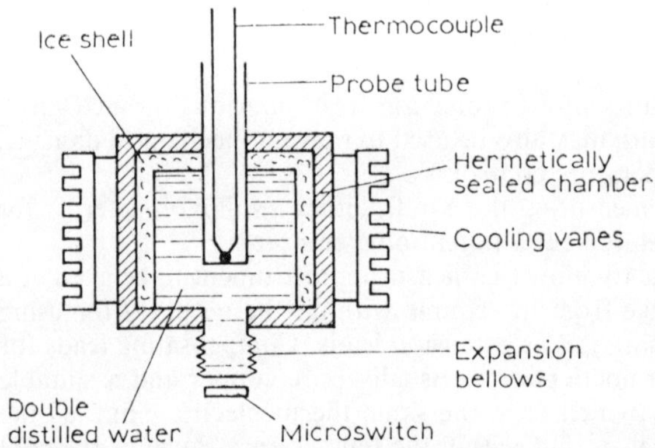

Fig. 11.12 ZEREF SOLID STATE COLD JUNCTION SYSTEM

Thermocouples in common use are listed in Table 11.8.

The e.m.f.'s generated in base metal couples are much greater than those generated in precious metal couples and they are thus more sensitive to small temperature changes. Because of their relative cheapness, thicker, stronger wires can be used (up to 3 mm) but they are unable to withstand high temperatures without melting and damage by gas attack. The tolerances for base metal thermocouples vary from ± 2°C at the lower part to $\pm\frac{3}{4}$°C at the upper part of their temperature ranges. Precious metal couples withstand high temperatures well and smaller wire diameters are used (0·5 mm or less) which helps to reduce cost and facilitates "point" measurement of temperature. Their tolerances range from ± 1°C at lower temperatures to ± 3°C at high temperatures.

11.16 Applications of Thermocouple Pyrometers

Because of its reliability and sensitivity the Pl./Pl.-10% Rh couple is used to define the International Practical Temperature Scale in the range 630·74°C to 1064°C, and in general because of their versatility thermocouples can be made to suit virtually any temperature measurement application, although it may not always be convenient or economical to do so. Apart from normal

TABLE 11.8 *Common Thermocouples and their Characteristics*

Type	*Thermocouple*	*Approximate sensitivity millivolts* (°C)[1]	*Normal temperature range* (°C)	*Maximum temperature* (°C)[5]	*Service characteristics*
Base metal	Copper/Constantan[2]	0·03	−200 + 400	+ 600	High resistance to condensed moisture corrosion.
Base metal	Iron/Constantan	0·05	−200 + 800	+ 1000	A low-cost couple withstanding oxidation and reducing atmospheres.
Base metal	Chromel[3]/Alumel[4]	0·04	−200 + 1200	+ 1350	Withstands oxidizing atmospheres.
Precious metal	Platinum/Platinum 90% /Rhodium 10%	0·006	0 + 1400	+ 1700	Withstands oxidation but e.m.f. is affected by the presence of metals, carbon or silicates in a reducing atmosphere.

[1] At lower temperatures, e.g. 20°C.
[2] 60% Cu. 40% Ni (Eureka).
[3] 90% Ni. 10% Cr.
[4] 95% Ni. 2% Al. 3% Mn.
[5] Short time immersion (seconds).

industrial applications, furnace temperature measurement, etc., thermocouples are manufactured small enough for such delicate applications as the temperature of nerves, tiny insects, transistors, and so on. Thermal conduction along the couple wires can be reduced almost to vanishing point by the use of very fine wires which may be sited in the most difficult and inaccessible places. Thermocouples are used for measuring the temperature of solids, liquids and gases, and the hot junction must be applied in such a way as to suit the characteristics of the substance measured.

In Fig. 11.13 *a* to *k* various methods of thermocouple application are indicated.

Fig. 11.13 *a* to *d* illustrate methods of surface measurement. The junction is welded to the surface (capacitance discharge welding offers an excellent method of attachment), or soldered. At *a* the wires run along the surface and in close contact with it for a short distance, which reduces or eliminates heat conduction away from the junction. At *b* an alternative method is shown where direct attachment is unsuitable. The junction is attached to a thin copper plate which is firmly held in position by clamping or strapping to ensure good heat transfer to the bead. At *c* the surface temperature may be estimated by extrapolation along X–X. Brick or concrete walls, etc., may be measured as shown at *d*, alternatively a piece of asbestos sheet may be used to cover the groove which contains the couple.

Diagram *e* illustrates temperature gradient measurement across a wall and at *f* pairs of couples are shown enabling average temperatures across a liquid medium to be measured. Care must be taken when measuring still liquids to ensure that the liquid is fully mixed to prevent local temperature variations, and the couples should not be sited in "dead" areas, e.g. corners. Heat transfer to the hot junction can be improved by immersing the couple in mercury or aluminium powder as illustrated at *g* and *h* but this method limits the application of the couple to the upright position. Liquid metals may be measured by direct insertion of the unjoined couple, but care must be taken to ensure that insertion is well below the surface slag (*j*). A disposable silicon protective sheath is used when measuring liquid steel, and the immersion time is only a few seconds.

Gas temperature measurement poses particular difficulties because of radiation interchange between the couple and the cooler enclosure or furnace walls; this means that the couple indicates a lower temperature than that of the gas in which it is immersed. The suction pyrometer is used to overcome this difficulty (*k*). Gas is drawn from the furnace and flows over the couple which is protected by a series of metal or refractory radiation shields depending on the gas temperature and conditions. The function of the shields is to maintain equal radiation between themselves and the couple, thus enabling the couple to measure the actual gas temperature.

Thermopiles

Series-connected thermocouples in which a number of hot junctions are laid side by side or in star formation are called thermopiles (Fig. 11.14). The total output is equal to the sum of the e.m.f.'s and a much more sensitive element

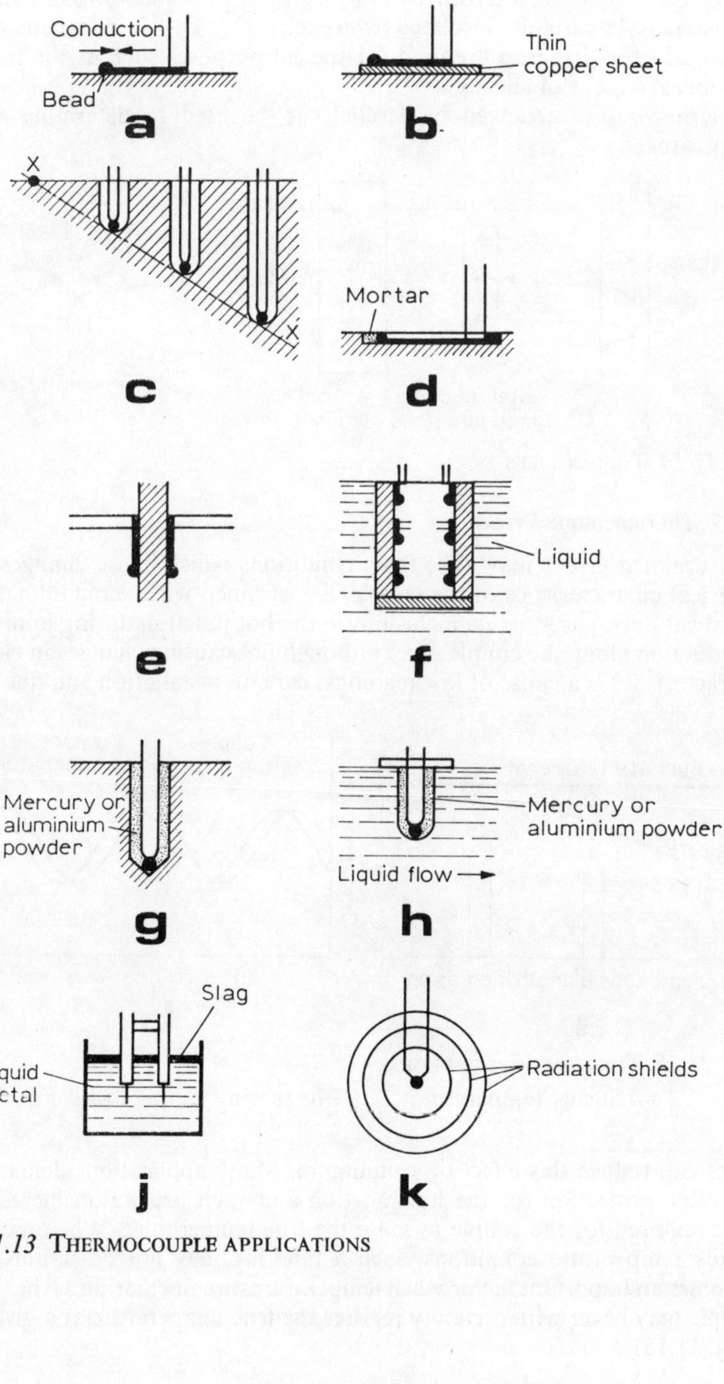

Fig. 11.13 THERMOCOUPLE APPLICATIONS

is obtained. A chromel/constantan thermopile having 14 couples gives about 1 mV/°C and achieves a sensitivity of 0·002°C. The hot junctions of individual couples must be carefully insulated from each other. Thermopiles consisting of semiconductors in series are used for special purposes such as the temperature measurement of sheet glass.

Thermocouples arranged in parallel can be used to determine average temperatures.

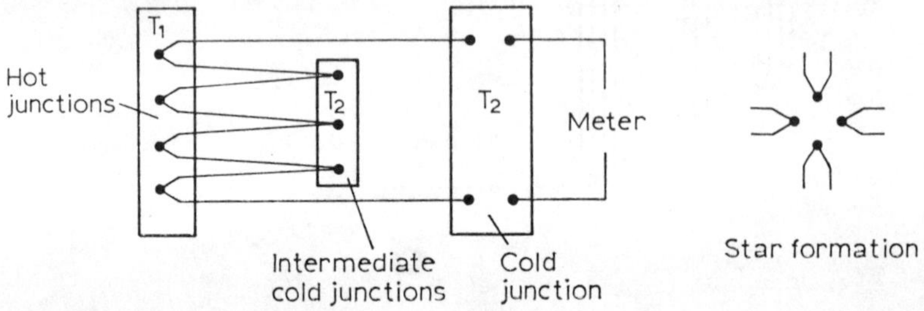

Fig. 11.14 THERMOPILES

11.17 Thermocouple Errors

Measurement errors may arise from conditions which cause changes in the electrical characteristics of the couple, e.g. strained wires, contamination by metal vapours, gases, oxide inclusions in the hot junction during joining, etc. Conduction along the couple wires although not causing changes in electrical characteristics is a cause of low readings; care on installation and fine couple

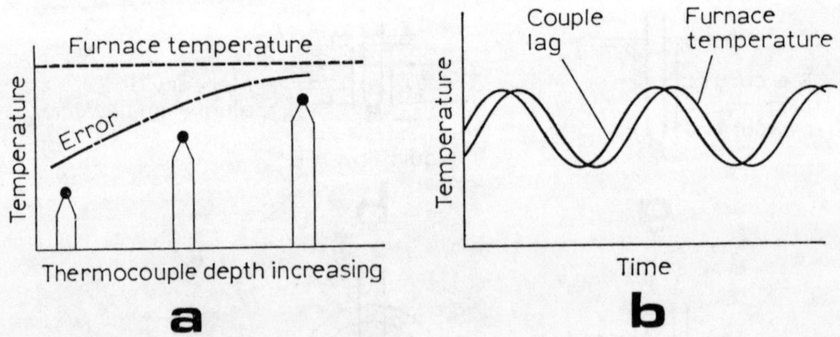

Fig. 11.15 THERMOCOUPLE ERRORS
a Steady temperature *b* Fluctuating temperature

wires can reduce this effect to a minimum. Many applications demand considerable protection for the hot junction and such protection increases the time required for the couple to sense the true temperature. When measuring steady temperature conditions, such a time lag may not be serious, but it becomes an important factor when temperatures are fluctuating as the thermocouple may never with certainty register the true temperature at a given time (Fig. 11.15).

A simple check for non-metallic inclusions in the couple circuit is to apply a flame to one of the wires in the circuit. If the circuit is sound no change of reading should be seen on the meter as the e.m.f. gradients developed balance each other (see also section 11.14).

11.18 Temperature Measurement by Detection of Radiation

A body radiating heat at a given intensity emits a range or spectrum of radiation of varying wavelengths and as temperature falls the radiation intensity also falls. The fall in radiation intensity is accompanied by a corresponding increase in radiation wavelength. If the temperature of the body is

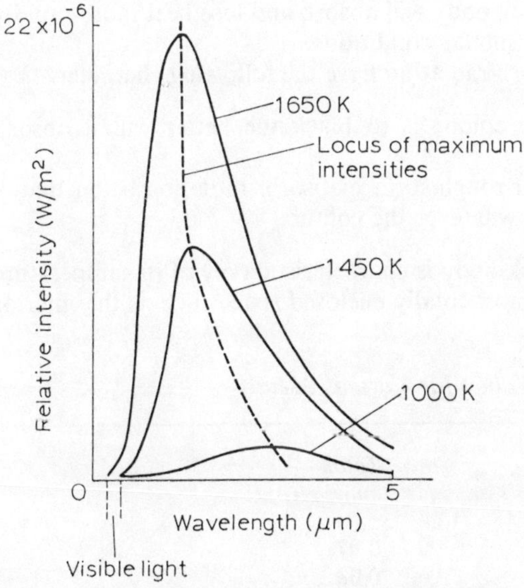

Fig. 11.16 BLACK BODY RADIATION INTENSITIES

increasing, then the intensity of radiation increases and the average radiation wavelength decreases. At about 500°C some of the radiation wavelengths are small enough to constitute visible light and the body begins to glow red; a further increase in temperature increases the proportion of light wavelengths emitted and when the body is white hot it is emitting both infra red (heat) and the whole range of visible (light) radiation. A small amount of ultra violet radiation which has an even shorter wavelength than visible light is also emitted at higher temperatures, the proportion of these radiations increasing with temperature (Fig. 11.16).

The Black Body

Heat is normally transferred by a combination of conduction, convection and radiation.

Transfer by conduction requires a material along which heat can flow, and transfer by convection requires the displacement of a heated material. Radiation requires no medium for heat transfer; thus if a radiating body were suspended in a vacuum all heat lost or gained by the body would be by radiation, as conduction and convection would be entirely prevented by the vacuum.

Likewise, if the surface of the body were a perfect reflector, then the quantity of heat in the body would not change and thus it would remain at constant temperature. It must be emphasized that such conditions do not arise in practice, although very highly reflective surfaces can be manufactured.

A body which is able to absorb all radiations falling upon it, reflecting and transmitting none, is called a "black body" and this ability depends on the condition of its surface. A black body will absorb and lose heat more rapidly than a non-black body under similar conditions.

Surfaces which readily absorb radiation have the following characteristics:

(1) Colour: the nearer the colour is to black the better will it absorb radiation.
(2) Surface finish: a matt or rough surface absorbs radiation better than a smooth, bright surface, whatever the colour.

The heat radiated by a black body is characteristic only of its temperature and in practical terms any almost totally enclosed space such as the interior of a muffled furnace fulfils the requirements of black body radiation. Thus if the radiation issuing from a small aperture in a furnace can be measured, the results will indicate the temperature of the interior of the furnace. Total radiation pyrometers are used for this purpose.

TABLE 11.9 *Total Emissivity Values for Various Metals*

Material	*Condition*	*Total emissivity*
Steel plate	Rough	0·97
Steel	Polished	0·08
Cast iron	Strongly oxidized	0·95
Cast iron	Polished	0·21
Aluminium	Oxidized	0·11
Aluminium	Polished	0·028

N.B. Emissivity values vary with temperature and tend to increase with temperature.

A body which is not radiating as a black body radiates less heat and the ratio of this to black body radiation is called the *total emissivity* of the surface. Thus a black body can be said to have an emissivity of one, whereas bodies which reflect or transmit some of the radiation falling upon them have an emissivity of less than one.

If E = radiation falling upon the body (W/m^2)
a = fraction of radiation power absorbed by the body
e = total emissivity

$$e = \frac{aE}{E} = \frac{\text{Radiation emitted from the body}}{\text{Radiation falling upon the body}}$$

A selection of total emissivity values for various materials is shown in Table 11.9.

The Stefan–Boltzmann Law

The total energy radiated by a black body is characteristic only of its absolute temperature and is proportional to the fourth power of that temperature.

If E = total energy radiated per unit area per unit time, then

$$E \propto K^4 \qquad \text{or} \qquad E = \sigma K^4$$

Stefans constant $\sigma = 5{\cdot}7 \times 10^{-8}\ WK^4/m^2$ approximately where K is the absolute temperature.

In practice, the pyrometer besides receiving heat from the radiating source also radiates heat to the source, thus the net radiation affecting the pyrometer will be:

$$E = \sigma(K_2^4 - K_1^4)$$

where K_1 is the pyrometer temperature and K_2 the temperature of the radiating source, but the fourth power of K_1 is small compared with that of K_2 and may be neglected.

If the emissivity of the body at the given temperature is e, then

$$E = e\sigma K^4 \qquad \text{(i)}$$

The temperature of a body radiating in the open can be found if the emissivity of its surface is known.

If K is the true temperature and K_a is the apparent temperature taken with a radiation pyrometer positioned so that the body fills the whole field of view, then

$$E = \sigma K_a^4 \qquad \text{(ii)}$$

The energy received by the pyrometer equals the quantity of energy which it would receive from a perfect black body radiating at temperature K_a, that is, at a lower temperature than the temperature of the body being measured. Now the energy emitted by a non-black body is expressed by equation (i).

Equating equations (i) and (ii) we have:

$$e\sigma K^4 = \sigma K_a^4 \qquad \text{i.e. } K^4 = K_a^4/e$$

Therefore

$$K = K_a/\sqrt[4]{e} \qquad \text{(iii)}$$

11.19 Total Radiation Pyrometers

Total radiation pyrometers are used mainly in the range 700–2000°C. The advantage of most types is that they are not in contact with the hot body. They measure all the radiations emitted from the hot body and are calibrated for black body conditions. The radiation emitted from the furnace is focused on to the hot junction of a thermocouple, a thermopile, or a platinum resistance element which are blackened to improve their ability to

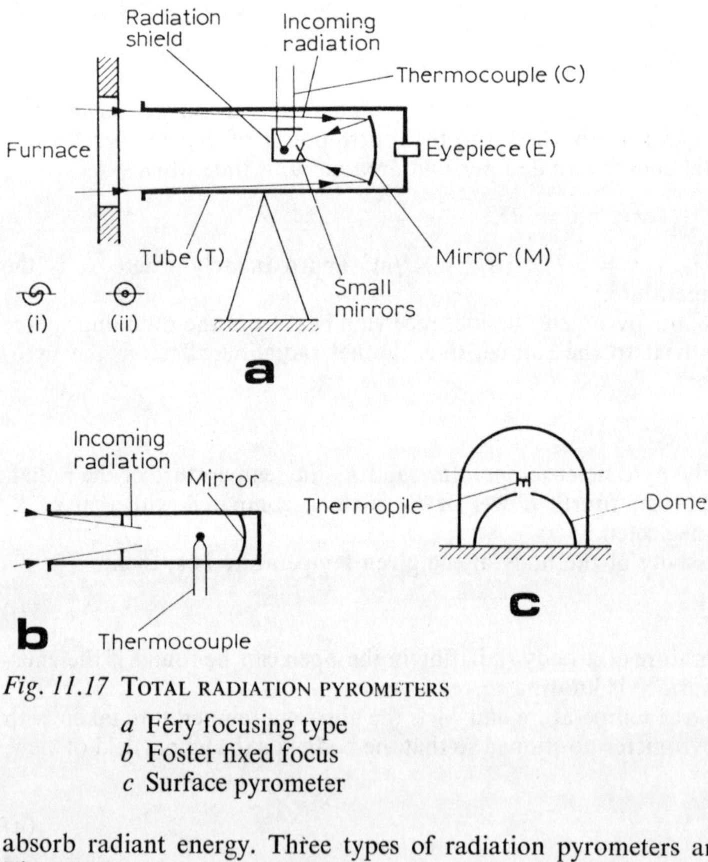

Fig. 11.17 TOTAL RADIATION PYROMETERS
a Féry focusing type
b Foster fixed focus
c Surface pyrometer

absorb radiant energy. Three types of radiation pyrometers are shown in Fig. 11.17.

Féry Total Radiation Pyrometer

The main features of a Féry pyrometer (11.17*a*) are:

(*a*) A blackened tube T open at one end to receive radiation and carrying an adjustable eyepiece E at the other.
(*b*) A thermocouple C shielded from incoming radiation and carrying a blackened copper target disc.

(*c*) A concave mirror M to provide maximum reflection of radiation on to the couple and centrally pierced to allow light to reach the eyepiece. The position of the mirror may be adjusted by means of a rack and pinion.
(*d*) Two small flat mirrors having semicircles cut out of them are mounted between the couple and the mirror M and inclined at a slight angle from the vertical plane. The resulting hole which is smaller than the target allows radiation from M to reach the couple.

The pyrometer is carefully positioned in line with the hole in the furnace door and as close as possible to reduce radiation absorption by the atmosphere. The eyepiece is adjusted to focus on the target and the concave mirror is adjusted to focus the radiations from the furnace on to the target. The function of the small mirrors is to aid focusing. When the radiation is not focused on to the target, the mirrors appear as shown at (i); when focusing is achieved they appear as at (ii).

The Foster Pyrometer

This is a fixed focus thermocouple instrument (Fig. 11.17*b*) arranged in such a way that provided the cone of radiation fills the tube the distance of the pyrometer from the source is unimportant.

The "Land" Surface Pyrometer

This pyrometer (Fig. 11.17*c*) is specially designed for measuring the temperature of surfaces in the open. It is also used for determining the total emissivity of a surface. The emissivity of a surface can be made to approach very closely to black body conditions if a gold-plated dome of high reflectivity is placed over the surface. In these circumstances virtually all the radiation both emitted and reflected is contained within the dome and the radiation passing through a small focusing lens on to a thermopile is virtually black body radiation. A black hemispherical insert can be fixed into the dome when required to determine the emissivity of a particular surface. Two readings are taken, one with the reflector exposed, and the other when the reflector is covered by the black insert. Thus the true and apparent temperatures of the surface are obtained and the emissivity can be determined.

Sources of Error

(1) The intervention of gas, flame, smoke or vapours between the radiating body and the pyrometer reduces the quantity of radiation received by the instrument.
(2) Radiation absorbed between the body and the instrument.
(3) Misalignment of the instrument with the radiating body.
(4) Failure of the radiating source to fill the whole field of view of the instrument. Radiation decreases as the square of the distance from the source but the area of the cone of radiation received by the instrument increases in the same proportion within the limits of size of the source (Fig. 11.18).

(5) Overheating of the pyrometer. Where the temperature is high because of operating conditions, e.g. when the pyrometer is permanently fixed to a furnace, then pyrometers are water cooled.

The accuracy of total radiation pyrometers is of the order of ±2% of the scale range.

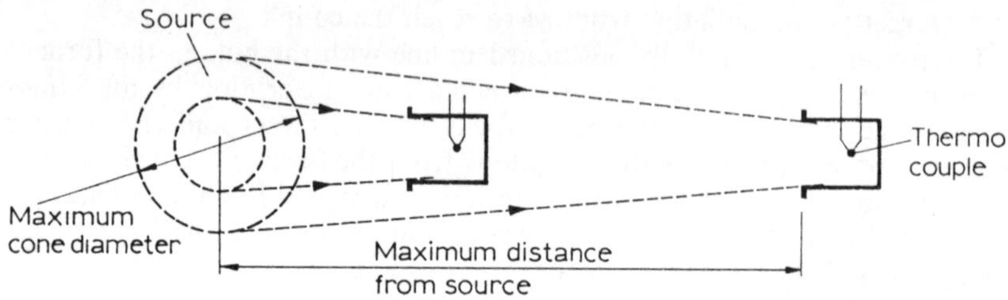

Fig. 11.18 CONE OF RADIATION

11.20 Optical Pyrometers

The curves shown in Fig. 11.16 indicate that a radiating body emits a spectrum of radiation wavelengths and that the wavelength at which maximum energy is emitted grows shorter as temperature increases. This is shown by the displacement of the peaks of the curves towards the left of the diagram. Further, the quantity of energy emitted at the maximum energy wavelength also increases as temperature increases.

If λ_m is the wavelength of maximum radiation and K is the absolute temperature, then

$$\lambda_m \propto 1/K \qquad \text{i.e. } \lambda_m \text{ is inversely proportional to } K$$

Therefore

$$\lambda_m K = \text{constant}$$

This is Wien's displacement law. Wien's second law states that

$$E_m \propto K^5 \qquad \text{i.e. } E_m = \text{constant} \times K^5$$

where E_m is the energy emitted at wavelength λ_m.

Formulae have also been evolved by Wien and by Planck which allow the energy emitted by a hot body at a given wavelength to be determined.

Optical pyrometers are based on comparison of the energy emitted by a hot body at a given wavelength with that of a black body calibrated lamp. They are used in the range 600–3000°C. The intensity of red light of approximate wavelength 0·65 μm is matched against a calibrated light source in the Féry disappearing filament pyrometer which is the instrument in most common use (Fig. 11.19). It should be remembered that light is simply the visible form of electromagnetic radiation and red light has the longest wavelength and is thus the first visible sign of radiation.

The pyrometer takes the form of a telescope. The lens O focuses radiation from the source at the plane of the filament F, the eyepiece E can also be focused at the same plane, and the removable red filter R prevents all radiation from the hot source except the red from reaching the eye. The battery, rheostat and ammeter carrying a temperature scale are all incorporated in the instrument. The current through the filament is adjusted by means of the rheostat. When the current passing through the filament is too low, the filament will emit radiation of lesser intensity than that of the source; it will thus appear dark against a bright background; when the current is too high it will appear brighter than the background; but when the correct current is

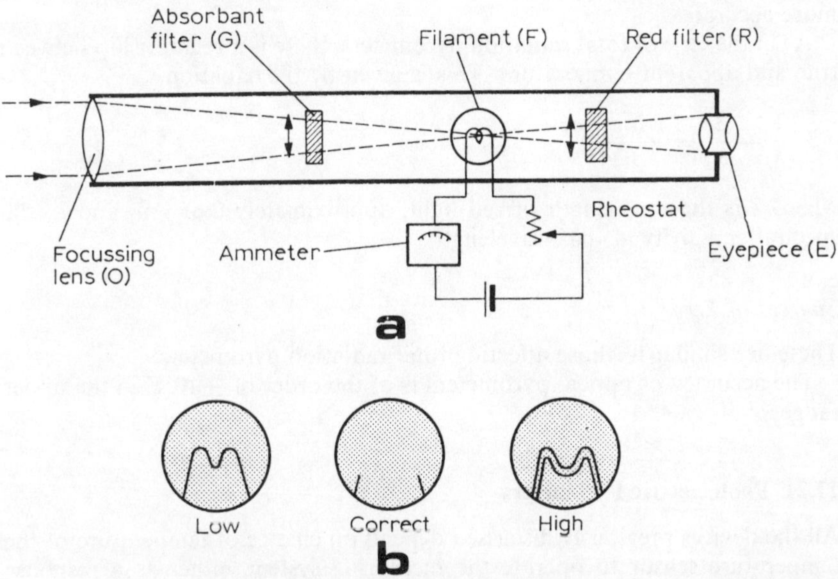

Fig. 11.19 DISAPPEARING FILAMENT PYROMETER

passing through the filament, the filament "disappears" into the background because it is radiating at the same intensity as the source (11.19*b*).

For temperatures up to 800°C the red filter may be removed. The ammeter has two overlapping scales, a lower temperature scale up to 1300°C and a higher temperature scale 1000–1800°C. When the higher scale is used an absorbant filter of known transmission factor is fitted between the lens O and the filament, thus enabling the same lamp to be used for higher temperatures without overloading the lamp filament.

Another type of optical pyrometer matches the "red" wavelength from the hot body with the radiation of similar wavelength emitted by an electric lamp which has been previously calibrated.

Spectral Emissivity

The colour of a body affects its ability to absorb radiation. A body of a particular colour, e.g. blue, absorbs all light radiation falling upon it except that which corresponds to the "blue" wavelength. This wavelength is reflected from the body and produces a "blue" sensation on the observer's brain. A body which only weakly absorbs a particular wavelength can only weakly emit it, thus the emissivity at that wavelength is less than one.

The ratio of the energy emitted by the body at a particular wavelength to the energy emitted by a black body at the same wavelength and temperature is called the spectral emissivity e_λ and an allowance must be made for this when using optical pyrometers. The correction required is usually less than for a total radiation pyrometer, and the temperature values obtained are more accurate.

As in the case of total radiation pyrometers, there is a relationship between true and apparent temperature. This is given by the equation

$$\frac{1}{K} - \frac{1}{K_a} = \frac{\lambda \log_{10} e_\lambda}{6245}$$

where λ is the wavelength of red light, approximately 0·65 μm, and e_λ the spectral emissivity at that wavelength.

Sources of Error

These are similar to those affecting total radiation pyrometers.

The accuracy of optical pyrometers is of the order of ± 10°C in the upper range.

11.21 Photoelectric Pyrometers

All the devices previously discussed depend on change of temperature of the temperature sensor to operate the measuring system, either as a response (bulb thermometers, thermocouples, etc.) or as a matching device (the filament pyrometer).

Photoelectric cells do not operate on this principle, but on the changes brought about by the effect of light falling on certain materials. Of the three types of photoelectric cells the photo-emissive, the photo-conductive and the photo-voltaic, it is the latter which is used for almost all temperature measurement. This is the silicon solar cell which is used to convert sunlight into electricity in space vehicles; it has also been proposed as a means of providing the energy to operate vehicles in hot sunny climates. It has the advantage of providing a current without a battery (Fig. 11.20).

The mechanism of the process is as follows: when light falls onto the cell, electrons are released in such a way as to set up an e.m.f. in a circuit connected to it; this e.m.f. is allowed to operate against a load resistance of about 500Ω and the current is measured by means of a deflection galvanometer or a self-balancing potentiometer.

Photoelectric cells do not respond to light of all wavelengths, but only to those of less than a certain critical value, this value depending on the material being irradiated. Below the critical wavelength the response improves as the wavelength decreases and also increases with the intensity of radiation.

The layout of photoelectric pyrometers is similar to that of optical and total radiation pyrometers, thermistors being used to compensate for changes in ambient temperature between zero and 80°C.

The advantages of this type of pyrometer include high accuracy (±0·2% of scale temperature), stability (i.e. little drift from calibration with time), and speed of response, which is of the order of one millisecond, making it very suitable for automatic control.

Spectral emissivity affects the accuracy of reading and allowance must be made for this. Photoelectric pyrometers are particularly suited to the measurement of small radiant sources and can be used effectively at a greater distance from the body than other pyrometers. They also have the advantage

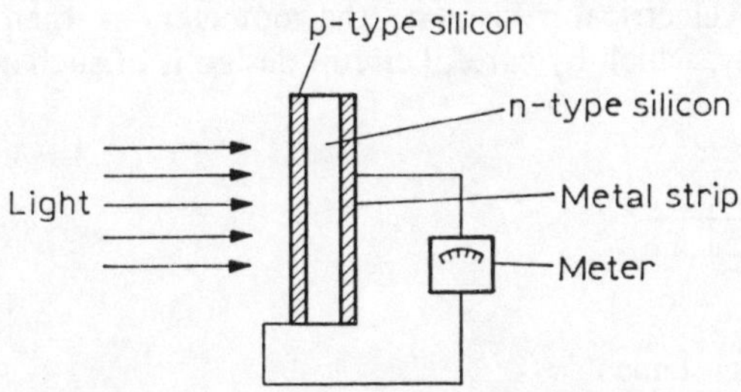

Fig. 11.20 SILICON SOLAR CELL

of good sensivity in the red or near infra-red, responding best to wavelengths of between 0·5 and 1 μm.

Sources of Error

The error sources are similar to those of other pyrometers but in addition errors may be introduced by light falling on the hot body from a window or other source. This is a particular danger when measurements are made in the open, especially if the surface of the material is highly reflective, and also when the temperature is low, as this increases the significance of any reflected light.

Temperatures upwards from 700°C are usually measured with these instruments although with precautions against reflected light, etc., lower temperatures are possible. There is no known upper temperature limit, but calibration difficulties arise at temperatures above 2000°C.

In addition to silicon, other materials are sometimes used which respond to different wavebands.

11.22 Temperature Measurement by Change of Electrical Resistance

The electrical resistance of conductors and semiconductors remains constant provided their physical and chemical conditions do not change. When the temperature of a conductor varies its resistance also varies.

Two kinds of electrical resistance thermometers are used: those using high purity metals, copper, nickel or platinum and having positive temperature coefficients of resistance, i.e. resistance increases with temperature; and those manufactured from mixtures of sintered metallic oxides, carbon, germanium alloys, etc., which have negative coefficients of resistance, i.e. resistance decreases with increase in temperature. Electrical resistance thermometers rely on resistance change in metals, often platinum, and thermistors (thermal resistors) rely on resistance change in semiconductors, usually germanium based alloys.

11.23 Electrical Resistance Thermometers

The most valuable feature of electrical resistance thermometers is their accuracy, sensitivity and stability, which by careful circuit design is of such a

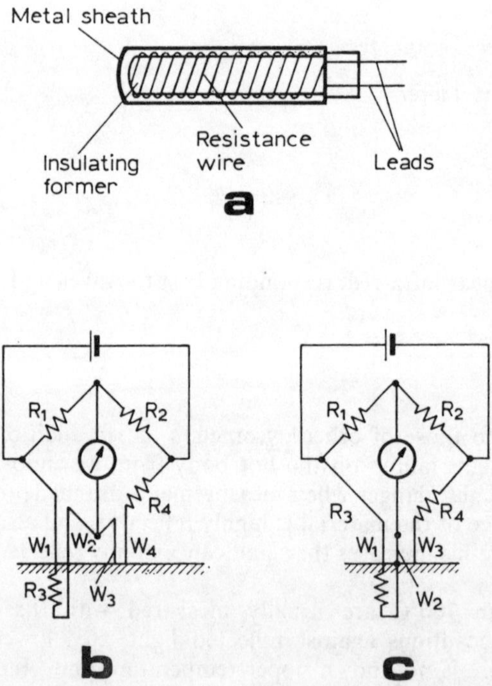

Fig. 11.21 ELECTRICAL RESISTANCE THERMOMETER

high order that they are suitable for IPTS determination. They are used for measurement in the range 10 K to 1300 K but industrially from 73 K to 873 K (−200 to 600°C) although largely in the upper part of the range. A coil of wire is wound non-inductively on to an insulting former which is inserted into a refractory protective sheath (Fig. 11.21*a*). Copper leads from the coil are connected into a Wheatstone bridge. When the leads from the thermometer are short they are made of large cross-section to reduce their resistance and thus remain largely unaffected by ambient temperature change, but when they are long, and to reduce cost, dummy leads are used and these take the form of three lead and four lead connections to the bridge (Fig. 11.21 *b* and *c*).

Three-lead Connection

The thermometer resistance coil is connected by three lead wires, W_1, W_2 and W_3 to the bridge circuit. If leads W_1 and W_3 are of similar material, length, diameter and condition, then any change of ambient temperature will cause an equal resistance change in both. Thus the arms of the bridge to which they are connected sustain equal resistance changes and the bridge measures only the resistance change of the coil.

Four-lead Connection

In this case two leads W_3 and W_4 are short-circuited near the resistance coil and will just compensate for W_1 and W_2 connected to the coil; again the bridge remains in balance and measures only the change in the coil.

When very high accuracy and sensitivity are required one of the forms of balanced Wheatstone bridge is used.

Applications

Temperature readings from widely separated points can easily be brought together on to a central console well removed from the measured body; thus resistance thermometers can be employed where the capillary lengths of liquid expansion thermometers would be excessive.

Resistance thermometers are available in a very wide variety of forms which include bonded wafers, ribbons and tapes for surface temperature determination, and bare wire types giving improved response for air and non-corrosive fluid measurement; they are also used for measuring gas temperatures. Special probes are available in the form of needle or skewer probes for soft materials such as rubber or meat. A number of thermometers may be connected in series into a bridge to give the average temperature of liquids, slurries or granulated solids. Temperature differences can be found by connecting two exactly similar resistance elements to two arms of the bridge. The bulbs are held at the same temperature and the bridge balanced; the scale may then be calibrated for temperature difference between the bulbs.

Resistance thermometer bulbs may be designed for almost any desired resistance change corresponding to the fundamental interval (ice to steam point) but industrial instruments usually have an interval of 50 or 100 ohms as this gives a large resistance change in the bulb compared with the lead wires. The current used for resistance thermometer bridges may be either ac or dc and will increase the temperature of the bridge circuit. However this effect is small and for most purposes may be neglected.

Sources of Error

(1) Impure resistance metal which can cause diffusion within the metal leading to segregation of phases and evaporation. The materials available are in fact of very high purity.
(2) A metal which has been strained will suffer a change in resistance characteristics, thus it must be annealed at a higher temperature than that at which it operates.
(3) Strains arising as a result of faulty installation, vibrations, etc., or strains developing from differential expansion between the measuring element and the solid to which the element is attached when measuring surface temperature.

Accuracy Electrical resistance thermometers have an accuracy of $\pm 0{\cdot}75\%$ of scale range at temperatures up to 600°C.

11.24 Thermistors

Metal atoms mutually share their valency electrons. Thus their valency electrons are in a continuous state of motion through the metal as they move from atom to atom: the electrons are in a sense "free". As temperature rises the vibration of the atoms in the metal lattices increases, the volume of space occupied by the atoms increases, and thus electron flow through the lattice becomes increasingly difficult.

The valency electrons in semiconductors (thermistors) are more firmly attached to their atoms and increase in temperature allows some of them to become detached so that they can flow; thus increase in temperature decreases electrical resistance by improving the conditions for flow. Thermistor elements used for temperature measurement have temperature coefficients of resistance many times greater than copper or platinum, e.g. at 25°C platinum $+0{\cdot}0036$, thermistors in general $-0{\cdot}045$. Resistivity is also much higher than that of any pure metal; they are thus highly sensitive to small changes in temperature, and respond very rapidly to such changes. Typical resistance ratio curves are shown in Fig. 11.22. Semiconductors have an exponential temperature–resistance relationship of the form

$$R = ae^{b/K}$$

where R = resistance (ohm)
K = absolute temperature
e = napierian base
a and b are constants over a small temperature range.

The ordinary Wheatstone bridge circuit has an output when unbalanced which is hyperbolic with respect to resistance variation of a particular arm. It is possible by carefully matching the non-linear characteristics of the thermistor and bridge to achieve a linear output at the scale over a small temperature range.

Among a variety of semiconductor materials germanium containing exact proportions of arsenic, gallium or antimony is in most wide use, because unlike many other semiconductors it is very stable and its drift from calibration is slight; it also shows a very large decrease in resistance with increasing temperature.

The temperature range measured with thermistors is −250°C to 650°C although both lower and higher temperatures are possible. Under carefully controlled laboratory conditions, precise measurements are made within 0·001 K near 4 K.

Battery and mains operated instruments, both portable and fixed, are obtainable having scale ranges as small as 10°C with 0·1°C subdivisions or

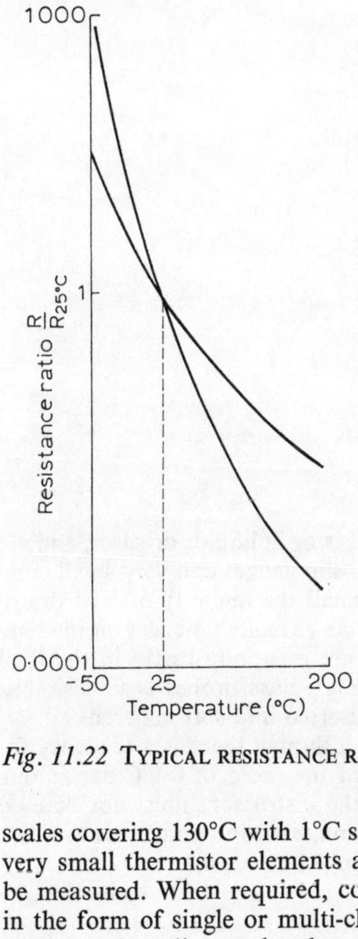

Fig. 11.22 TYPICAL RESISTANCE RATIO CURVES

scales covering 130°C with 1°C subdivisions. Because of their high resistivity very small thermistor elements are possible allowing point temperatures to be measured. When required, continuous reading instruments may be used in the form of single or multi-channel units. Thermistors take the form of beads, probes, discs and rods, and are sometimes used bare but are usually glass coated or positioned under a thin metal cap; discs and rods are used more as time delay elements, temperature compensators, and for voltage and power control in electronic circuits (Fig. 11.23).

Glass and metal probes less than 2 mm diameter are used for measuring metal surface temperatures, gases and liquids.

Glass Probes

These are made of specially toughened glass and are calibrated on metal surfaces or in liquids. They are particularly suited to materials or liquids which are poor heat conductors and have low heat capacity, e.g. insulating materials, powders, plastics, textiles, lacquered or painted surfaces. They are used for checking small components such as transistors, windings, brush gear and live cables up to 450 V providing proper precautions are taken.

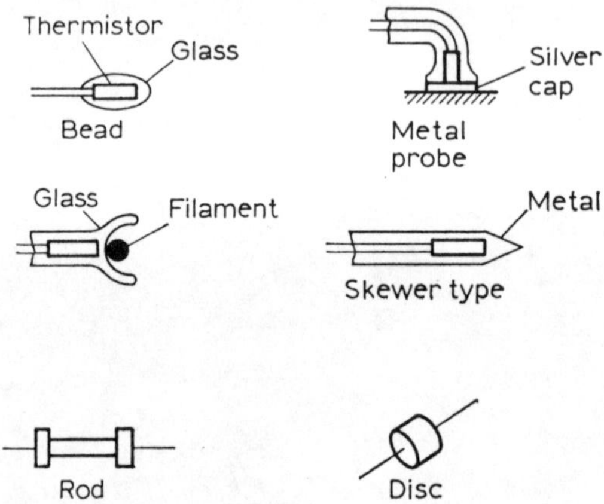

Fig. 11.23 THERMISTORS

Metal Probes

These are calibrated directly on metal surfaces or in liquids or gases, and are used for checking surface temperatures, e.g. slip gauges and very small components. Because probe diameters are so small the quantity of heat drawn from the tested surface is insignificant and has virtually no effect on the temperature of the component. Metal probes are especially useful in the food industry where contamination by breakage of glass probes is undesirable. Special stainless steel needle types can be inserted into soft materials.

Germanium transistors are highly stable although there is a tendency for calibration to drift with time but stability of the order of 0·1°C per annum can be obtained. To maintain calibration the instrument must not be used beyond the limits of its temperature range and vibration and stress avoided.

Special probes are used for lightly vibrating components.

11.25 Calibration

The need to establish fixed points on which temperature scales are based is discussed in sections 11.4, 11.5 and 11.6. Instruments may be calibrated directly by reference to the equilibrium temperatures of substances (primary calibration) or by comparison with standard instruments which have been calibrated in this way (secondary calibration).

Primary Calibration

The temperature sensor, e.g. a thermocouple or resistance thermometer, in the form in which it is to be used and connected in circuit is immersed in fully molten slowly cooling pure metal (see Table 11.4). At the onset of freezing, i.e. when solid begins to form, the fall in temperature of the substance is arrested and remains stationary until it is completely solid. A direct or an inverse cooling curve indicates when the freezing temperature has been reached. A direct cooling curve is obtained by plotting meter readings at regular

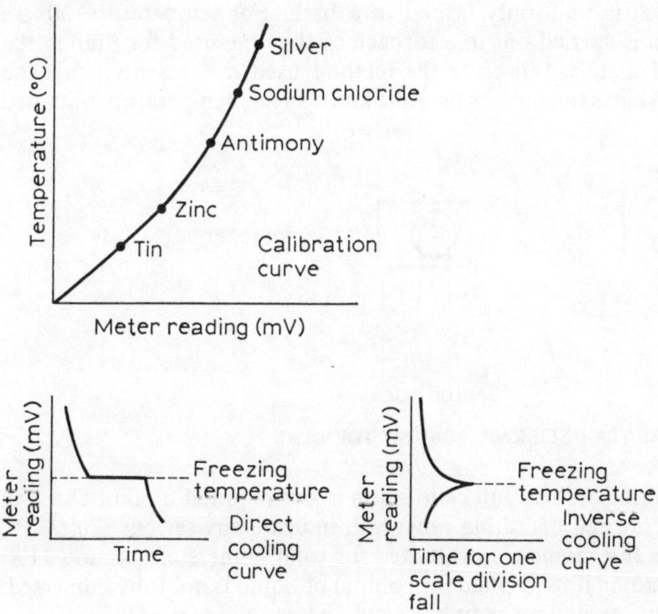

Fig. 11.24

intervals against a time base during cooling and an inverse cooling curve by plotting the meter reading against the time required for it to fall through one scale division. The combined freezing points for a number of substances can then be plotted to form a full calibration curve for the instrument (Fig. 11.24). It is of course essential to ensure that all sources of error in the system be eliminated before any attempt at calibration is made.

Rare metal thermocouples can be calibrated at the gold point (1064·43°C) by fixing a small piece of gold wire between the free ends of the thermocouple hot junction. This method makes use of the law of intermediate metals (see section 11.14). The thermocouple is heated very slowly in the region of the melting point while meter readings are observed. When the temperature is sufficient to melt the wire the thermocouple circuit breaks and the meter reading at breakage is taken to be the gold point. This technique can also be used to determine the palladium point (1554°C) and, with modifications, the platinum point (1772°C), but it is not suitable for metals which oxidize rapidly unless special precautions are taken.

Secondary Calibration

The simplest form of secondary calibration involves direct comparison with a previously calibrated instrument and this is the normal method used for temperatures up to the gold point.

For temperatures up to 550°C water, oil, salt or metal baths incorporating a stirrer and heating system are used or the instruments may be inserted in a metal block which is uniformly heated in a bath. For temperatures above 550°C calibration is carried out in a furnace of the type used for finding the freezing point of metals. Whatever the method used it is essential that the instruments be as close together as possible and a steady temperature obtained

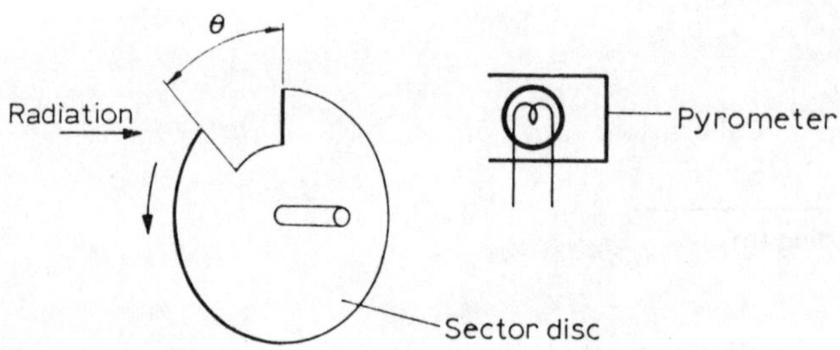

Fig. 11.25 PYROMETER CALIBRATION BY SECTOR DISC

prior to each reading. Where full calibration is not required a point check in the instrument's normal operating range will indicate any serious fault.

Liquid in glass thermometers calibrated for total immersion (section 11.8) will give a low reading if used when the column of liquid is not fully immersed. Emergent column corrections may be calculated as follows:

$$\text{Correction} = N \times (t_2 - t_1) \times \text{a constant}$$

where N = number of degrees shown on the emergent column
t_1 = average temperature of the emergent column
t_2 = the emergent column temperature reading.

Constant = the coefficient of apparent expansion of the liquid in glass, e.g. mercury 0·000 155, alcohol 0·001 04. An emergent column should be kept as short as possible.

Optical Pyrometers

Standard optical pyrometers are calibrated by sighting on a black body immersed in freezing gold and the lamp current adjusted until matching is obtained.

For temperatures below the gold point sector discs are rotated between the pyrometer and the source. The angle of the sector removed from the disc is known and thus a known proportion of the total radiation previously received passes the disc to the instrument (Fig. 11.25). Different discs may be used to determine other temperatures. The temperature indicated on the instrument by this technique is calculated from the formula

$$\log \frac{J_2}{J_1} = \frac{C_2}{\lambda}\left(\frac{1}{1336} - \frac{1}{K}\right)$$

where C_2 = 14·32 mm degrees
λ = wavelength (mm)
K = absolute temperature, and λK is less than 3 mm degrees
J_2 = intensity of monochromatic light radiated at the temperature K of wavelength λ
J_1 = intensity of monochromatic light of the same wavelength radiated by a black body at the gold point.

Temperatures above the gold point are determined by rotating sector discs between the pyrometer and a tungsten filament lamp. The brightness of the filament lamp is adjusted until the pyrometer is receiving the equivalent of gold point radiation, the radiation proportion is then known and the temperature of the filament lamp calculated as before. The disc is removed and the pyrometer filament matched to the lamp filament.

Bibliography

(1) Ambrosius, Fellow and Brickman, *Mechanical Measurement and Instrumentation* (Ronald Press, 1966)
(2) P. S. Barna, *Fluid Mechanics for Engineers* (Butterworth, 1969)
(3) H. G. Bass, *Introduction to Engineering Measurements* (McGraw-Hill, 1971)
(4) T. G. Beckwith and N. Lewis Buck, *Mechanical Measurements* (Addison-Wesley, 1961)
(5) R. P. Benedict, *Fundamentals of Temperature, Pressure and Flow Measurement* (Wiley, 1969)
(6) T. Bradshaw, *Experimental Fluid Mechanics* (Pergamon Press, 1952)
(7) J. J. Brophy, *Semi-Conductor Devices* (Allen and Unwin, 1966)
(8) A. Dinsdale and F. Moore, *Viscosity and its Measurement* (Chapman and Hall, 1962)
(9) E. O. Doebelin, *Measurement Systems: Application and Design* (McGraw-Hill, 1966)
(10) A. W. Hendry, *Elements of Experimental Stress Analysis* (Pergamon Press)
(11) R. A. Higgins, *Engineering Metallurgy*, Vol. 1, 3rd edition (E.U.P., 1968)
(12) G. S. Holister, *Experimental Stress Analysis* (C.U.P., 1967)
(13) J. P. Holman, *Experimental Methods for Engineers* (McGraw-Hill)
(14) K. J. Hume, *Engineering Metrology* (Macdonald, 1966)
(15) E. B. Jones, *Instrument Technology*, 2nd edition (Butterworths, 1965)
(16) R. Kingslake, *Applied Optics and Optical Engineering*, Vol. 1 (Academic Press Inc. (London), 1965)
(17) A. Linford, *Flow Measurement and Meters* (Spon, 1961)
(18) B. B. Low, *Strength of Materials* (Longman)
(19) A. C. Merrington, *Viscometry* (Arnold, 1949)
(20) M. B. Moore, *Theory and Application of Mechanical Engineering Measurement* (Van Nostrand, 1960)
(21) J. T. Miller, *Revised Course in Instrument Technology* (United Trade Press, 1964)
(22) M. Nelkon and P. Parker, *Advanced Level Physics* (Heinemann, 1962)
(23) H. K. P. Neubert, *Strain Gauges: Kinds and Uses* (Macmillan, 1968)
(24) P. J. O'Higgins, *Basic Instrumentation, Industrial Measurement* (McGraw-Hill, 1966)
(25) E. B. Pearson, *Technology of Measurement* (E.U.P., 1957)

(26) C. C. Perry and H. R. Lissner, *The Strain Gauge Primer* (McGraw-Hill, 1962)
(27) L. Prandtl, *Essentials of Fluid Mechanics* (Blackie, 1960)
(28) F. H. Rolt, *Use of Light Waves for Controlling the Accuracy of Block Gauges* (Hilger and Watts)
(29) G. H. Ryder, *Strength of Materials* (S.I. edition) (Macmillan, 1968)
(30) F. C. Tyson, *Industrial Instrumentation* (Prentice-Hall, 1961)

Journals

(1) *Technographs*, Engineering Measurement Series, 1967
(2) Science and Industry, Vol. 5, *Proceedings of the Instrument Society of America* (Reinhold Pub. Corp.)
(3) N.P.L. Notes on Applied Science, No. 12, *Calibration of Temperature Measurement Instruments*
(4) *Metrologia*, Vol. 5, No. 2, 1969 (Springer)
(5) R. C. Spragg and D. J. Whitehouse, *Proceedings of the Institution of Mechanical Engineers*, Vol. 185, 47/71

British Standard Specifications

B.SS: DD6: 1972	Methods for Calibration of Bonded Electrical Resistance Strain Gauges
817: 1972	Surface Plates and Tables
818: 1963	Cast Iron Straight Edges (Bow Section and I Section)
863: 1939	Steel Straightedges of Rectangular Section
869: 1939	Toolmakers Flats and High Precision Surface Plates
888: 1950	Slip (or Block) Gauges and Their Accessories
1041: ——	Temperature Measurement
1042: 1964	Part I Flow Measurement in Pipes
1134: 1972	Method for the Assessment of Surface Texture
1780: 1960	Pressure Gauges
1790: 1961	Length Bars and Their Accessories
2634: ——	Roughness Comparison Specimens
3064: 1969	Sine Bars and Sine Tables (Excluding Compound Tables)
4311: 1968	Metric Gauge Blocks

Institute of Petroleum Standard IP 73/53

Index